Foundation Maths

Essential Maths for Students

Series Editors: Anthony Croft, Department of Mathematical Sciences, Loughborough University, and Robert Davison, Department of Mathematical Sciences, De Montfort University, Leicester

Also available

Engineering Maths
Mustoe

Maths for Computing and Information Technology
Giannasi and Low

Maths and Statistics for Business
Lawson, Hubbard and Pugh

Engineering Systems: Modelling and Control
Hargreaves

Improve your Maths!: a refresher course
Bancroft and Fletcher

Essential Maths for Students

Foundation Maths

Second edition

Anthony Croft
Loughborough University

Robert Davison
De Montfort University

Prentice
Hall

An imprint of Pearson Education

Harlow, England · London · New York · Reading, Massachusetts · San Francisco · Toronto · Don Mills, Ontario · Sydney
Tokyo · Singapore · Hong Kong · Seoul · Taipei · Cape Town · Madrid · Mexico City · Amsterdam · Munich · Paris · Milan

510

© **Addison Wesley Longman 1997**

Pearson Education Limited
Edinburgh Gate, Harlow
Essex CM20 2JE, England
and Associated Companies throughout the world

Visit us on the World Wide Web at:
http://www.pearsoneduc.com

Illustrated and typeset by 16
Printed and bound in Great Britain by Henry Ling Ltd,
at the Dorset Press, Dorchester, Dorset.

First edition published 1995
This edition first printed 1997.

ISBN 0–201–17804–4

British Library Cataloguing-in-Publication Data
A catalogue record for this book is available from the British Library

Library of Congress Cataloging-in-Publication Data
Croft, Tony 1957–
  Foundation maths / Anthony Croft, Robert Davison. – 2nd ed.
    p.   cm. – (Essential maths for students)
  Includes index.
    ISBN 0–201–17804–4 (alk. paper)
    1. Mathematics.   I. Davison, Robert,   II. Title.   III. Series.
QA37 2.C74   1997
510–dc21   97-34029
                                                                CIP

10   9   8   7   6   5   4
04   03   02   01   00

Contents

Preface to the second edition

Since the first edition of *Foundation Maths* was published in October 1994, we have received numerous comments about the book. Before embarking on the second edition we considered all of these.

The style of the first edition appears to have been popular; readers commented favourably on the numerous worked examples, the highlighted key points and the jargon-free language. All of these features have been retained in the second edition. There are, of course, some changes.

The first chapter of the book has been re-written in order to make it more accessible to those students who have returned only recently to studying mathematics and who need thorough revision of basic arithmetical techniques. Particular care has been taken with the chapter on Fractions since it is our experience that students who master the arithmetic of fractions are better placed to handle algebraic techniques later in their studies.

The numerical work in the early chapters benefits from a new chapter on Decimals. Some of the algebraic development has been restructured and new sections added where appropriate, including a section on solving simultaneous equations graphically. The chapter on Functions now includes a section on composite functions. The Trigonometry has had a major overhaul and now covers three chapters. Finally, numerous extra exercises have been added throughout.

We hope this second edition will help you in your study of Mathematics.

A. Croft, R. Davison
August 1997

Preface to the first edition

Today, a huge variety of disciplines require their students to have knowledge of certain mathematical tools in order to appreciate quantitative aspects of their subjects. At the same time, higher education institutions have widened access so that there is much greater variety in the mathematical profile of the student body. Many of these students are returning to education after many years in the workplace or at home bringing up families. One consequence has been that lecturers can no longer assume that a firm bedrock of mathematical knowledge and competence has been established. In addition, students are finding themselves inadequately prepared for the mathematical demands placed upon them. In some institutions, a remedy has been access or foundation course provision which prepares students to make the transition to a higher education course. In others, numeracy and mathematics learning centres have been established to provide support and encouragement.

Foundation Maths has been written to pave the way into higher education for those students who have not specialized in mathematics at *A* level. It is intended for non-specialists who need some but not a great deal of mathematics as they embark upon their courses of higher education. It takes students from around the lower levels of GCSE to a standard which will enable them to participate fully in a degree or diploma course. It is ideally suited for foundation and access courses in mathematics and for those who wish to enter a wide range of courses such as marketing, business studies, management, science, engineering, social science, geography, combined studies and design. It will also be useful for those who lack confidence and need careful, steady guidance in mathematical methods. Even for those whose mathematical expertise is already established, the book will be a helpful revision and reference guide. The style of the book also makes it suitable for those who wish to engage in self-study or distance learning.

We have tried throughout to adopt an informal, user-friendly approach and have described mathematical processes in everyday language. Mathematical ideas are usually developed by example rather than by formal proof. This reflects our experience that students learn better from examples than from abstract development. Where appropriate, the examples contain a great deal of detail so that the student is not left wondering how one stage of a calculation leads to the next. In *Foundation*

Maths, objectives are clearly stated at the beginning of each chapter, and key points and formulae are highlighted throughout the book. Self-assessment questions are provided at the end of most sections. These test understanding of important features in the section and answers can be checked by referring back. These are followed by exercises; it is essential that these are attempted as the only way to develop competence and understanding is through practice. Detailed solutions to these exercises are given at the back of the book and should be consulted only after the exercises have been attempted. A further set of test and assignment exercises is given at the end of each chapter. These are provided so that the tutor can set regular assignments or tests throughout the course. Solutions to these are not provided.

In order to keep the size of the book reasonable, we have endeavoured to include topics which we think are most important, cause the most problems for students and have the widest applicability. Inevitably compromise has been necessary. However, recent experience has shown us that the root of many of the problems occurring in mathematical education today is the lack of knowledge of algebraic manipulation. So many staff exclaim that if only their students could master algebra, all of the other techniques would fall into place. A large proportion of students now entering higher education have had little opportunity to develop algebraic skills to the standard needed in applications. With this in mind we have placed great emphasis on algebra in the book, with new work being gradually introduced over several chapters.

The best strategy for those using the book would be to read through each section, carefully studying all of the worked examples and solutions. Many of these solutions develop important results needed later in the book. It is then a good idea to cover up the solution and try to work the example again for yourself. It is only by actually doing the calculation for yourself that the necessary techniques will be mastered. At the end of each section attempt the self-assessment questions. If you cannot answer these, read the previous few pages again to find the answers in the text. Then attempt the exercises, regularly checking your solutions with those given at the back of the book.

In conclusion, remember that learning mathematics takes time and effort. Carrying out a large number of exercises allows you to experience a greater variety of problems, thus building your expertise and confidence. Armed with these you will be able to tackle more unfamiliar and demanding problems that arise in other aspects of your course. We hope you find *Foundation Maths* useful and wish you the very best of luck.

A. Croft and R. Davison

Mathematical symbols

$+$	plus
$-$	minus
$\pm$	plus or minus
$\times$	multiply by
$\cdot$	multiply by
$\div$	divide by
$=$	is equal to
$\equiv$	is identically equal to
$\approx$	is approximately equal to
$\neq$	is not equal to
$>$	is greater than
$\geqslant$	is greater than or equal to
$<$	is less than
$\leqslant$	is less than or equal to
$\in$	is a member of set
$\mathbb{R}$	set of real numbers
$\therefore$	therefore
∞	infinity
e	the base of natural logarithms (2.718...)
$\ln$	natural logarithm
$\log$	logarithm to base 10
$\sum$	sum of terms
$\int$	integral
$\frac{dy}{dx}$	derivative of y with respect to x
π	'pi' ≈ 3.14159

1 Arithmetic

Objectives	This chapter
	• explains the rules for adding, subtracting, multiplying and dividing positive and negative numbers
	• explains what is meant by an integer
	• explains what is meant by a prime number
	• explains what is meant by a factor
	• explains how to prime factorize an integer
	• explains the terms 'highest common factor' and 'lowest common multiple'

1.1 Addition, subtraction, multiplication and division

Arithmetic is the study of numbers and their manipulation. A clear and firm understanding of the rules of arithmetic is essential for tackling everyday calculations. Arithmetic also serves as a springboard for tackling more abstract mathematics such as algebra and calculus.

The calculations in this chapter will involve mainly whole numbers, or **integers** as they are often called. The **positive integers** are the numbers

$$1, 2, 3, 4, 5 \ldots$$

and the negative integers are the numbers

$$\ldots -5, -4, -3, -2, -1$$

The number 0 is also an integer but is neither positive nor negative.

To find the **sum** of two or more numbers, the numbers are added together. To find the **difference** of two numbers, the second is subtracted from the first. The **product** of two numbers is found by multiplying the numbers together. Finally, the **quotient** of two numbers is found by dividing the first number by the second.

Worked example

1.1   (a) Find the sum of 3, 6 and 4.
(b) Find the difference of 6 and 4.
(c) Find the product of 7 and 2.
(d) Find the quotient of 20 and 4.

Solution   (a) The sum of 3, 6 and 4 is
$$3 + 6 + 4 = 13$$
(b) The difference of 6 and 4 is
$$6 - 4 = 2$$
(c) The product of 7 and 2 is
$$7 \times 2 = 14$$
(d) The quotient of 20 and 4 is $\frac{20}{4}$, that is 5.

When writing products we sometimes replace the sign $\times$ by '.' or even omit it completely. For example, $3 \times 6 \times 9$ could be written as $3.6.9$ or $(3)(6)(9)$.

On occasions it is necessary to perform calculations involving negative numbers. To understand how these are added and subtracted consider Figure 1.1 which shows a number line.

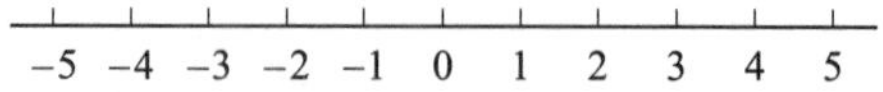

Figure 1.1
The number line

$$
\begin{array}{ccccccccccc}
-5 & -4 & -3 & -2 & -1 & 0 & 1 & 2 & 3 & 4 & 5
\end{array}
$$

Any number can be represented by a point on the line. Positive numbers are on the right-hand side of the line and negative numbers are on the left. From any given point on the line, we can add a positive number by moving that number of places to the right. For example to find the sum $5 + 3$, start at the point 5 and move 3 places to the right, to arrive at 8. This is shown in Figure 1.2.

Figure 1.2
To add a positive number, move that number of places to the right

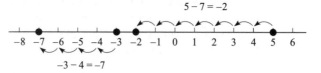

To subtract a positive number, we move that number of places to the left. For example to find the difference $5 - 7$, start at the point 5 and move 7 places to the left to arrive at -2. Thus $5 - 7 = -2$. This is shown in Figure 1.3. The result of finding $-3 - 4$ is also shown to be -7.

Figure 1.3
To subtract a positive number, move that number of places to the left

To add or subtract a negative number, the motions just described are reversed. So, to add a negative number, we move to the left. To subtract a negative number we move to the right. The result of finding $2 + (-3)$ is shown in Figure 1.4.

Figure 1.4
Adding a negative number involves moving to the left

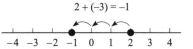

We see that $2 + (-3) = -1$. Note that this is the same as the result of finding $2 - 3$, so that adding a negative number is equivalent to subtracting a positive number.

The result of finding $5 - (-3)$ is shown in Figure 1.5.

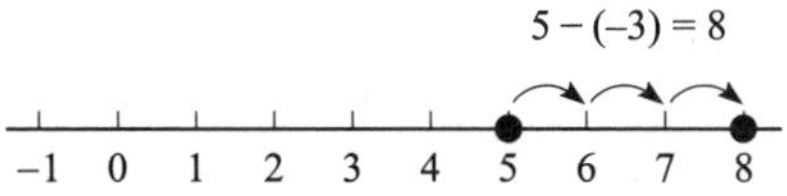

Figure 1.5
Subtracting a negative number involves moving to the right

We see that $5 - (-3) = 8$. This is the same as the result of finding $5 + 3$, so subtracting a negative number is equivalent to adding a positive number.

> **KEY POINT**
>
> Adding a negative number is equivalent to subtracting a positive number.
>
> Subtracting a negative number is equivalent to adding a positive number.

Worked example

1.2   Evaluate (a) $8 + (-4)$, (b) $-15 + (-3)$, (c) $-15 - (-4)$

Solution   (a) $8 + (-4)$ is equivalent to $8 - 4$, that is 4.
(b) Because adding a negative number is equivalent to subtracting a positive number we find $-15 + (-3)$ is equivalent to $-15 - 3$, that is -18.
(c) $-15 - (-4)$ is equivalent to $-15 + 4$, that is -11.

When we need to multiply or divide negative numbers, care must be taken with the **sign** of the answer, that is, whether the result is positive or negative. The following rules apply for determining the sign of the answer when multiplying or dividing positive and negative numbers.

KEY POINT

$(\text{positive}) \times (\text{positive}) = \text{positive}$ and $\dfrac{\text{positive}}{\text{positive}} = \text{positive}$

$(\text{positive}) \times (\text{negative}) = \text{negative}$

$(\text{negative}) \times (\text{positive}) = \text{negative}$ $\dfrac{\text{positive}}{\text{negative}} = \text{negative}$

$(\text{negative}) \times (\text{negative}) = \text{positive}$

$\dfrac{\text{negative}}{\text{positive}} = \text{negative}$

$\dfrac{\text{negative}}{\text{negative}} = \text{positive}$

Worked example

1.3   Evaluate

(a) $3 \times (-2)$ (b) $(-1) \times 7$ (c) $(-2) \times (-4)$ (d) $\dfrac{12}{(-4)}$ (e) $\dfrac{-8}{4}$ (f) $\dfrac{-6}{-2}$

Solution   (a) $3 \times (-2) = -6$ (b) $(-1) \times 7 = -7$ (c) $(-2) \times (-4) = 8$

(d) $\dfrac{12}{-4} = -3$ (e) $\dfrac{-8}{4} = -2$ (f) $\dfrac{-6}{-2} = 3$

Self-assessment questions 1.1

1. Explain what is meant by an integer, a positive integer, and a negative integer.
2. Explain the terms sum, difference, product and quotient.
3. State the sign of the result obtained after performing the following calculations:
   (a) $(-5) \times (-3)$ (b) $(-4) \times 2$ (c) $\frac{7}{-2}$ (d) $\frac{-8}{-4}$.

Exercise 1.1

1. Without using a calculator, evaluate each of the following:

   (a) $6 + (-3)$ (b) $6 - (-3)$ (c) $16 + (-5)$
   (d) $16 - (-5)$ (e) $27 - (-3)$ (f) $27 - (-29)$
   (g) $-16 + 3$ (h) $-16 + (-3)$ (i) $-16 - 3$
   (j) $-16 - (-3)$ (k) $-23 + 52$
   (l) $-23 + (-52)$ (m) $-23 - 52$
   (n) $-23 - (-52)$

2. Without using a calculator, evaluate

   (a) $3 \times (-8)$ (b) $(-4) \times 8$ (c) $15 \times (-2)$
   (d) $(-2) \times (-8)$ (e) $14 \times (-3)$

3. Without using a calculator, evaluate

   (a) $\frac{15}{-3}$ (b) $\frac{21}{7}$ (c) $\frac{-21}{7}$ (d) $\frac{-21}{-7}$ (e) $\frac{21}{-7}$
   (f) $\frac{-12}{2}$ (g) $\frac{-12}{-2}$ (h) $\frac{12}{-2}$

4. Find the sum and product of (a) 3 and 6, (b) 10 and 7, (c) 2, 3 and 6.

5. Find the difference and quotient of (a) 18 and 9, (b) 20 and 5, (c) 100 and 20.

1.2   The BODMAS rule

When evaluating numerical expressions we need to know the order in which addition, subtraction, multiplication and division are carried out. As a simple example, consider evaluating $2 + 3 \times 4$. If the addition is carried out first we get $2 + 3 \times 4 = 5 \times 4 = 20$. If the multiplication is carried out first we get $2 + 3 \times 4 = 2 + 12 = 14$. Clearly the order of carrying out numerical operations is important. The BODMAS rule tells us the order in which we must carry out the operations of addition, subtraction, multiplication and division.

KEY POINT

BODMAS stands for

Brackets ( )	First priority
Of $\times$	Second priority
Division $\div$	Second priority
Multiplication $\times$	Second priority
Addition $+$	Third priority
Subtraction $-$	Third priority

This is the order of carrying out arithmetical operations, with bracketed expressions having highest priority and subtraction and addition having the lowest priority. Note that 'Of', 'Division' and 'Multiplication' have equal priority, as do 'Addition' and 'Subtraction'. 'Of' is used to show multiplication when dealing with fractions, for example, find $\frac{1}{2}$ of 6 means $\frac{1}{2} \times 6$.

If an expression contains only multiplication and division, we evaluate by working from left to right. Similarly, if an expression contains only addition and subtraction, we also evaluate by working from left to right.

Worked examples

1.4   Evaluate
(a) $2 + 3 \times 4$ (b) $(2 + 3) \times 4$

Solution   (a) Using the BODMAS rule we see that multiplication is carried out first. So
$$2 + 3 \times 4 = 2 + 12 = 14$$
(b) Using the BODMAS rule we see that the bracketed expression takes priority over all else. Hence
$$(2 + 3) \times 4 = 5 \times 4 = 20$$

1.5   Evaluate
(a) $4 - 2 \div 2$ (b) $1 - 3 + 2 \times 2$

Solution   (a) Division is carried out before subtraction, and so

$$4 - 2 \div 2 = 4 - \frac{2}{2} = 3$$

(b) Multiplication is carried out before subtraction or addition,
$$1 - 3 + 2 \times 2 = 1 - 3 + 4 = 2$$

1.6   Evaluate
(a) $(12 \div 4) \times 3$ (b) $12 \div (4 \times 3)$

Solution   Recall that bracketed expressions are evaluated first.

(a) $(12 \div 4) \times 3 = \left(\dfrac{12}{4}\right) \times 3 = 3 \times 3 = 9$

(b) $12 \div (4 \times 3) = 12 \div 12 = 1$

Example 1.6 shows the importance of the position of brackets in an expression.

Self-assessment questions 1.2

1. State the BODMAS rule used to evaluate expressions.
2. The position of brackets in an expression is unimportant. True or false?

Exercise 1.2

1. Evaluate the following expressions:

   (a) $6 - 2 \times 2$ (b) $(6 - 2) \times 2$
   (c) $6 \div 2 - 2$ (d) $(6 \div 2) - 2$
   (e) $6 - 2 + 3 \times 2$ (f) $6 - (2 + 3) \times 2$
   (g) $(6 - 2) + 3 \times 2$ (h) $\dfrac{16}{-2}$ (i) $\dfrac{-24}{-3}$
   (j) $(-6) \times (-2)$ (k) $(-2)(-3)(-4)$

2. Place brackets in the following expressions to make them correct:

   (a) $6 \times 12 - 3 + 1 = 55$
   (b) $6 \times 12 - 3 + 1 = 68$
   (c) $6 \times 12 - 3 + 1 = 60$
   (d $5 \times 4 - 3 + 2 = 7$
   (e) $5 \times 4 - 3 + 2 = 15$
   (f) $5 \times 4 - 3 + 2 = -5$

1.3   Prime numbers and factorization

A **prime number** is a positive integer which cannot be expressed as the product of two smaller positive integers. To put it another way, a prime number is one which can be divided exactly only by 1 and itself. For example, $6 = 2 \times 3$, so 6 can be expressed as a product of smaller numbers and hence 6 is not a prime number. However, 7 is prime. Examples of prime numbers are 1, 2, 3, 5, 7, 11, 13, 17, 19, 23. Note that 2 is the only even prime.

Factorize means 'write as a product'. By writing 12 as 3×4 we have factorized 12. We say 3 is a **factor** of 12 and 4 is also a factor of 12. The way in which a number is factorized is not unique, for example 12 may be expressed as 3×4 or 2×6. Note that 2 and 6 are also factors of 12.

When a number is written as a product of prime numbers we say the number has been **prime factorized**.

To prime factorize a number, consider the technique used in the following examples.

Worked examples

1.7   Prime factorize the following numbers:
(a) 12 (b) 42 (c) 40 (d) 70

Solution   (a)   We begin with 2 and see if this is a factor of 12. Clearly it is, so we write

$$12 = 2 \times 6$$

Now we consider 6. Again 2 is a factor so we write

$$12 = 2 \times 2 \times 3$$

All the factors are now prime, that is the prime factorization of 12 is $2 \times 2 \times 3$.

(b)   We begin with 2 and see if this is a factor of 42. Clearly it is and so we can write

$$42 = 2 \times 21$$

Now we consider 21. Now 2 is not a factor of 21, so we examine the next prime, 3. Clearly 3 is a factor of 21 and so we can write

$$42 = 2 \times 3 \times 7$$

All the factors are now prime, and so the prime factorization of 42 is $2 \times 3 \times 7$.

(c)   Clearly 2 is a factor of 40,

$$40 = 2 \times 20$$

Clearly 2 is a factor of 20,

$$40 = 2 \times 2 \times 10$$

Again 2 is a factor of 10,

$$40 = 2 \times 2 \times 2 \times 5$$

All the factors are now prime. The prime factorization of 40 is $2 \times 2 \times 2 \times 5$.

(d) Clearly 2 is a factor of 70,

$$70 = 2 \times 35$$

We consider 35: 2 is not a factor, 3 is not a factor, but 5 is.

$$70 = 2 \times 5 \times 7$$

All the factors are prime. The prime factorization of 70 is $2 \times 5 \times 7$.

1.8   Prime factorize 2299.

Solution   We note that 2 is not a factor and so we try 3. Again 3 is not a factor and so we try 5. This process continues until we find the first prime factor. It is 11,

$$2299 = 11 \times 209$$

We now consider 209. The first prime factor is 11,

$$2299 = 11 \times 11 \times 19$$

All the factors are prime. The prime factorization of 2299 is $11 \times 11 \times 19$.

Self-assessment questions 1.3

1. Explain what is meant by a prime number.
2. List the first ten prime numbers.
3. Explain why all even numbers other than 2 cannot be prime.

Exercise 1.3

1. State which of the following numbers are prime numbers

   (a) 13 (b) 1000 (c) 2 (d) 29 (e) $\frac{1}{2}$

2. Prime factorize the following numbers:

   (a) 26 (b) 100 (c) 27 (d) 71 (e) 64 (f) 87 (g) 437 (h) 899

3. Prime factorize the two numbers 30 and 42. List any prime factors which are common to both numbers.

Highest common factor and lowest common multiple

Highest common factor

Suppose we prime factorize 12. This gives $12 = 2 \times 2 \times 3$. From this prime factorization we can deduce all the factors of 12.

> 2 is a factor of 12
> 3 is a factor of 12
> $2 \times 2 = 4$ is a factor of 12
> $2 \times 3 = 6$ is a factor of 12

Hence 12 has 4 factors 2, 3, 4 and 6, in addition to the obvious factors of 1 and 12.

Similarly we could prime factorize 18 to obtain $18 = 2 \times 3 \times 3$. From this we can list the factors of 18.

> 2 is a factor of 18
> 3 is a factor of 18
> $2 \times 3 = 6$ is a factor of 18
> $3 \times 3 = 9$ is a factor of 18

The factors of 18 are 1, 2, 3, 6, 9 and 18. Some factors are common to both 12 and 18. These are 2, 3 and 6. These are **common factors** of 12 and 18. The highest common factor of 12 and 18 is 6.

The highest common factor of 12 and 18 can be obtained directly from their prime factorization. We simply note all the primes common to both factorizations:

$$12 = 2 \times 2 \times 3 \qquad 18 = 2 \times 3 \times 3$$

Common to both is 2×3. Thus the highest common factor is $2 \times 3 = 6$. Thus, 6 is the highest number which divides exactly into both 12 and 18.

KEY POINT

Given two or more numbers the **highest common factor** (h.c.f.) is the largest (highest) number which is a factor of all the given numbers.

Worked examples

1.9   Find the h.c.f. of 12 and 27.

Solution   We prime factorize 12 and 27,

$$12 = 2 \times 2 \times 3 \qquad 27 = 3 \times 3 \times 3$$

Common to both is 3. Thus, 3 is the h.c.f. of 12 and 27. This means that 3 is the highest number which divides both 12 and 27.

1.10   Find the h.c.f. of 28 and 210.

Solution   The numbers are prime factorized,

$$28 = 2 \times 2 \times 7$$

$$210 = 2 \times 3 \times 5 \times 7$$

The factors which are common are identified: a 2 is common to both and a 7 is common to both. Hence both numbers are divisible by $2 \times 7 = 14$. Since this number contains all the common factors it is the highest common factor.

1.11   Find the h.c.f. of 90 and 108.

Solution   The numbers are prime factorized,

$$90 = 2 \times 3 \times 3 \times 5$$

$$108 = 2 \times 2 \times 3 \times 3 \times 3$$

The common factors are 2, 3 and 3 and so the h.c.f. is $2 \times 3 \times 3$, that is 18. This is the highest number which divides both 90 and 108.

1.12   Find the h.c.f. of 12, 18 and 20.

Solution   Prime factorization yields

$$12 = 2 \times 2 \times 3 \qquad 18 = 2 \times 3 \times 3 \qquad 20 = 2 \times 2 \times 5$$

There is only one factor common to all three numbers; it is 2. Hence 2 is the h.c.f. of 12, 18 and 20.

Lowest common multiple

Suppose we are given two or more numbers and wish to find numbers into which all the given numbers will divide. For example, given 4 and 6 we see that they both divide exactly into 12, 24, 36, 48, 60 and so on. The smallest number into which they both divide is 12. We say 12 is the **lowest common multiple** of 4 and 6.

KEY POINT	The lowest common multiple (l.c.m.) of a set of numbers is the smallest (lowest) number into which all the given numbers will divide exactly.

Worked example

1.13   Find the l.c.m. of 6 and 10.

Solution   We seek the smallest number into which both 6 and 10 will divide exactly. There are many numbers into which 6 and 10 will divide, for example 60, 120, 600, but we are seeking the smallest such number. By inspection, the smallest such number is 30. Thus, the l.c.m. of 6 and 10 is 30.

A more systematic method of finding the l.c.m. involves the use of prime factorization.

Worked examples

1.14   Find the l.c.m. of 15 and 20.

Solution   The numbers are prime factorized,

$$15 = 3 \times 5 \qquad 20 = 2 \times 2 \times 5$$

We require the smallest number into which both of the given numbers will divide. Consider the number of 2s required in the l.c.m. The highest number of 2s occurs in the factorization of 20. Hence the l.c.m. requires 2 factors of 2. Consider the number of 3s required. The highest number of 3s occurs in the factorization of 15. Hence the l.c.m. requires 1 factor of 3. Consider the number of 5s required. The highest number of 5s is 1 and so the l.c.m. requires 1 factor of 5. Hence the l.c.m. is $2 \times 2 \times 3 \times 5 = 60$.

1.15   Find the l.c.m. of 20, 24 and 25.

Solution   The numbers are prime factorized,

$$20 = 2 \times 2 \times 5 \quad 24 = 2 \times 2 \times 2 \times 3 \quad 25 = 5 \times 5$$

Consider the number of 2s required. The highest number of 2s required is 3 from factorizing 24. The highest number of 3s required is 1, again from factorizing 24. The highest number of 5s required is 2, found from factorizing 25. Hence the l.c.m. is given by

$$\text{l.c.m.} = 2 \times 2 \times 2 \times 3 \times 5 \times 5 = 600$$

Hence 600 is the smallest number into which 20, 24 and 25 will all divide exactly.

Self-assessment questions 1.4

1. Explain what is meant by the h.c.f. of a set of numbers.
2. Explain what is meant by the l.c.m. of a set of numbers.

Exercise 1.4

1. Calculate the h.c.f. of the following sets of numbers:
   (a) 12, 15, 21 (b) 16, 24, 40 (c) 28, 70, 120, 160 (d) 35, 38, 42 (e) 96, 120, 144

2. Calculate the l.c.m. of the following sets of numbers:
   (a) 5, 6, 8 (b) 20, 30 (c) 7, 9, 12 (d) 100, 150, 235 (e) 96, 120, 144

Test and assignment exercises 1

1. Evaluate
   (a) $6 \div 2 + 1$ (b) $6 \div (2 + 1)$
   (c) $12 + 4 \div 4$ (d) $(12 + 4) \div 4$
   (e) $3 \times 2 + 1$ (f) $3 \times (2 + 1)$
   (g) $6 - 2 + 4 \div 2$ (h) $(6 - 2 + 4) \div 2$
   (i) $6 - (2 + 4 \div 2)$ (j) $6 - (2 + 4) \div 2$
   (k) $2 \times 4 - 1$ (l) $2 \times (4 - 1)$
   (m) $2 \times 6 \div (3 - 1)$ (n) $2 \times (6 \div 3) - 1$
   (o) $2 \times (6 \div 3 - 1)$

2. Prime factorize (a) 56, (b) 39, (c) 74.

3. Find the h.c.f. of
   (a) 8, 12, 14
   (b) 18, 42, 66
   (c) 20, 24, 30
   (d) 16, 24, 32, 160

4. Find the l.c.m. of
   (a) 10, 15
   (b) 11, 13
   (c) 8, 14, 16
   (d) 15, 24, 30

2 Fractions

Objectives

This chapter

- explains what is meant by a fraction
- defines the terms 'improper fraction', 'proper fraction', and 'mixed fraction'
- explains how to write fractions in different but equivalent forms
- explains how to simplify fractions by cancelling common factors
- explains how to add, subtract, multiply and divide fractions

2.1 Introduction

The arithmetic of fractions is very important groundwork which must be mastered before topics in algebra such as formulas and equations can be understood. The same techniques which are used to manipulate fractions are used in these more advanced topics. You should use this chapter to ensure that you are confident at handling fractions before moving on to algebra. In all the examples and exercises it is important that you should carry out the calculations without the use of a calculator.

Fractions are numbers like $\frac{1}{2}, \frac{3}{4}, \frac{11}{8}$ and so on. In general a fraction is a number of the form $\frac{p}{q}$ where the letters p and q represent whole numbers or integers. The integer q can never be zero because it is never possible to divide by zero.

In any fraction $\frac{p}{q}$, the number p is called the **numerator** and the number q is called the **denominator**.

KEY POINT

$$\text{fraction} = \frac{\text{numerator}}{\text{denominator}} = \frac{p}{q}$$

Suppose that p and q are both positive numbers. If p is less than q, the fraction is said to be a **proper fraction**. So, $\frac{1}{2}$ and $\frac{3}{4}$ are proper fractions since the numerator is less than the denominator. If p is greater than or equal to q, the fraction is said to be **improper**. So, $\frac{11}{8}$, $\frac{7}{4}$ and $\frac{3}{3}$ are all improper fractions.

If either of p or q is negative, we simply ignore the negative sign when determining whether the fraction is proper or improper. So, $-\frac{3}{5}$, $\frac{-7}{21}$ and $\frac{4}{-21}$ are proper fractions, but $\frac{3}{-3}$, $\frac{-8}{2}$ and $-\frac{11}{2}$ are improper.

Note that all proper fractions have a value less than one.

Self-assessment questions 2.1

1. Explain the terms (a) fraction, (b) improper fraction, (c) proper fraction. In each case give an example of your own.
2. Explain the terms (a) numerator, (b) denominator.

Exercise 2.1

1. Classify each of the following as proper or improper:

   (a) $\frac{9}{17}$ (b) $\frac{-9}{17}$ (c) $\frac{8}{8}$ (d) $-\frac{7}{8}$ (e) $\frac{110}{77}$

2.2 Expressing a fraction in equivalent forms

Given a fraction, we may be able to express it in a different form. For example you will know that $\frac{1}{2}$ is equivalent to $\frac{2}{4}$. Note that multiplying both numerator and denominator by the same number leaves the value of the fraction unchanged. So, for example,

$$\frac{1}{2} = \frac{1 \times 2}{2 \times 2} = \frac{2}{4}$$

We say that $\frac{1}{2}$ and $\frac{2}{4}$ are **equivalent fractions**. Whilst they might look different, they have the same value.

Similarly, given the fraction $\frac{8}{12}$ we can divide both numerator and denominator by 4 to obtain

$$\frac{8}{12} = \frac{8/4}{12/4} = \frac{2}{3}$$

so $\frac{8}{12}$ and $\frac{2}{3}$ have the same value and are equivalent fractions.

| **KEY POINT** | Multiplying or dividing both numerator and denominator of a fraction by the same number produces a fraction having the same value, called an equivalent fraction. |

A fraction is in its **simplest form** when there are no factors common to both numerator and denominator. For example $\frac{5}{12}$ is in its simplest form, but $\frac{3}{6}$ is not since 3 is a factor common to both numerator and denominator. Its simplest form is the equivalent fraction $\frac{1}{2}$.

To express a fraction in its simplest form we look for factors which are common to both the numerator and denominator. This is done by prime factorizing both of these. Dividing both the numerator and denominator by any common factors removes them but leaves an equivalent fraction. This is equivalent to cancelling any common factors. For example, to simplify $\frac{4}{6}$ we prime factorize to produce

$$\frac{4}{6} = \frac{2 \times 2}{2 \times 3}$$

Dividing both numerator and denominator by 2 leaves $\frac{2}{3}$. This is equivalent to cancelling the common factor of 2.

Worked examples

2.1 Express $\frac{24}{36}$ in its simplest form.

Solution We seek factors common to both numerator and denominator. To do this we prime factorize 24 and 36.

Prime factorization has been described in §1.3.

$$24 = 2 \times 2 \times 2 \times 3 \qquad 36 = 2 \times 2 \times 3 \times 3$$

The factors $2 \times 2 \times 3$ are common to both 24 and 36 and so these may be cancelled. Note that only common factors may be cancelled when simplifying a fraction. Hence

Finding the highest common factor (h.c.f.) of two numbers is detailed in §1.4.

$$\frac{24}{36} = \frac{\cancel{2} \times \cancel{2} \times 2 \times \cancel{3}}{\cancel{2} \times \cancel{2} \times \cancel{3} \times 3} = \frac{2}{3}$$

In its simplest form $\frac{24}{36}$ is $\frac{2}{3}$. In effect we have divided 24 and 36 by 12, which is their h.c.f.

2.2 Express $\frac{49}{21}$ in its simplest form.

Solution Prime factorizing 49 and 21 gives

$$49 = 7 \times 7 \qquad 21 = 3 \times 7$$

Their h.c.f. is 7. Dividing 49 and 21 by 7 gives

$$\frac{49}{21} = \frac{7}{3}$$

Hence the simplest form of $\frac{49}{21}$ is $\frac{7}{3}$.

Before we can start to add and subtract fractions it is necessary to be able to convert fractions into a variety of equivalent forms. Work through the following examples.

Worked examples

2.3   Express $\frac{3}{4}$ as an equivalent fraction having a denominator of 20.

Solution   To achieve a denominator of 20, the existing denominator must be multiplied by 5. To produce an equivalent fraction both numerator and denominator must be multiplied by 5, so

$$\frac{3}{4} = \frac{3 \times 5}{4 \times 5} = \frac{15}{20}$$

2.4   Express 7 as an equivalent fraction with a denominator of 3.

Solution   Note that 7 is the same as the fraction $\frac{7}{1}$. To achieve a denominator of 3, the existing denominator must be multiplied by 3. To produce an equivalent fraction both numerator and denominator must be multiplied by 3, so

$$7 = \frac{7}{1} = \frac{7 \times 3}{1 \times 3} = \frac{21}{3}$$

Self-assessment questions 2.2

1. All integers can be thought of as fractions. True or false?
2. Explain the use of h.c.f. in the simplification of fractions.
3. Give an example of 3 fractions which are equivalent.

Exercise 2.2

1. Express the following fractions in their simplest form:

   (a) $\frac{18}{27}$ (b) $\frac{12}{20}$ (c) $\frac{15}{45}$ (d) $\frac{25}{80}$ (e) $\frac{15}{60}$ (f) $\frac{90}{200}$

   (g) $\frac{15}{20}$ (h) $\frac{2}{18}$ (i) $\frac{16}{24}$ (j) $\frac{30}{65}$ (k) $\frac{12}{21}$ (l) $\frac{100}{45}$

   (m) $\frac{6}{9}$ (n) $\frac{12}{16}$ (o) $\frac{13}{42}$ (p) $\frac{13}{39}$ (q) $\frac{11}{33}$ (r) $\frac{14}{30}$

   (s) $-\frac{12}{16}$ (t) $\frac{11}{-33}$ (u) $\frac{-14}{-30}$

2. Express $\frac{3}{4}$ as an equivalent fraction having a denominator 28.

3. Express 4 as an equivalent fraction with a denominator of 5.

4. Express $\frac{5}{12}$ as an equivalent fraction having a denominator of 36.

5. Express 2 as an equivalent fraction with a denominator of 4.

6. Express 6 as an equivalent fraction with a denominator of 3.

7. Express each of the fractions $\frac{2}{3}, \frac{5}{4}$ and $\frac{5}{6}$ as an equivalent fraction with a denominator of 12.

8. Express each of the fractions $\frac{4}{9}, \frac{1}{2}$ and $\frac{5}{6}$ as an equivalent fraction with a denominator of 18.

9. Express each of the following numbers as an equivalent fraction with a denominator of 12.

   (a) $\frac{1}{2}$ (b) $\frac{3}{4}$ (c) $\frac{5}{2}$ (d) 5 (e) 4 (f) 12

2.3 | **Addition and subtraction of fractions**

To add and subtract fractions we first rewrite each fraction so that they all have the same denominator. This is known as the **common denominator**. The denominator is chosen to be the lowest common multiple of the original denominators. Then, the numerators only are added or subtracted as appropriate, and the result is divided by the common denominator.

Worked examples

2.5   Find $\frac{2}{3} + \frac{5}{4}$.

Solution   The denominators are 3 and 4. The l.c.m. of 3 and 4 is 12. We need to express both fractions with a denominator of 12.

Finding the lowest common multiple (l.c.m.) is detailed in §1.4.

To express $\frac{2}{3}$ with a denominator of 12 we multiply both numerator and denominator by 4. Hence $\frac{2}{3}$ is the same as $\frac{8}{12}$. To express $\frac{5}{4}$ with a denominator of 12 we multiply both numerator and denominator by 3. Hence $\frac{5}{4}$ is the same as $\frac{15}{12}$. So

$$\frac{2}{3} + \frac{5}{4} = \frac{8}{12} + \frac{15}{12} = \frac{8+15}{12} = \frac{23}{12}$$

2.6   Find $\frac{4}{9} - \frac{1}{2} + \frac{5}{6}$.

Solution   The denominators are 9, 2 and 6. Their l.c.m. is 18. Each fraction is expressed with 18 as the denominator

$$\frac{4}{9} = \frac{8}{18} \qquad \frac{1}{2} = \frac{9}{18} \qquad \frac{5}{6} = \frac{15}{18}$$

Then

$$\frac{4}{9} - \frac{1}{2} + \frac{5}{6} = \frac{8}{18} - \frac{9}{18} + \frac{15}{18} = \frac{8-9+15}{18} = \frac{14}{18}$$

The fraction $\frac{14}{18}$ can be simplified to $\frac{7}{9}$. Hence

$$\frac{4}{9} - \frac{1}{2} + \frac{5}{6} = \frac{7}{9}$$

2.7   Find $\frac{1}{4} - \frac{5}{9}$.

Solution   The l.c.m. of 4 and 9 is 36. Each fraction is expressed with a denominator of 36. Thus

$$\frac{1}{4} = \frac{9}{36}, \qquad \text{and} \qquad \frac{5}{9} = \frac{20}{36}$$

Then

$$\begin{aligned}
\frac{1}{4} - \frac{5}{9} &= \frac{9}{36} - \frac{20}{36} \\
&= \frac{9 - 20}{36} \\
&= \frac{-11}{36} \\
&= -\frac{11}{36}
\end{aligned}$$

Consider the number $2\frac{3}{4}$. This is referred to as a **mixed fraction** because it contains a whole number part, 2, and a fractional part $\frac{3}{4}$. We can convert this mixed fraction into an improper fraction as follows. Recognize that 2 is equivalent to $\frac{8}{4}$, and so $2\frac{3}{4}$ is $\frac{8}{4} + \frac{3}{4} = \frac{11}{4}$.

The reverse of this process is to convert an improper fraction into a mixed fraction. Consider the improper fraction $\frac{11}{4}$. Now 4 divides into 11 twice leaving a remainder of 3; so $\frac{11}{4} = 2$ remainder 3, which we write as $2\frac{3}{4}$.

Worked example

2.8   (a) Express $4\frac{2}{5}$ as an improper fraction.
(b) Find $4\frac{2}{5} + \frac{1}{3}$.

Solution   (a) $4\frac{2}{5}$ is a mixed fraction. Note that $4\frac{2}{5}$ is equal to $4 + \frac{2}{5}$. We can write 4 as the equivalent fraction $\frac{20}{5}$. Therefore

$$\begin{aligned}
4\frac{2}{5} &= \frac{20}{5} + \frac{2}{5} \\
&= \frac{22}{5}
\end{aligned}$$

(b)
$$\begin{aligned}
4\frac{2}{5} + \frac{1}{3} &= \frac{22}{5} + \frac{1}{3} \\
&= \frac{66}{15} + \frac{5}{15} \\
&= \frac{71}{15}
\end{aligned}$$

Self-assessment questions 2.3

1. Explain the use of l.c.m. when adding and subtracting fractions.

Exercise 2.3

1. Find

   (a) $\dfrac{1}{4}+\dfrac{2}{3}$ (b) $\dfrac{3}{5}+\dfrac{5}{3}$ (c) $\dfrac{12}{14}-\dfrac{2}{7}$

   (d) $\dfrac{3}{7}-\dfrac{1}{2}+\dfrac{2}{21}$ (e) $1\dfrac{1}{2}+\dfrac{4}{9}$

   (f) $2\dfrac{1}{4}-1\dfrac{1}{3}+\dfrac{1}{2}$ (g) $\dfrac{10}{15}-1\dfrac{2}{5}+\dfrac{8}{3}$

   (h) $\dfrac{9}{10}-\dfrac{7}{16}+\dfrac{1}{2}-\dfrac{2}{5}$

2. Find

   (a) $\dfrac{7}{8}+\dfrac{1}{3}$ (b) $\dfrac{1}{2}-\dfrac{3}{4}$ (c) $\dfrac{3}{5}+\dfrac{2}{3}+\dfrac{1}{2}$

   (d) $\dfrac{3}{8}+\dfrac{1}{3}+\dfrac{1}{4}$ (e) $\dfrac{2}{3}-\dfrac{4}{7}$

   (f) $\dfrac{1}{11}-\dfrac{1}{2}$ (g) $\dfrac{3}{11}-\dfrac{5}{8}$

3. Express as improper fractions:

   (a) $2\dfrac{1}{2}$ (b) $3\dfrac{2}{3}$ (c) $10\dfrac{1}{4}$ (d) $5\dfrac{2}{7}$

   (e) $6\dfrac{2}{9}$ (f) $11\dfrac{1}{3}$ (g) $15\dfrac{1}{2}$ (h) $13\dfrac{3}{4}$

   (i) $12\dfrac{1}{11}$ (j) $13\dfrac{2}{3}$ (k) $56\dfrac{1}{2}$

4. Without using a calculator express these improper fractions as mixed fractions:

   (a) $\dfrac{10}{3}$ (b) $\dfrac{7}{2}$ (c) $\dfrac{15}{4}$ (d) $\dfrac{25}{6}$

2.4   Multiplication of fractions

The product of two or more fractions is found by multiplying their numerators to form a new numerator, and then multiplying their denominators to form a new denominator.

Worked examples

2.9   Find $\dfrac{4}{9}\times\dfrac{3}{8}$.

Solution   The numerators are multiplied: $4\times3=12$. The denominators are multiplied: $9\times8=72$. Hence

$$\frac{4}{9}\times\frac{3}{8}=\frac{12}{72}$$

This may now be expressed in its simplest form,

$$\frac{12}{72}=\frac{1}{6}$$

Hence

$$\frac{4}{9} \times \frac{3}{8} = \frac{1}{6}$$

An alternative, but equivalent, method is to cancel any factors common to both numerator and denominator at the outset:

$$\frac{4}{9} \times \frac{3}{8} = \frac{4 \times 3}{9 \times 8}$$

A factor of 4 is common to the 4 and the 8. Hence

$$\frac{4 \times 3}{9 \times 8} = \frac{1 \times 3}{9 \times 2}$$

A factor of 3 is common to the 3 and the 9. Hence

$$\frac{1 \times 3}{9 \times 2} = \frac{1 \times 1}{3 \times 2} = \frac{1}{6}$$

2.10   Find $\frac{12}{25} \times \frac{2}{7} \times \frac{10}{9}$.

Solution   We cancel factors common to both numerator and denominator. A factor of 5 is common to 10 and 25. Cancelling this gives

$$\frac{12}{25} \times \frac{2}{7} \times \frac{10}{9} = \frac{12}{5} \times \frac{2}{7} \times \frac{2}{9}$$

A factor of 3 is common to 12 and 9. Cancelling this gives

$$\frac{12}{5} \times \frac{2}{7} \times \frac{2}{9} = \frac{4}{5} \times \frac{2}{7} \times \frac{2}{3}$$

There are no more common factors. Hence

$$\frac{12}{25} \times \frac{2}{7} \times \frac{10}{9} = \frac{4}{5} \times \frac{2}{7} \times \frac{2}{3} = \frac{16}{105}$$

2.11   Find $\frac{3}{4}$ of $\frac{5}{9}$.

Recall that 'of' means multiply

Solution $\frac{3}{4}$ of $\frac{5}{9}$ is the same as $\frac{3}{4} \times \frac{5}{9}$. Cancelling a factor of 3 from numerator and denominator gives $\frac{1}{4} \times \frac{5}{3}$, that is $\frac{5}{12}$. Hence $\frac{3}{4}$ of $\frac{5}{9}$ is $\frac{5}{12}$.

2.12   Find $\frac{5}{6}$ of 70.

Solution   We can write 70 as $\frac{70}{1}$. So

$$\frac{5}{6} \text{ of } 70 = \frac{5}{6} \times \frac{70}{1} = \frac{5}{3} \times \frac{35}{1} = \frac{175}{3} = 58\frac{1}{3}$$

2.13 Find $2\frac{7}{8} \times \frac{2}{3}$.

Solution In this example the first fraction is a mixed fraction. We convert it to an improper fraction before performing the multiplication. Note that $2\frac{7}{8} = \frac{23}{8}$. Then

$$\frac{23}{8} \times \frac{2}{3} = \frac{23}{4} \times \frac{1}{3}$$
$$= \frac{23}{12}$$
$$= 1\frac{11}{12}$$

Self-assessment questions 2.4

1. Describe how to multiply fractions together.

Exercise 2.4

1. Evaluate

   (a) $\frac{2}{3} \times \frac{6}{7}$ (b) $\frac{8}{15} \times \frac{25}{32}$ (c) $\frac{1}{4} \times \frac{8}{9}$

   (d) $\frac{16}{17} \times \frac{34}{48}$ (e) $2 \times \frac{3}{5} \times \frac{5}{12}$ (f) $2\frac{1}{3} \times 1\frac{1}{4}$

   (g) $1\frac{3}{4} \times 2\frac{1}{2}$ (h) $\frac{3}{4} \times 1\frac{1}{2} \times 3\frac{1}{2}$

2. Evaluate

   (a) $\frac{2}{3}$ of $\frac{3}{4}$ (b) $\frac{4}{7}$ of $\frac{21}{30}$

   (c) $\frac{9}{10}$ of 80 (d) $\frac{6}{7}$ of 42

3. Is $\frac{3}{4}$ of $\frac{12}{15}$ the same as $\frac{12}{15}$ of $\frac{3}{4}$?

4. Find

   (a) $-\frac{1}{3} \times \frac{5}{7}$ (b) $\frac{3}{4} \times -\frac{1}{2}$

   (c) $\left(-\frac{5}{8}\right) \times \frac{8}{11}$ (d) $\left(-\frac{2}{3}\right) \times \left(-\frac{15}{7}\right)$

5. Find

   (a) $5\frac{1}{2} \times \frac{1}{2}$ (b) $3\frac{3}{4} \times \frac{1}{3}$

   (c) $\frac{2}{3} \times 5\frac{1}{9}$ (d) $\frac{3}{4} \times 11\frac{1}{2}$

6. Find

   (a) $\frac{3}{5}$ of $11\frac{1}{4}$ (b) $\frac{2}{3}$ of $15\frac{1}{2}$

   (c) $\frac{1}{4}$ of $-8\frac{1}{3}$

| 2.5 | **Division by a fraction** |

To divide one fraction by another fraction, we invert the second fraction and then multiply.

Worked examples

2.14   Find $\frac{6}{25} \div \frac{2}{5}$.

Solution   We invert $\frac{2}{5}$ to obtain $\frac{5}{2}$. Multiplication is then performed. So

$$\frac{6}{25} \div \frac{2}{5} = \frac{6}{25} \times \frac{5}{2} = \frac{3}{25} \times \frac{5}{1} = \frac{3}{5} \times \frac{1}{1} = \frac{3}{5}$$

2.15   Evaluate (a) $1\frac{1}{3} \div \frac{8}{3}$ (b) $\frac{20}{21} \div \frac{5}{7}$.

Solution   (a) Firstly we express $1\frac{1}{3}$ as an improper fraction.

$$1\frac{1}{3} = 1 + \frac{1}{3} = \frac{3}{3} + \frac{1}{3} = \frac{4}{3}$$

So we calculate $\frac{4}{3} \div \frac{8}{3} = \frac{4}{3} \times \frac{3}{8} = \frac{4}{8} = \frac{1}{2}$

Hence $1\frac{1}{3} \div \frac{8}{3} = \frac{1}{2}$

(b) $\frac{20}{21} \div \frac{5}{7} = \frac{20}{21} \times \frac{7}{5} = \frac{4}{21} \times \frac{7}{1} = \frac{4}{3}$

Self-assessment questions 2.5

1. Explain the process of division by a fraction.

Exercise 2.5

1. Evaluate

(a) $\frac{3}{4} \div \frac{1}{8}$ (b) $\frac{8}{9} \div \frac{4}{3}$ (c) $\frac{-2}{7} \div \frac{4}{21}$

(d) $\frac{9}{4} \div 1\frac{1}{2}$ (e) $\frac{5}{6} \div \frac{5}{12}$ (f) $\frac{99}{100} \div 1\frac{4}{5}$

(g) $3\frac{1}{4} \div 1\frac{1}{8}$ (h) $\left(2\frac{1}{4} \div \frac{3}{4}\right) \times 2$

(i) $2\frac{1}{4} \div \left(\frac{3}{4} \times 2\right)$ (j) $6\frac{1}{4} \div 2\frac{1}{2} + 5$

(k) $6\frac{1}{4} \div \left(2\frac{1}{2} + 5\right)$

Test and assignment exercises 2

1. Evaluate

   (a) $\dfrac{3}{4} + \dfrac{1}{6}$ (b) $\dfrac{2}{3} + \dfrac{3}{5} - \dfrac{1}{6}$

   (c) $\dfrac{5}{7} - \dfrac{2}{3}$ (d) $2\dfrac{1}{3} - \dfrac{9}{10}$

   (e) $5\dfrac{1}{4} + 3\dfrac{1}{6}$ (f) $\dfrac{9}{8} - \dfrac{7}{6} + 1$

   (g) $\dfrac{5}{6} - \dfrac{5}{3} + \dfrac{5}{4}$ (h) $\dfrac{4}{5} + \dfrac{1}{3} - \dfrac{3}{4}$

2. Evaluate

   (a) $\dfrac{4}{7} \times \dfrac{21}{32}$ (b) $\dfrac{5}{6} \times \dfrac{8}{15}$

   (c) $\dfrac{3}{11} \times \dfrac{20}{21}$ (d) $\dfrac{9}{14} \times \dfrac{8}{18}$

   (e) $\dfrac{5}{4} \div \dfrac{10}{13}$ (f) $\dfrac{7}{16} \div \dfrac{21}{32}$

   (g) $\dfrac{-24}{25} \div \dfrac{51}{50}$ (h) $\dfrac{45}{81} \div \dfrac{25}{27}$

3. Evaluate the following expressions using the BODMAS rule.

   (a) $\dfrac{1}{2} + \dfrac{1}{3} \times 2$ (b) $\dfrac{3}{4} \times \dfrac{2}{3} + \dfrac{1}{4}$

   (c) $\dfrac{5}{6} \div \dfrac{2}{3} + \dfrac{3}{4}$ (d) $\left(\dfrac{2}{3} + \dfrac{1}{4}\right) \div 4 + \dfrac{3}{5}$

   (e) $\left(\dfrac{4}{3} - \dfrac{2}{5} \times \dfrac{1}{3}\right) \times \dfrac{1}{4} + \dfrac{1}{2}$

   (f) $\dfrac{3}{4}$ of $\left(1 + \dfrac{2}{3}\right)$

   (g) $\dfrac{2}{3}$ of $\dfrac{1}{2} + 1$ (h) $\dfrac{1}{5} \times \dfrac{2}{3} + \dfrac{2}{5} \div \dfrac{4}{5}$

4. Express in their simplest form:

   (a) $\dfrac{21}{84}$ (b) $\dfrac{6}{80}$ (c) $\dfrac{34}{85}$ (d) $\dfrac{22}{143}$ (e) $\dfrac{69}{253}$

3 Decimal fractions

Objectives

This chapter

- revises the decimal number system
- shows how to write a number to a given number of significant figures
- shows how to write a number to a given number of decimal places

3.1 Decimal numbers

Consider the whole number 478. We can regard it as the sum

$$400 + 70 + 8$$

In this way we see that in the number 478, the 8 represents 8 ones, or 8 units, the 7 represents 7 tens, or 70, and the number 4 represents 4 hundreds or 400. Thus, we have the system of hundreds, tens and units familiar from early years in school. All whole numbers can be thought of in this way.

When we wish to deal with proper fractions and mixed fractions, we extend the hundreds, tens and units system as follows. A **decimal point**, '.', marks the end of the whole number part, and the numbers which follow it, to the right, form the fractional part.

A number immediately to the right of the decimal point, that is, in the **first decimal place**, represents tenths, so

$$.1 = \frac{1}{10}$$
$$.2 = \frac{2}{10} \text{ or } \frac{1}{5}$$
$$.3 = \frac{3}{10} \qquad \text{and so on}$$

When there are no whole numbers involved it is usual to write a zero in front of the decimal point, thus, .2 would be written 0.2.

Worked example

3.1 Express the following decimal numbers as proper fractions in their simplest form.

(a) 0.4   (b) 0.5   (c) 0.6

Solution The first number after the decimal point represents tenths.

(a) $0.4 = \frac{4}{10}$ which simplifies to $\frac{2}{5}$

(b) $0.5 = \frac{5}{10}$ or simply $\frac{1}{2}$

(c) $0.6 = \frac{6}{10} = \frac{3}{5}$

Frequently we will deal with numbers having a whole number part and a fractional part. Thus

$$5.2 = 5 \text{ units } + 2 \text{ tenths}$$
$$= 5 + \frac{2}{10}$$
$$= 5 + \frac{1}{5}$$
$$= 5\frac{1}{5}$$

Similarly,

$$175.8 = 175\frac{8}{10} = 175\frac{4}{5}$$

Numbers in the second position after the decimal point, or the **second decimal place**, represent hundredths so

$$0.01 = \frac{1}{100}$$
$$0.02 = \frac{2}{100} \quad \text{or} \quad \frac{1}{50}$$
$$0.03 = \frac{3}{100} \quad \text{and so on}$$

In fact we can regard any numbers occupying the first two decimal places as hundredths so that

$$0.25 = \frac{25}{100} \qquad \text{or simply} \qquad \frac{1}{4}$$

$$0.50 = \frac{50}{100} \qquad \text{or } \frac{1}{2}$$

$$0.75 = \frac{75}{100} = \frac{3}{4}$$

Worked examples

3.2   Express the following decimal numbers as proper fractions in their simplest form:

(a) 0.35   (b) 0.56   (c) 0.68

Solution   The first two decimal places represent hundredths:

(a) $0.35 = \frac{35}{100} = \frac{7}{20}$

(b) $0.56 = \frac{56}{100} = \frac{14}{25}$

(b) $0.68 = \frac{68}{100} = \frac{17}{25}$

3.3   Express 37.25 as a mixed fraction in its simplest form.

Solution
$$37.25 = 37 + 0.25$$
$$= 37 + \frac{25}{100}$$
$$= 37 + \frac{1}{4}$$
$$= 37\frac{1}{4}$$

Numbers in the third position after the decimal point, or **third decimal place**, represent thousandths so

$$0.001 = \frac{1}{1000}$$

$$0.002 = \frac{2}{1000} \qquad \text{or } \frac{1}{500}$$

$$0.003 = \frac{3}{1000} \qquad \text{and so on}$$

In fact we can regard any numbers occupying the first three positions after the decimal point as thousandths so that

$$0.356 = \frac{356}{1000} \quad \text{or} \quad \frac{89}{250}$$

$$0.015 = \frac{15}{1000} \quad \text{or} \quad \frac{3}{200}$$

$$0.075 = \frac{75}{1000} = \frac{3}{40}$$

Worked example

3.4 Write each of the following as a decimal number:

(a) $\frac{3}{10} + \frac{7}{100}$ (b) $\frac{8}{10} + \frac{3}{1000}$

Solution (a) $\frac{3}{10} + \frac{7}{100} = 0.37$

(b) $\frac{8}{10} + \frac{3}{1000} = 0.803$

You will normally use a calculator to add, subtract, multiply and divide decimal numbers. Generally the more decimal places used, the more accurately we can state a number. This idea is developed in the next section.

Self-assessment questions 3.1

1. State which is the largest and which is the smallest of the following numbers:
   23.001, 23.0, 23.00001, 23.0008, 23.01
2. Which is the largest of the following numbers?
   0.1, 0.02, 0.003, 0.0004, 0.00005

Exercise 3.1

1. Express the following decimal numbers as proper fractions in their simplest form:

   (a) 0.7 (b) 0.8 (c) 0.9

2. Express the following decimal numbers as proper fractions in their simplest form:

   (a) 0.55 (b) 0.158 (c) 0.98 (d) 0.099

3. Express each of the following as a mixed fraction in its simplest form:

   (a) 4.6 (b) 5.2 (c) 8.05 (d) 11.59
   (e) 121.09

4. Write each of the following as a decimal number:

   (a) $\frac{6}{10} + \frac{9}{100} + \frac{7}{1000}$ (b) $\frac{8}{100} + \frac{3}{1000}$

   (c) $\frac{17}{1000} + \frac{5}{10}$

| 3.2 | **Significant figures and decimal places** |

The accuracy with which we state a number often depends upon the context in which the number is being used. The volume of a petrol tank is usually given to the nearest litre. It is of no practical use to give such a volume to the nearest cubic centimetre.

When writing a number we often give the accuracy by stating the **number of significant figures** or the **number of decimal places** used. These terms are now explained.

Significant figures

Suppose we are asked to write down the number nearest to 857 using at most two non-zero digits, or numbers. We would write 860. This number is nearer to 857 than any other number with two non-zero digits. We say that 857 to 2 **significant figures** is 860. The words 'significant figures' are usually abbreviated to s.f. Because 860 is larger than 857 we say that the 857 has been **rounded up** to 860.

To write a number to three significant figures we can use no more than three non-zero digits. For example, the number closest to 1784 which has no more than three non-zero digits is 1780. We say that 1784 to 3 significant figures is 1780. In this case, because 1780 is less than 1784 we say that 1784 has been **rounded down** to 1780.

Worked examples

3.5   Write down the number nearest to 86 using only one non-zero digit. Has 86 been rounded up or down?

Solution   The number 86 written to one significant figure is 90. This number is nearer to 86 than any other number having only one non-zero digit. 86 has been rounded up to 90.

3.6   Write down the number nearest to 999 which uses only one non-zero digit.

Solution   The number 999 to one significant figure is 1000. This number is nearer to 999 than any other number having only one non-zero digit.

We now explain the process of writing to a given number of significant figures.

When asked to write a number to, say, three significant figures, 3 s.f., the first step is to look at the first four digits. If asked to write a number to two significant figures we look at the first three digits and so on. We always look at one more digit than the number of significant figures required.

For example, to write 6543.19 to 2 s.f. we would consider the number 6540.00; the digits 3, 1 and 9 are effectively ignored. The next step is to round up or down. If the final digit is a 5 or more then we round up by increasing the previous digit by 1. If the final digit is 4 or less we round down by leaving the previous digit unchanged. Hence when considering 6543.19 to 2 s.f., the 4 in the third place means that we round down to 6500.

To write 23865 to 3 s.f. we would consider the number 23860. The next step is to increase the 8 to a 9. Thus 23865 is rounded up to 23900.

Zeros at the beginning of a number are ignored. To write 0.004693 to 2 s.f. we would firstly consider the number 0.00469. Note that the zeros at the beginning of the number have not been counted. We then round the 6 to a 7, producing 0.0047.

The following examples illustrate the process.

Worked examples

3.7   Write 36.482 to 3 s.f.

Solution   We consider the first 4 digits, that is, 36.48. The final digit is 8 and so we round up 36.48 to 36.5. To 3 s.f. 36.482 is 36.5.

3.8   Write 1.0049 to 4 s.f.

Solution   To write to 4 s.f. we consider the first 5 digits, that is, 1.0049. The final digit is a 9 and so 1.0049 is rounded up to 1.005.

3.9   Write 695.3 to 2 s.f.

Solution   We consider 695. The final digit is a 5 and so we round up. We cannot round up the 9 to a 10 and so the 69 is rounded up to 70. Hence to 2 s.f. the number is 700.

3.10   Write 0.0473 to 1 s.f.

Solution   We do not count the initial zeros and consider 0.047. The final digit tells us to round up. Hence to 1 s.f. we have 0.05.

3.11   A number is given to 2 s.f. as 67.

(a) What is the maximum value the number could have?

(b) What is the minimum value the number could have?

Solution   (a) To 2 s.f. 67.5 is 68. Any number just below 67.5, for example 67.49 or 67.499, to 2 s.f. is 67. Hence the maximum value of the number is 67.4999....

(b) To 2 s.f. 66.4999... is 66. However, 66.5 to 2 s.f. is 67. The minimum value of the number is thus 66.5.

Decimal places

When asked to write a number to 3 decimal places (3 d.p.) we consider the first 4 decimal places, that is, numbers after the decimal point. If asked to write to 2 d.p. we consider the first 3 decimal places and so on. If the final digit is 5 or more we round up, otherwise we round down.

Worked examples

3.12   Write 63.4261 to 2 d.p.

Solution   We consider the number to 3 d.p., that is, 63.426. The final digit is 6 and so we round up 63.426 to 63.43. Hence 63.4261 to 2 d.p. is 63.43.

3.13   Write 1.97 to 1 d.p.

Solution   In order to write to 1 d.p. we consider the number to 2 d.p., that is, we consider 1.97. The final digit is a 7 and so we round up. The 9 cannot be rounded up and so we look at 1.9. This can be rounded up to 2.0. Hence 1.97 to 1 d.p. is 2.0. Note that it is crucial to write 2.0 and not simply 2, as this shows that the number is written to 1 d.p.

3.14   Write −6.0439 to 2 d.p.

Solution   We consider −6.043. As the final digit is a 3 the number is rounded down to −6.04.

Self-assessment questions 3.2

1. Explain the meaning of 'significant figures'.
2. Explain the process of writing a number to so many decimal places.

Exercise 3.2

1. Write to 3 s.f.

   (a) 6962 (b) 70.406 (c) 0.0123
   (d) 0.010991 (e) 45.607 (f) 2345

2. Write 65.999 to

   (a) 4 s.f. (b) 3 s.f. (c) 2 s.f. (d) 1 s.f.
   (e) 2 d.p. (f) 1 d.p.

3. Write 9.99 to

   (a) 1 s.f. (b) 1 d.p.

4. Write 65.4555 to

   (a) 3 d.p. (b) 2 d.p. (c) 1 d.p. (d) 5 s.f.

   (e) 4 s.f. (f) 3 s.f. (g) 2 s.f. (h) 1 s.f.

Test and assignment exercises 3

1. Express the following numbers as proper fractions in their simplest form:

   (a) 0.74 (b) 0.96 (c) 0.05 (d) 0.25

2. Express each of the following as a mixed fraction in its simplest form:

   (a) 2.5 (b) 3.25 (c) 3.125 (d) 6.875

3. Write each of the following as a decimal number:

   (a) $\frac{3}{10} + \frac{1}{100} + \frac{7}{1000}$

   (b) $\frac{5}{1000} + \frac{9}{100}$

   (c) $\frac{4}{1000} + \frac{9}{10}$

4. Write 0.09846 to (a) 1 d.p, (b) 2 s.f., (c) 1 s.f.

5. Write 9.513 to (a) 3 s.f., (b) 2 s.f., (c) 1 s.f.

6. Write 19.96 to (a) 1 d.p., (b) 2 s.f., (c) 1 s.f.

4 Percentage and ratio

Objectives

This chapter

- explains the terms 'percentage' and 'ratio'
- shows how to perform calculations using percentages and ratios

4.1 Percentage

A **percentage** is a fraction whose denominator is 100. We use the per cent symbol, %, to represent a percentage, for example $\frac{17}{100}$ may be written as 17%.

Worked example

4.1 Express $\frac{19}{100}$ and $\frac{35}{100}$ as percentages.

Solution Both of these fractions have a denominator of 100. So, we can write simply

$$\frac{19}{100} = 19\%, \quad \text{and} \quad \frac{35}{100} = 35\%$$

Sometimes it is necessary to convert a fraction whose denominator is not 100 into a percentage. This can be done in two ways. The first way is to express the fraction as an equivalent fraction having a denominator of 100. Thus the fraction $\frac{2}{5}$ can be written as

$$\frac{2}{5} = \frac{2 \times 20}{5 \times 20} = \frac{40}{100}$$

$\frac{2}{5}$ and $\frac{40}{100}$ are equivalent fractions. We see that $\frac{2}{5} = 40\%$.

The second way in which a fraction can be converted into a percentage is to multiply the fraction by 100 and then label the result as a percentage. Thus to convert $\frac{2}{5}$ we write

$$\frac{2}{5} \times 100 = \frac{200}{5} = 40$$

and so $\frac{2}{5} = 40\%$.

KEY POINT

> To convert a fraction into a percentage, multiply by 100 and label the result as a percentage.

Worked examples

4.2 Convert $\frac{5}{8}$ into a percentage.

Solution We multiply $\frac{5}{8}$ by 100 to obtain

$$\frac{5}{8} \times 100 = \frac{500}{8} = 62.5$$

The result is labelled as a percentage. Thus $\frac{5}{8} = 62.5\%$.

4.3 Bill scores $\frac{13}{17}$ in a test. Mary scores $\frac{14}{19}$. Express the scores as percentages.

Solution To convert $\frac{13}{17}$ into a percentage we multiply by 100 and label the result as a percentage. Thus

$$\frac{13}{17} \times 100 = \frac{1300}{17} = 76.5 \,(1\text{d.p.})$$

So, $\frac{13}{17} = 76.5\%$.

Similarly,

$$\frac{14}{19} \times 100 = \frac{1400}{19} = 73.7 \,(1\text{d.p.})$$

So $\frac{14}{19} = 73.7\%$.

In this form it is easier to make a comparison of the two marks.

4.4 Express 30% as a decimal.

Solution
$$30\% = \frac{30}{100} = 0.3$$

Hence 30% is equivalent to 0.3.

Some common percentage–fraction conversions are given here for reference.

$10\% = \frac{1}{10}$, $25\% = \frac{1}{4}$, $50\% = \frac{1}{2}$, $75\% = \frac{3}{4}$

Recall from §1.2 that 'of' means multiply.

We are often asked to calculate a percentage of a quantity. For example, find 50% of 160, or find 10% of 95. Such calculations frequently arise when calculating discounts on prices etc. Since $50\% = \frac{1}{2}$ we find

$$50\% \text{ of } 160 = \frac{1}{2} \times 160 = 80$$

$$10\% \text{ of } 95 = \frac{1}{10} \times 95 = 9.5$$

Worked examples

4.5   Calculate 27% of 90.

Solution   Recall that 'of' means multiply.

$$27\% \text{ of } 90 = \frac{27}{100} \times 90 = \frac{27}{10} \times 9 = 24.3$$

From example 4.6 we note that 100% of a number is simply the number itself.

4.6   Calculate 100% of 6.

Solution $$100\% \text{ of } 6 = \frac{100}{100} \times 6 = 1 \times 6 = 6$$

4.7   A deposit of £750 increases by 9%. Calculate the resulting deposit.

Solution   We calculate 9% of 750.

$$9\% \text{ of } 750 = \frac{9}{100} \times 750 = 67.5$$

The deposit has increased by £67.5. The resulting deposit is 750 + 67.5 = £817.5.

Alternatively we may perform the calculation as follows. The original deposit represents 100%. The deposit increases by 9% to 109% of the original. So the resulting deposit is 109% of £750.

$$109\% \text{ of } 750 = \frac{109}{100} \times 750 = 817.5$$

4.8   A television set is advertised at £315. The retailer offers a 10% discount. How much do you pay for the television?

Solution

$$10\% \text{ of } 315 = \frac{10}{100} \times 315 = 31.5$$

The discount is £31.5 and so the cost is $315 - 31.5 = £283.5$.

Using a calculator

Many calculators have the facility to calculate percentages of quantities directly although care must be taken with the order in which you enter the numbers. For example, to find 27% of 90, you might type

$$\boxed{9}\boxed{0}\boxed{\times}\boxed{2}\boxed{7}\boxed{\%}$$

and obtain the answer 24.3. Check that you can use your calculator to find percentages, and ask your tutor for help if necessary.

Self-assessment questions 4.1

1. Give one reason why it is sometimes useful to express fractions as percentages.

Exercise 4.1

1. Calculate 23% of 124.

2. Express the following as percentages:
   (a) $\frac{9}{11}$ (b) $\frac{15}{20}$ (c) $\frac{9}{10}$ (d) $\frac{45}{50}$ (e) $\frac{75}{90}$

3. Express $\frac{13}{12}$ as a percentage.

4. Calculate 217% of 500.

5. A worker earns £400 a week. She receives a 6% increase. Calculate her new weekly wage.

6. A debt of £1200 is decreased by 17%. Calculate the remaining debt.

7. Express the following percentages as decimals:
   (a) 50% (b) 36% (c) 75% (d) 100% (e) 12.5%

8. A compact disc player normally priced at £256 is reduced in a sale by 20%. Calculate the sale price.

9. A bank deposit earns 7.5% interest in one year. Calculate the interest earned on a deposit of £15 000.

| 4.2 | **Ratio** |

Ratios are simply an alternative way of expressing fractions. Consider the problem of dividing £200 between 2 people, Ann and Bill, in the ratio of 7:3. This means that Ann receives £7 for every £3 that Bill receives. So, every £10 is divided as £7 to Ann and £3 to Bill. Thus Ann receives $\frac{7}{10}$ of the money. Now $\frac{7}{10}$ of £200 is $\frac{7}{10} \times 200 = £140$. So Ann receives £140 and Bill receives £60.

Worked example

4.9   Divide 170 in the ratio 3:2.

Solution   A ratio of 3:2 means that every 5 parts are split as 3 and 2, that is, the first number is $\frac{3}{5}$ of the total; the second number is $\frac{2}{5}$ of the total.

$$\frac{3}{5} \text{ of } 170 = \frac{3}{5} \times 170 = 102$$

$$\frac{2}{5} \text{ of } 170 = \frac{2}{5} \times 170 = 68$$

The number is divided into 102 and 68.

Note from Example 4.9 that to split a number in a given ratio we first find the total number of parts. The total number of parts is found by adding the numbers in the ratio. For example, if the ratio is given as $m : n$, the total number of parts is $m + n$. Then, these $m + n$ parts are split into two with the first number being $\frac{m}{m+n}$ of the total, and the second number being $\frac{n}{m+n}$ of the total. Compare this with Example 4.9.

Worked example

4.10   Divide 250 cm in the ratio 1:3:4.

Solution   Every 8 cm is divided into 1 cm, 3 cm and 4 cm. Thus, the first length is $\frac{1}{8}$ of the total, the second length is $\frac{3}{8}$ of the total and the third length is $\frac{4}{8}$ of the total.

$$\frac{1}{8} \text{ of } 250 = \frac{1}{8} \times 250 = 31.25$$

$$\frac{3}{8} \text{ of } 250 = \frac{3}{8} \times 250 = 93.75$$

$$\frac{4}{8} \text{ of } 250 = \frac{4}{8} \times 250 = 125$$

The 250 cm length is divided into 31.25 cm, 93.75 cm and 125 cm.

Ratios can be written in different ways. The ratio 3:2 can also be written as 6:4. This is clear if we note that 6:4 is a total of ten parts split as $\frac{6}{10}$ and $\frac{4}{10}$ of the total. Since $\frac{6}{10}$ is equivalent to $\frac{3}{5}$, and $\frac{4}{10}$ is equivalent to $\frac{2}{5}$, we see that 6:4 is equivalent to 3:2.

Generally, any ratio can be expressed as an equivalent ratio by multiplying or dividing each term in the ratio by the same number. So, for example

$$5:3 \text{ is equivalent to } 15:9$$

and

$$\frac{3}{4}:2 \text{ is equivalent to } 3:8$$

Worked examples

4.11   Divide a mass of 380 kg in the ratio $\frac{3}{4}:\frac{1}{5}$.

Solution   It is simpler to work with whole numbers, so first of all we produce an equivalent ratio by multiplying each term, first by 4, and then by 5, to give

$$\frac{3}{4}:\frac{1}{5} = 3:\frac{4}{5} = 15:4$$

Note that this is equivalent to multiplying through by the lowest common multiple of 4 and 5.

So dividing 380 kg in the ratio $\frac{3}{4}:\frac{1}{5}$ is equivalent to dividing it in the ratio 15:4.

Now the total number of parts is 19 and so we split the 380 kg mass as

$$\frac{15}{19} \times 380 = 300$$

and

$$\frac{4}{19} \times 380 = 80$$

The total mass is split into 300 kg and 80 kg.

4.12   Bell metal, which is a form of bronze, is used for casting bells. It is an alloy of copper and tin. To manufacture bell metal requires 17 parts of copper to every 3 parts of tin.

(a) Express this requirement as a ratio.

(b) Express the amount of tin required as a percentage of the total.

(c) If the total amount of tin in a particular casting is 150 kg, find the amount of copper.

Solution   (a) Copper and tin are needed in the ratio 17:3.

(b) $\frac{3}{20}$ of the alloy is tin. Since $\frac{3}{20} = 15\%$ we find that 15% of the alloy is tin.

(c) A mass of 150 kg of tin makes up 15% of the total. So 1% of the total would have a mass of 10 kg. Copper which makes up 85% will have a mass of 850 kg.

Self-assessment questions 4.2

1. Dividing a number in the ratio 2:3 is the same as dividing it in the ratio of 10:15. True or false?

Exercise 4.2

1. Divide 180 in the ratio 8:1:3.

2. Divide 930 cm in the ratio 1:1:3.

3. A 6 m length of wood is cut in the ratio 2:3:4. Calculate the length of each piece.

4. Divide 1200 in the ratio 1:2:3:4.

5. A sum of £2600 is divided between Alan, Bill and Claire in the ratio of $2\frac{3}{4}:1\frac{1}{2}:2\frac{1}{4}$. Calculate the amount that each receives.

6. A mass of 40 kg is divided into three portions in the ratio 3:4:8. Calculate the mass of each portion.

7. Express the following ratios in their simplest forms:

   (a) 12:24 (b) 3:6 (c) 3:6:12 (d) $\frac{1}{3}:7$

8. A box contains two sizes of nails. The ratio of long nails to short nails is 2:7. Calculate the number of each type if the total number of nails is 108.

Test and assignment exercises 4

1. Express as decimals

   (a) 8% (b) 18% (c) 65%

2. Express as percentages

   (a) $\frac{3}{8}$ (b) $\frac{79}{100}$ (c) $\frac{56}{118}$

3. Calculate 27.3% of 1496.

4. Calculate 125% of 125.

5. Calculate 85% of 0.25.

6. Divide 0.5 in the ratio of 2:4:9.

7. A bill totals £234.5 to which is added tax at 17.5%. Calculate the amount of tax to be paid.

8. An inheritance is divided between 3 people in the ratio 4:7:2. If the least amount received is £2300 calculate how much the other two people received.

9. Divide 70 in the ratio of 0.5:1.3:2.1.

10. Divide 50% in the ratio 2:3.

5 Algebra

Objectives

This chapter

- explains what is meant by 'algebra'
- introduces important algebraic notations
- explains what is meant by a 'power' or 'index'
- explains what is meant by a 'formula'

5.1 What is algebra?

In order to extend the techniques of arithmetic so that they can be more useful in applications we introduce letters or **symbols** to represent quantities of interest. For example, we may choose the capital letter I to stand for the *interest rate* in a business calculation, or the lower case letter t to stand for the *time* in a scientific calculation, and so on. The choice of which letter to use for which quantity is largely up to the user although some conventions have been developed. Very often the letters x and y are used to stand for arbitrary quantities. **Algebra** is the body of mathematical knowledge which has been developed to manipulate symbols. Some symbols take fixed and unchanging values, and these are known as **constants**. For example, suppose we let the symbol b stand for the boiling point of water. This is fixed at 100 °C and so b is a constant. Some symbols represent quantities which can vary, and these are called **variables.** For example, the velocity of a car might be represented by the symbol v, and might vary from 0 to 100 kilometres per hour.

Algebraic notation

In algebraic work particular attention must be paid to the type of symbol used, so that, for example, the symbol T is quite different from the

Table 5.1
The Greek alphabet

A	α	alpha	I	ι	iota	P	ρ	rho
B	β	beta	K	κ	kappa	Σ	σ	sigma
Γ	γ	gamma	Λ	λ	lambda	T	τ	tau
Δ	δ	delta	M	μ	mu	Y	υ	upsilon
E	ϵ	epsilon	N	ν	nu	Φ	ϕ	phi
Z	ζ	zeta	Ξ	ξ	xi	X	χ	chi
H	η	eta	O	o	omicron	Ψ	ψ	psi
Θ	θ	theta	Π	π	pi	Ω	ω	omega

Your scientific calculator is preprogrammed with the value of π. Check that you can use it.

symbol t. Usually the symbols chosen are letters from the English alphabet although we frequently meet Greek letters. You may already be aware that the Greek letter 'pi' which has the symbol π is used in the formula for the area of a circle, and is equal to the constant $3.14159\ldots$. In many calculations π can be approximated by $\frac{22}{7}$. For reference the full Greek alphabet is given in Table 5.1.

Another important feature is the position of a symbol in relation to other symbols. As we shall see in this chapter, the quantities xy, x^y, y^x and x_y all can mean quite different things. When a symbol is placed to the right and slightly higher than another symbol it is referred to as a **superscript**. So, the quantity x^y contains the superscript y. Likewise, if a symbol is placed to the right and slightly lower than another symbol it is called a **subscript**. The quantity x_1 contains the subscript 1.

The arithmetic of symbols

Addition $(+)$ If the letters x and y stand for two numbers, their **sum** is written as $x + y$. Note that $x + y$ is the same as $y + x$ just as $4 + 7$ is the same as $7 + 4$.

Subtraction $(-)$ The quantity $x - y$ is called the **difference** of x and y, and means the number y subtracted from the number x. Note that $x - y$ is not the same as $y - x$, in the same way that $5 - 3$ is different from $3 - 5$.

Multiplication $(\times)$ Five times the number x is written $5 \times x$, although when multiplying the $\times$ sign is sometimes replaced with '.', or is even left out altogether. This means that $5 \times x$, $5.x$ and $5x$ all mean five times the number x. Similarly $x \times y$ can be written $x.y$ or simply xy. When multiplying, the order of the symbols is not important so that xy

is the same as yx just as 5×4 is the same as 4×5. The quantity xy is also known as the **product** of x and y.

Division ($\div$) $x \div y$ means the number x divided by the number y. This is also written x/y. Here the order is important and x/y is quite different from y/x. An expression involving one symbol divided by another is known as an **algebraic fraction.** The top line is called the **numerator** and the bottom line is called the **denominator.** The quantity x/y is known as the **quotient** of x and y.

A quantity made up of symbols together with $+$, $-$, $\times$ or $\div$ is called an **algebraic expression.** When evaluating an algebraic expression the BODMAS rule given in Chapter 1 applies. This rule reminds us of the correct order in which to evaluate an expression.

Self-assessment questions 5.1

1. Explain what you understand by the term 'algebra'.
2. If m and n are two numbers, explain what is meant by mn.
3. What is an algebraic fraction? Explain the meaning of the terms 'numerator' and 'denominator'.
4. What is the distinction between a superscript and a subscript?
5. What is the distinction between a variable and a constant?

5.2   Powers, or indices

Frequently we shall need to multiply a number by itself several times. For example $3 \times 3 \times 3$, or $a \times a \times a \times a$.

To abbreviate such quantities a new notation is introduced. $a \times a \times a$ is written a^3, pronounced 'a cubed'. The superscript 3 is called a **power** or **index** and the letter a is called the **base.** Similarly $a \times a$ is written a^2, pronounced 'a squared' or 'a raised to the power 2'.

The power is the number of times that the base is multiplied by itself. Most calculators have a button marked x^y which can be used to evaluate expressions such as 2^8, 3^{11} and so on. Check to see if your calculator can do these by verifying that $2^8 = 256$ and $3^{11} = 177147$. Note that the plural of index is **indices.**

As a^2 means $a \times a$, and a^3 means $a \times a \times a$, then we interpret a^1 as simply a. That is, any number raised to the power 1 is itself.

The calculator button x^y is used to find powers of numbers.

KEY POINT

Any number raised to the power 1 is itself, that is $a^1 = a$.

Worked examples

5.1   In the expression 3^8 identify the index and the base.

Solution   In the expression 3^8, the index is 8 and the base is 3.

5.2   Explain what is meant by y^5.

Solution y^5 means $y \times y \times y \times y \times y$.

5.3   Explain what is meant by $x^2 y^3$.

Solution x^2 means $x \times x$; y^3 means $y \times y \times y$. Therefore $x^2 y^3$ means
$x \times x \times y \times y \times y$.

5.4   Evaluate 2^3 and 3^4.

Solution 2^3 means $2 \times 2 \times 2$, that is, 8. Similarly 3^4 means $3 \times 3 \times 3 \times 3$, that is 81.

5.5   Explain what is meant by 7^1.

Solution   Any number to the power 1 is itself, that is 7^1 is simply 7.

5.6   Evaluate 10^2 and 10^3.

Solution 10^2 means 10×10 or 100. Similarly 10^3 means $10 \times 10 \times 10$ or 1000.

5.7   Use indices to write the expression $a \times a \times b \times b \times b$ more compactly.

Solution $a \times a$ can be written a^2; $b \times b \times b$ can be written b^3. Therefore
$a \times a \times b \times b \times b$ can be written as $a^2 \times b^3$ or simply $a^2 b^3$.

5.8   Write out fully $z^3 y^2$.

Solution $z^3 y^2$ means $z \times z \times z \times y \times y$. Note we could also write this as $zzzyy$.

We now consider how to deal with expressions involving not only powers but other operations as well. Recall from §5.1 that the BODMAS rule tells us the order in which operations should be carried out, but the rule makes no reference to powers. In fact, powers should be given higher priority than any other operation and evaluated first. Consider the expression -4^2. Because the power must be evaluated first -4^2 is equal to -16. On the other hand $(-4)^2$ means $(-4) \times (-4)$ which is equal to $+16$.

Worked examples

5.9   Simplify (a) -5^2, (b) $(-5)^2$.

Solution   (a) The power is evaluated first. Noting that $5^2 = 25$, we see that $-5^2 = -25$.

Recall that when a negative number is multiplied by another negative number the result is positive.

(b) $(-5)^2$ means $(-5) \times (-5) = +25$.

Note how the brackets can significantly change the meaning of an expression.

5.10   Explain the meanings of $-x^2$ and $(-x)^2$. Are these different?

Solution   In the expression $-x^2$ it is the quantity x which is squared, so that $-x^2 = -(x \times x)$. On the other hand $(-x)^2$ means $(-x) \times (-x)$ which equals $+x^2$. The two expressions are not the same.

Following the previous two examples we emphasize again the importance of the position of brackets in an expression.

Self-assessment questions 5.2

1. Explain the meaning of the terms 'power' and 'base'.
2. What is meant by an index?
3. Explain the distinction between $(xyz)^2$ and xyz^2.
4. Explain the distinction between $(-3)^4$ and -3^4.

Exercise 5.2

1. Evaluate the following without using a calculator: 2^4, $\left(\frac{1}{2}\right)^2$, 1^8, 3^5 and 0^3.

2. Evaluate 10^4, 10^5 and 10^6 without using a calculator.

3. Use a calculator to evaluate 11^4, 16^8, 39^4 and 1.5^7.

4. Write out fully (a) $a^4 b^2 c$ and (b) $xy^2 z^4$.

5. Write the following expressions compactly using indices:

   (a) $xxxyyx$ (b) $xxyyzzz$
   (c) $xyzxyz$ (d) $abccba$

6. Using a calculator, evaluate

   (a) 7^4 (b) 7^5 (c) $7^4 \times 7^5$ (d) 7^9

   (e) 8^3 (f) 8^7 (g) $8^3 \times 8^7$ (h) 8^{10}

Can you spot a rule for multiplying numbers with powers?

7. Without using a calculator, find $(-3)^3$, $(-2)^2$, $(-1)^7$ and $(-1)^4$.

8. Use a calculator to find $(-16.5)^3$, $(-18)^2$ and $(-0.5)^5$.

9. Without using a calculator find

   (a) $(-6)^2$ (b) $(-3)^2$ (c) $(-4)^3$
   (d) $(-2)^3$.

   Carefully compare your answers with the results of finding, -6^2, -3^2, -4^3 and -2^3.

5.3   Substitution and formulae

Substitution means replacing letters by actual numerical values.

Worked examples

5.11 Find the value of a^4 when $a = 3$.

Solution a^4 means $a \times a \times a \times a$. When we **substitute** the number 3 in place of the letter a we find 3^4 or $3 \times 3 \times 3 \times 3$, that is, 81.

5.12 Find the value of $a + 7b + 3c$ when $a = 1$, $b = 2$ and $c = 3$.

Solution Letting $b = 2$ we note that $7b = 14$. Letting $c = 3$ we note that $3c = 9$. Therefore, with $a = 1$,

$$a + 7b + 3c = 1 + 14 + 9 = 24$$

5.13 If $x = 4$, find the value of (a) $8x^3$ and (b) $(8x)^3$.

Solution (a) Substituting $x = 4$ into $8x^3$ we find $8 \times 4^3 = 8 \times 64 = 512$.
(b) Substituting $x = 4$ into $(8x)^3$ we obtain $(32)^3 = 32768$. Note that the use of brackets makes a significant difference to the result.

5.14 Evaluate mk, mn and nk when $m = 5$, $n = -4$ and $k = 3$.

Solution $mk = 5 \times 3 = 15$. Similarly $mn = 5 \times (-4) = -20$ and $nk = (-4) \times 3 = -12$.

5.15 Find the value of $-7x$ when (a) $x = 2$ and (b) $x = -2$.

Solution (a) Substituting $x = 2$ into $-7x$ we find -7×2 which equals -14.
(b) Substituting $x = -2$ into $-7x$ we find -7×-2 which equals 14.

5.16 Find the value of x^2 when $x = -3$.

Solution Because x^2 means $x \times x$, its value when $x = -3$ is -3×-3, that is $+9$.

5.17 Find the value of $-x^2$ when $x = -3$.

Solution Recall that a power is evaluated first. So $-x^2$ means $-(x \times x)$. When $x = -3$, this evaluates to $-(-3 \times -3) = -9$.

5.18 Find the value of $x^2 + 3x$ when (a) $x = 2$, (b) $x = -2$.

Solution (a) Letting $x = 2$ we find

$$x^2 + 3x = (2)^2 + 3(2) = 4 + 6 = 10$$

(b) Letting $x = -2$ we find

$$x^2 + 3x = (-2)^2 + 3(-2) = 4 - 6 = -2$$

5.19 Find the value of $\frac{3x^2}{4} + 5x$ when $x = 2$.

Solution Letting $x = 2$ we find

$$\frac{3x^2}{4} + 5x = \frac{3(2)^2}{4} + 5(2)$$
$$= \frac{12}{4} + 10$$
$$= 13$$

5.20 Find the value of $\frac{x^3}{4}$ when $x = 0.5$.

Solution When $x = 0.5$ we find

$$\frac{x^3}{4} = \frac{0.5^3}{4} = 0.03125$$

A **formula** is used to relate two or more quantities. You may already be familiar with the common formula used to find the area of a rectangle:

area = length × breadth

In symbols, writing A for area, l for length and b for breadth we have

$$A = l \times b \qquad \text{or simply } A = lb$$

If we are now given particular numerical values for l and b we can use this formula to find A.

Worked examples

5.21 Use the formula $A = lb$ to find A when $l = 10$ and $b = 2.5$.

Solution Substituting the values $l = 10$ and $b = 2.5$ into the formula $A = lb$ we find $A = 10 \times 2.5 = 25$.

5.22 The formula $V = IR$ is used by electrical engineers. Find the value of V when $I = 12$ and $R = 7$.

Solution Substituting $I = 12$ and $R = 7$ in $V = IR$ we find $V = 12 \times 7 = 84$.

5.23   Use the formula $y = x^2 + 3x + 4$ to find y when $x = -2$.

Solution   Substituting $x = -2$ into the formula gives

$$y = (-2)^2 + 3(-2) + 4 = 4 - 6 + 4 = 2$$

Self-assessment questions 5.3

1. What is the distinction between an algebraic expression and a formula?

Exercise 5.3

1. Evaluate $3x^2y$ when $x = 2$ and $y = 5$.

2. Evaluate $8x + 17y - 2z$ when $x = 6$, $y = 1$ and $z = -2$.

3. The area A of a circle is found from the formula $A = \pi r^2$ where r is the length of the radius. Taking π to be 3.142 find the areas of the circles whose radii, in centimetres, are (a) $r = 10$, (b) $r = 3$, (c) $r = 0.2$.

4. Evaluate $3x^2$ and $(3x)^2$ when $x = 4$.

5. Evaluate $5x^2$ and $(5x)^2$ when $x = -2$.

6. If $y = 4.85$ find

   (a) $7y$ (b) y^2 (c) $5y + 2.5$ (d) $y^3 - y$

7. If $a = 12.8$, $b = 3.6$ and $c = 9.1$ find

   (a) $a + b + c$ (b) ab (c) bc (d) abc

8. If $C = \frac{5}{9}(F - 32)$, find C when $F = 100$.

9. Evaluate (a) x^2, (b) $-x^2$ and (c) $(-x)^2$, when $x = 7$.

10. Evaluate the following when $x = -2$:

    (a) x^2 (b) $(-x)^2$ (c) $-x^2$ (d) $3x^2$
    (e) $-3x^2$ (f) $(-3x)^2$

11. Evaluate the following when $x = -3$:

    (a) $\frac{x^2}{3}$ (b) $(-x)^2$ (c) $-\left(\frac{x}{3}\right)^2$ (d) $4x^2$
    (e) $-4x^2$ (f) $(-4x)^2$

12. Evaluate $x^2 - 7x + 2$ when $x = -9$.

13. Evaluate $2x^2 + 3x - 11$ when $x = -3$.

14. Evaluate $-x^2 + 3x - 5$ when $x = -1$.

15. Evaluate $-9x^2 + 2x$ when $x = 0$.

16. Evaluate $5x^2 + x + 1$ when (a) $x = 3$, (b) $x = -3$, (c) $x = 0$, (d) $x = -1$.

17. Evaluate $\frac{2x^2}{3} - \frac{x}{2}$ when

    (a) $x = 6$ (b) $x = -6$ (c) $x = 0$ (d) $x = 1$

18. Evaluate $\frac{4x^2}{5} + 3$ when

    (a) $x = 0$ (b) $x = 1$ (c) $x = 5$ (d) $x = -5$

19. Evaluate $\frac{x^3}{2}$ when

    (a) $x = -1$ (b) $x = 2$ (c) $x = 4$

20. Use the formula $y = \frac{x^3}{2} + 3x^2$ to find y when

    (a) $x = 0$ (b) $x = 2$ (c) $x = 3$ (d) $x = -1$

21. If $g = 2t^2 - 1$, find g when

    (a) $t = 3$ (b) $t = 0.5$ (c) $t = -2$

22. In business calculations, the simple interest earned on an investment, I, is calculated from the formula $I = Prn$ where P is the amount invested, r is the interest rate and n is the number of time periods. Evaluate I when

    (a) $P = 15\,000$, $r = 0.08$ and $n = 5$
    (b) $P = 12\,500$, $r = 0.075$ and $n = 3$.

23. An investment earning 'compound interest' has a value, S given by $S = P(1 + r)^n$, where P is the amount invested, r is the interest rate and n is the number of time periods. Calculate S when

    (a) $P = 8\,250$, $r = 0.05$ and $n = 15$
    (b) $P = 125\,000$, $r = 0.075$ and $n = 11$.

Test and assignment exercises 5

1. Using a calculator, evaluate 44^3, 0.44^2 and 32.5^3.

2. Write the following compactly using indices:

    (a) $xxxyyyy$ (b) $\dfrac{xxx}{yyyy}$ (c) a^2baab

3. Evaluate the expression $4x^3yz^2$ when $x = 2$, $y = 5$ and $z = 3$.

4. The circumference C of a circle which has a radius of length r is given by the formula $C = 2\pi r$. Find the circumference of the circle with radius 0.5 cm. Take $\pi = 3.142$.

5. Find (a) $21^2 - 16^2$, (b) $(21 - 16)^2$. Comment upon the result.

6. If $x = 4$ and $y = -3$, evaluate

    (a) xy (b) $\dfrac{x}{y}$ (c) $\dfrac{x^2}{y^2}$ (d) $\left(\dfrac{x}{y}\right)^2$

7. Evaluate $2x(x + 4)$ when $x = 7$.

8. Evaluate $4x^2 + 7x$ when $x = 9$.

9. Evaluate $3x^2 - 7x + 12$ when $x = -2$.

10. Evaluate $-x^2 - 11x + 1$ when $x = -3$.

11. The formula $I = V/R$ is used by engineers. Find I when $V = 10$ and $R = 0.01$.

12. Given the formula $A = 1/x$, find A when (a) $x = 1$, (b) $x = 2$, (c) $x = 3$.

13. From the formula $y = 1/(x^2 + x)$ find y when (a) $x = 1$, (b) $x = -1$, (c) $x = 3$.

14. Find the value of $(-1)^n$ (a) when n is an even natural number and (b) when n is an odd natural number. (A natural number is a positive whole number.)

15. Find the value of $(-1)^{n+1}$ (a) when n is an even natural number and (b) when n is an odd natural number.

6 Indices

Objectives

This chapter

- states three laws used for manipulating indices
- shows how expressions involving indices can be simplified using the three laws
- explains the use of negative powers
- explains square roots, cube roots and fractional powers
- revises multiplication and division by powers of ten
- explains 'scientific notation' for representing very large and very small numbers

6.1 The laws of indices

Recall from Chapter 5 that an index is simply a power and that the plural of index is indices. Expressions involving indices can often be simplified if use is made of the **laws of indices**.

The first law

$$a^m \times a^n = a^{m+n}$$

In words, this states that if two numbers involving the same base but possibly different indices are to be multiplied together, their indices are added. Note that this law can only be applied if both bases are the same.

KEY POINT

The first law: $a^m \times a^n = a^{m+n}$.

Worked examples

6.1   Use the first law of indices to simplify $a^4 \times a^3$.

Solution   Using the first law we have $a^4 \times a^3 = a^{4+3} = a^7$. Note that the same result could be obtained by actually writing out all the terms:
$a^4 \times a^3 = (a \times a \times a \times a) \times (a \times a \times a) = a^7$.

6.2   Use the first law of indices to simplify $3^4 \times 3^5$.

Solution   From the first law $3^4 \times 3^5 = 3^{4+5} = 3^9$.

6.3   Simplify $a^4 a^7 b^2 b^4$.

Solution $a^4 a^7 b^2 b^4 = a^{4+7} b^{2+4} = a^{11} b^6$. Note that only those quantities with the same base can be combined using the first law.

The second law

$$\frac{a^m}{a^n} = a^{m-n}$$

In words, this states that if two numbers involving the same base but possibly different indices are to be divided, their indices are subtracted.

KEY POINT

The second law: $\dfrac{a^m}{a^n} = a^{m-n}$.

Worked examples

6.4   Use the second law of indices to simplify $\dfrac{a^5}{a^3}$.

Solution   The second law states that we subtract the indices, that is,
$\dfrac{a^5}{a^3} = a^{5-3} = a^2$.

6.5   Use the second law of indices to simplify $\dfrac{3^7}{3^4}$.

Solution   From the second law $\dfrac{3^7}{3^4} = 3^{7-4} = 3^3$.

6.6   Using the second law of indices, simplify $\dfrac{x^3}{x^3}$.

Solution   Using the second law of indices we have $\dfrac{x^3}{x^3} = x^{3-3} = x^0$

However, note that any expression divided by itself equals 1, and so $\frac{x^3}{x^3}$ must equal 1. We can conclude from this that any number raised to the power 0 equals 1.

KEY POINT

Any number raised to the power 0 equals 1, that is, $a^0 = 1$.

Worked example

6.7 Evaluate (a) 14^0 (b) 0.5^0.

Solution (a) Any number to the power zero equals 1 and so $14^0 = 1$.

(b) Similarly, $0.5^0 = 1$.

The third law

$$(a^m)^n = a^{mn}$$

If a number is raised to a power, and the result is itself raised to a power, then the two powers are multiplied together.

KEY POINT

The third law: $(a^m)^n = a^{mn}$.

Worked examples

6.8 Simplify $(3^2)^4$.

Solution The third law states that the two powers are multiplied:
$(3^2)^4 = 3^{2\times4} = 3^8$.

6.9 Simplify $(x^4)^3$.

Solution Using the third law $(x^4)^3 = x^{4\times3} = x^{12}$.

6.10 Remove the brackets from the expression $(2a^2)^3$.

Solution $(2a^2)^3$ means $(2a^2) \times (2a^2) \times (2a^2)$. We can write this as

$$2 \times 2 \times 2 \times a^2 \times a^2 \times a^2$$

or simply $8a^6$. We could obtain the same result by noting that both terms in the bracket, that is, the 2 and the a^2, must be raised to the power 3, that is,

$$(2a^2)^3 = 2^3(a^2)^3 = 8a^6$$

The result of the previous example can be generalized to any term of the form $(a^m b^n)^k$. To simplify such an expression we make use of the formula $(a^m b^n)^k = a^{mk} b^{nk}$

KEY POINT

$$(a^m b^n)^k = a^{mk} b^{nk}$$

Worked example

6.11   Remove the brackets from the expression $(x^2 y^3)^4$.

Solution   Using the previous result we find

$$(x^2 y^3)^4 = x^8 y^{12}$$

Self-assessment questions 6.1

1. State the three laws of indices.
2. Explain what is meant by a^0.
3. Explain what is meant by x^1.

Exercise 6.1

1. Simplify

   (a) $5^7 \times 5^{13}$ (b) $9^8 \times 9^5$
   (c) $11^2 \times 11^3 \times 11^4$

2. Simplify

   (a) $\dfrac{15^3}{15^2}$ (b) $\dfrac{4^{18}}{4^9}$ (c) $\dfrac{5^{20}}{5^{19}}$

3. Simplify

   (a) $a^7 a^3$ (b) $a^4 a^5$ (c) $b^{11} b^{10} b$

4. Simplify

   (a) $x^7 \times x^8$ (b) $y^4 \times y^8 \times y^9$

5. Explain why the laws of indices cannot be used to simplify $19^8 \times 17^8$.

6. Simplify

   (a) $(7^3)^2$ (b) $(4^2)^8$ (c) $(7^9)^2$

7. Simplify $\dfrac{1}{(5^3)^8}$

8. Simplify

   (a) $(x^2 y^3)(x^3 y^2)$ (b) $(a^2 bc^2)(b^2 ca)$

9. Remove the brackets from

   (a) $(x^2 y^4)^5$ (b) $(9x^3)^2$ (c) $(-3x)^3$
   (d) $(-x^2 y^3)^4$

10. Simplify

    (a) $\dfrac{(z^2)^3}{z^3}$ (b) $\dfrac{(y^3)^2}{(y^2)^2}$ (c) $\dfrac{(x^3)^2}{(x^2)^3}$

6.2	**Negative powers**

Sometimes a number is raised to a negative power. This is interpreted as follows:

$$a^{-m} = \frac{1}{a^m}$$

This can also be rearranged and expressed in the form

$$a^m = \frac{1}{a^{-m}}$$

KEY POINT

$$a^{-m} = \frac{1}{a^m}, \qquad a^m = \frac{1}{a^{-m}}$$

For example,

$$3^{-2} \text{ means } \frac{1}{3^2}, \text{ that is, } \frac{1}{9}$$

Similarly,

the number $\dfrac{1}{5^{-2}}$ can be written 5^2, or simply 25

To see the justification for this, note that because any number raised to the power 0 equals 1 we can write

$$\frac{1}{a^m} = \frac{a^0}{a^m}$$

Using the second law of indices to simplify the right-hand side we obtain $\frac{a^0}{a^m} = a^{0-m} = a^{-m}$ so that $\frac{1}{a^m}$ is the same as a^{-m}.

Worked examples

6.12 Evaluate

(a) 2^{-5} (b) $\dfrac{1}{3^{-4}}$

Solution (a) $2^{-5} = \dfrac{1}{2^5} = \dfrac{1}{32}$ (b) $\dfrac{1}{3^{-4}} = 3^4$ or simply 81

6.13 Evaluate (a) 10^{-1} (b) 10^{-2}

Solution (a) 10^{-1} means $\frac{1}{10^1}$, or simply $\frac{1}{10}$. It is important to recognize that 10^{-1} is therefore the same as 0.1.

(b) 10^{-2} means $\frac{1}{10^2}$ or $\frac{1}{100}$. So 10^{-2} is therefore the same as 0.01.

6.14 Rewrite each of the following expressions using only positive powers:

(a) 7^{-3} (b) x^{-5}

Solution (a) 7^{-3} means the same as $\frac{1}{7^3}$. The expression has now been written using a positive power.

(b) $x^{-5} = \frac{1}{x^5}$.

6.15 Rewrite each of the following expressions using only positive powers:

(a) $\dfrac{1}{x^{-9}}$ (b) $\dfrac{1}{a^{-4}}$

Solution (a) $\dfrac{1}{x^{-9}} = x^9$ (b) $\dfrac{1}{a^{-4}} = a^4$

6.16 Rewrite each of the following using only negative powers:

(a) 6^8 (b) x^5 (c) z^a

Solution (a) $6^8 = \dfrac{1}{6^{-8}}$ (b) $x^5 = \dfrac{1}{x^{-5}}$ (c) $z^a = \dfrac{1}{z^{-a}}$

6.17 Simplify (a) $x^{-2}x^7$ (b) $\dfrac{x^{-3}}{x^{-5}}$

Solution (a) To simplify $x^{-2}x^7$ we can use the first law of indices to write it as $x^{-2+7} = x^5$.

(b) To simplify $\frac{x^{-3}}{x^{-5}}$ we can use the second law of indices to write it as $x^{-3-(-5)} = x^{-3+5} = x^2$.

6.18 Simplify (a) $(x^{-3})^5$ (b) $\dfrac{1}{(x^{-2})^2}$

Solution (a) To simplify $(x^{-3})^5$ we can use the third law of indices and write it as $x^{-3\times5} = x^{-15}$. The answer could also be written as $\frac{1}{x^{15}}$.

(b) Note that $(x^{-2})^2 = x^{-4}$ using the third law. So $\frac{1}{(x^{-2})^2} = \frac{1}{x^{-4}}$. This could also be written as x^4.

Self-assessment questions 6.2

1. Explain how the negative power in a^{-m} is interpreted.

Exercise 6.2

1. Without using a calculator express each of the following as a proper fraction:

   (a) 2^{-2} (b) 2^{-3} (c) 3^{-2} (d) 3^{-3} (e) 5^{-2}

   (f) 4^{-2} (g) 9^{-1} (h) 11^{-2} (i) 7^{-1}

2. Express each of the following as decimal fractions:

   (a) 10^{-1} (b) 10^{-2} (c) 10^{-6}

   (d) $\frac{1}{10^2}$ (e) $\frac{1}{10^3}$ (f) $\frac{1}{10^4}$

3. Write each of the following using only a positive power:

   (a) x^{-4} (b) $\frac{1}{x^{-5}}$ (c) x^{-7} (d) y^{-2} (e) $\frac{1}{y^{-1}}$

   (f) y^{-1} (g) y^{-2} (h) z^{-1} (i) $\frac{1}{z^{-1}}$

4. Simplify the following using the laws of indices and write your results using only positive powers:

   (a) $x^{-2}x^{-1}$ (b) $x^{-3}x^{-2}$ (c) x^3x^{-4}

   (d) $x^{-4}x^9$ (e) $\frac{x^{-2}}{x^{11}}$ (f) $(x^{-4})^2$

   (g) $(x^{-3})^3$ (h) $(x^2)^{-2}$

5. Simplify

   (a) $a^{13}a^{-2}$ (b) $x^{-9}x^{-7}$ (c) $x^{-21}x^2x$

   (d) $(4^{-3})^2$

6. Evaluate

   (a) 10^{-3} (b) 10^{-4} (c) 10^{-5}

7. Evaluate $4^{-8}/4^{-6}$ and $3^{-5}/3^{-8}$ without using a calculator.

6.3　Square roots, cube roots and fractional powers

Square roots

Consider the relationship between the numbers 5 and 25. We know that $5^2 = 25$ and so 25 is the square of 5. Equivalently we say that 5 is a **square root** of 25. The symbol $\sqrt[2]{}$, or simply $\sqrt{}$ is used to denote a square root and we write

$$5 = \sqrt{25}$$

We can picture this as follows:

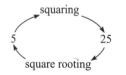

From this we see that taking the square root can be thought of as reversing the process of squaring.

We also note that

$$(-5) \times (-5) = (-5)^2$$
$$= 25$$

and so -5 is also a square root of 25. Hence we can write

$$-5 = \sqrt{25}$$

We can write both results together by using the 'plus or minus' sign $\pm$. We write

$$\sqrt{25} = \pm 5$$

In general, a **square root** of a number is a number which when squared gives the original number. Note that there are two square roots of any positive number but negative numbers possess no square roots.

Most calculators enable you to find square roots although only the positive value is normally given. Look for a $\sqrt{\ }$ button on your calculator.

Worked example

6.19 (a) Use your calculator to find $\sqrt{79}$ correct to four decimal places.

(b) Check your answers are correct by squaring them.

Solution (a) Using the $\sqrt{\ }$ button on the calculator you should verify that

$$\sqrt{79} = 8.8882 \text{ (to four decimal places)}$$

The second square root is -8.8882. Thus we can write
$$\sqrt{79} = \pm 8.8882.$$

(b) Squaring either of the numbers ± 8.8882 we recover the original number, 79.

Cube roots

The **cube root** of a number is a number which when cubed gives the original number. The symbol for a cube root is $\sqrt[3]{\ }$. So, for example, since $2^3 = 8$ we can write $\sqrt[3]{8} = 2$.

We can picture this as follows:

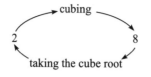

We can think of taking the cube root as reversing the process of cubing. As another example we note that $(-2)^3 = -8$ and hence $\sqrt[3]{-8} = -2$. All numbers, both positive and negative, possess a single cube root.

Your calculator may enable you to find a cube root. Look for a button marked $\sqrt[3]{}$. If so, check that you can use it correctly by verifying that

$$\sqrt[3]{46} = 3.5830$$

Fourth, fifth and other roots are defined in a similar way. For example, since

$$8^5 = 32768$$

we can write

$$\sqrt[5]{32768} = 8$$

Fractional powers

Sometimes fractional powers are used. The following example helps us to interpret a fractional power.

Worked example

6.20 Simplify (a) $x^{\frac{1}{2}}x^{\frac{1}{2}}$ (b) $x^{\frac{1}{3}}x^{\frac{1}{3}}x^{\frac{1}{3}}$

Use your results to interpret the fractional powers $\frac{1}{2}$ and $\frac{1}{3}$.

Solution   (a) Using the first law we can write

$$x^{\frac{1}{2}}x^{\frac{1}{2}} = x^{\frac{1}{2}+\frac{1}{2}} = x^1 = x$$

(b) Similarly,

$$x^{\frac{1}{3}}x^{\frac{1}{3}}x^{\frac{1}{3}} = x^{\frac{1}{3}+\frac{1}{3}+\frac{1}{3}} = x^1 = x$$

From (a) we see that

$$(x^{\frac{1}{2}})^2 = x$$

So when $x^{\frac{1}{2}}$ is squared, the result is x. Thus $x^{\frac{1}{2}}$ is simply the square root of x, that is

$$x^{\frac{1}{2}} = \sqrt{x}$$

Similarly, from (b)

$$(x^{\frac{1}{3}})^3 = x$$

and so $x^{\frac{1}{3}}$ is the cube root of x, that is

$$x^{\frac{1}{3}} = \sqrt[3]{x}$$

KEY POINT

$$x^{\frac{1}{2}} = \sqrt{x}, \quad x^{\frac{1}{3}} = \sqrt[3]{x}$$

More generally we have the following result:

KEY POINT

$$x^{\frac{1}{n}} = \sqrt[n]{x}$$

Your scientific calculator will probably be able to find fractional powers. The button may be marked $x^{1/y}$. Check that you can use it correctly by working through the following example.

Worked examples

6.21   Evaluate to three decimal places, using a calculator

(a) $3^{\frac{1}{4}}$ (b) $15^{1/5}$

Solution   Use your calculator to obtain the following solutions:

(a) 1.316 (b) 1.719

Note in part (a) that although the calculator gives just a single fourth root, there is another, -1.316.

6.22   Evaluate $(81)^{1/2}$.

Solution $(81)^{1/2} = \sqrt{81} = \pm 9$.

6.23   Explain what is meant by the number $27^{1/3}$.

Solution $27^{1/3}$ can be written $\sqrt[3]{27}$, that is, the cube root of 27. The cube root of 27 is 3, since $3 \times 3 \times 3 = 27$, and so $27^{1/3} = 3$. Note also that since $27 = 3^3$ we can write

$$(27)^{1/3} = (3^3)^{1/3} = 3^{(3 \times 1/3)} \quad \text{using the third law}$$
$$= 3^1 \quad = 3$$

The following worked example shows how we deal with negative fractional powers.

6.24   Explain what is meant by the number $(81)^{-1/2}$.

Solution   Recall from our work on negative powers that $a^{-m} = 1/a^m$. Therefore we can write $(81)^{-1/2}$ as $1/(81)^{1/2}$. Now $81^{1/2} = \sqrt{81} = \pm 9$ and so

$$(81)^{-1/2} = \frac{1}{\pm 9} = \pm \frac{1}{9}$$

6.25   Write each of the following using a single index.

(a) $(5^2)^{\frac{1}{3}}$ (b) $(5^{-2})^{\frac{1}{3}}$

Solution   (a) Using the third law of indices we find

$$(5^2)^{\frac{1}{3}} = 5^{2 \times \frac{1}{3}} = 5^{\frac{2}{3}}$$

Note that $(5^2)^{\frac{1}{3}}$ is the cube root of 5^2, that is $\sqrt[3]{25}$, or 2.9240.

(b) Using the third law of indices we find

$$(5^{-2})^{\frac{1}{3}} = 5^{-2 \times \frac{1}{3}} = 5^{-\frac{2}{3}}$$

Note that there is a variety of equivalent ways in which this can be expressed, for example, $\sqrt[3]{\frac{1}{5^2}}$ or $\sqrt[3]{\frac{1}{25}}$, or as $\frac{1}{5^{2/3}}$.

6.26   Write each of the following using a single index.

(a) $\sqrt{x^3}$ (b) $(\sqrt{x})^3$

Solution   (a) Because the square root of a number can be expressed as that number raised to the power $\frac{1}{2}$ we can write

$$\sqrt{x^3} = (x^3)^{\frac{1}{2}}$$
$$= x^{3 \times \frac{1}{2}} \qquad \text{using the third law}$$
$$= x^{\frac{3}{2}}$$

(b) $(\sqrt{x})^3 = (x^{\frac{1}{2}})^3$

$\qquad\qquad = x^{\frac{3}{2}}$ $\qquad\qquad$ using the third law

Note from this example that $\sqrt{x^3} = (\sqrt{x})^3$.

Note that by generalizing the results of the two previous worked examples we have the following:

KEY POINT $\quad \boxed{a^{\frac{m}{n}} = \sqrt[n]{a^m} = (\sqrt[n]{a})^m}$

Self-assessment questions 6.3

1. Explain the meaning of the fractional powers $x^{1/2}$ and $x^{1/3}$.
2. What are the square roots of 100? Explain why the number -100 does not have any square roots.

Exercise 6.3

1. Evaluate

    (a) $64^{1/3}$ (b) $144^{1/2}$ (c) $16^{-1/4}$ (d) $25^{-1/2}$

    (e) $\dfrac{1}{32^{-1/5}}$

2. Simplify and then evaluate

    (a) $(3^{-1/2})^4$ (b) $(8^{1/3})^{-1}$

3. Write each of the following using a single index:

    (a) $\sqrt{8}$ (b) $\sqrt[3]{12}$ (c) $\sqrt[4]{16}$ (d) $\sqrt{13^3}$

    (e) $\sqrt[3]{4^7}$

4. Write each of the following using a single index:

    (a) $\sqrt{x}$ (b) $\sqrt[3]{y}$ (c) $\sqrt[2]{x^5}$ (d) $\sqrt[3]{5^7}$

6.4 $\qquad$ **Multiplication and division by powers of 10**

To multiply and divide decimal fractions by powers of 10 is particularly simple. For example, to multiply 256.875 by 10 the decimal point is moved one place to the right, that is

$\qquad 256.875 \times 10 = 2568.75$

To multiply by 100 the decimal point is moved two places to the right. So

$$256.875 \times 100 = 25687.5$$

To divide a number by 10, the decimal point is moved one place to the left. This is equivalent to multiplying by 10^{-1}. To divide by 100, the decimal point is moved two places to the left. This is equivalent to multiplying by 10^{-2}.

In general, to multiply a number by 10^a, the decimal point is moved a places to the right if a is a positive integer, and a places to the left if a is a negative integer. If necessary, additional zeros are inserted to make up the required number of digits. Consider the following example.

Worked example

6.27 Without the use of a calculator, write down

(a) 75.45×10^3 (b) 0.056×10^{-2} (c) 96.3×10^{-3} (d) 0.00743×10^5

Solution (a) The decimal point is moved three places to the right: $75.45 \times 10^3 = 75450$. It has been necessary to include an additional zero to make up the required number of digits.

(b) The decimal point is moved two places to the left. $0.056 \times 10^{-2} = 0.00056$.

(c) $96.3 \times 10^{-3} = 0.0963$.

(d) $0.00743 \times 10^5 = 743$.

Exercise 6.4

1. Without the use of a calculator write down:

(a) 7.43×10^2 (b) 7.43×10^4
(c) 0.007×10^4 (d) 0.07×10^{-2}

2. Write each of the following as a multiple of 10^2.

(a) 300 (b) 356 (c) 32 (d) 0.57

| 6.5 | **Scientific notation** |

It is often necessary to use very large numbers such as 65000000000 or very small numbers such as 0.000000001. **Scientific notation** can be used to express such numbers in a more concise form which avoids writing very lengthy strings of numbers. Each number is written in the form

$$a \times 10^n$$

where a is usually a number between 1 and 10. We also make use of the fact that

$$10 = 10^1, \quad 100 = 10^2, \quad 1000 = 10^3 \text{ and so on}$$

and also that

$$10^{-1} = \frac{1}{10} = 0.1, \quad 10^{-2} = \frac{1}{100} = 0.01 \text{ and so on}$$

Then, for example,

the number 4000 can be written $4 \times 1000 = 4 \times 10^3$

Similarly

the number 68,000 can be written $6.8 \times 10,000 = 6.8 \times 10^4$

and

the number 0.09 can be written $9 \times 0.01 = 9 \times 10^{-2}$

Note that all three numbers have been written in the form $a \times 10^n$ where a lies between 1 and 10.

Worked examples

6.28 Express the following numbers in scientific notation:

(a) 54 (b) −276 (c) 0.3

Solution (a) 54 can be written as 5.4×10, so in scientific notation we have 5.4×10^1.

(b) Negative numbers cause no problem: $-276 = -2.76 \times 10^2$.

(c) We can write 0.3 as 3×0.1 or 3×10^{-1}.

6.29 Write out fully the following numbers:

(a) 2.7×10^{-1} (b) 9.6×10^5 (c) -8.2×10^2

Solution (a) $2.7 \times 10^{-1} = 0.27$.

(b) $9.6 \times 10^5 = 9.6 \times 100000 = 960000$.

(c) $-8.2 \times 10^2 = -8.2 \times 100 = -820$.

6.30   Simplify the expression $(3 \times 10^2) \times (5 \times 10^3)$.

Solution   The order in which the numbers are written down does not matter and so we can write

$$(3 \times 10^2) \times (5 \times 10^3) = 3 \times 5 \times 10^2 \times 10^3 = 15 \times 10^5$$

In the standard form of scientific notation we would write this as 1.5×10^6.

Self-assessment questions 6.5

1. What is the purpose of using scientific notation?

Exercise 6.5

1. Express each of the following numbers in scientific notation:

   (a) 45 (b) 45000 (c) −450
   (d) 90000000 (e) 0.15 (f) 0.00036
   (g) 3.5 (h) −13.2 (i) 1000000
   (j) 0.0975 (k) 45.34

2. Write out fully the following numbers:

   (a) 3.75×10^2 (b) 3.97×10^1
   (c) 1.875×10^{-1} (d) -8.75×10^{-3}

Test and assignment exercises 6

1. Simplify

   (a) $\dfrac{z^5}{z^{-5}}$ (b) z^0 (c) $\dfrac{z^8 z^6}{z^{14}}$

2. Evaluate

   (a) $0.25^{1/2}$ (b) $(4096)^{1/3}$ (c) $(2601)^{1/2}$
   (d) $16^{-1/2}$

3. Simplify $\dfrac{x^8 x^{-3}}{x^{-5} x^2}$

4. Find the value of $(1/7)^0$.

5. Remove the brackets from

   (a) $(abc^2)^2$ (b) $(xy^2 z^3)^2$ (c) $(8x^2)^{-3}$

6. Express each of the following numbers in scientific notation:

   (a) 5792 (b) 98.4 (c) 0.001 (d) −66.667

7 Simplifying algebraic expressions

Objectives

This chapter

- describes a number of ways in which complicated algebraic expressions can be simplified

7.1 Addition and subtraction of like terms

Like terms are multiples of the same quantity. For example $3y$, $72y$ and $0.5y$ are all multiples of y and so are like terms. Similarly, $5x^2$, $-3x^2$ and $\frac{1}{2}x^2$ are all multiples of x^2 and so are like terms. xy, $17xy$ and $-91xy$ are all multiples of xy and are therefore like terms. Like terms can be collected together and added or subtracted in order to simplify them.

Worked examples

7.1 Simplify $3x + 7x - 2x$.

Solution All three terms are multiples of x and so are like terms. Therefore
$3x + 7x - 2x = 8x$.

7.2 Simplify $3x + 2y$.

Solution $3x$ and $2y$ are not like terms. One is a multiple of x and the other is a multiple of y. The expression $3x + 2y$ cannot be simplified.

7.3 Simplify $x + 7x + x^2$.

Solution The like terms are x and $7x$. These can be simplified to $8x$. Then
$x + 7x + x^2 = 8x + x^2$. Note that $8x$ and x^2 are not like terms and so this expression cannot be simplified further.

7.4 Simplify $ab + a^2 - 7b^2 + 9ab + 8b^2$.

Solution The terms ab and $9ab$ are like terms. Similarly the terms $-7b^2$ and $8b^2$ are like terms. These can be collected together and then added or subtracted as appropriate. Thus

$$ab + a^2 - 7b^2 + 9ab + 8b^2 = ab + 9ab + a^2 - 7b^2 + 8b^2 = 10ab + a^2 + b^2$$

Exercise 7.1

1. Simplify, if possible,

   (a) $5p - 10p + 11q + 8q$
   (b) $-7r - 13s + 2r + z$
   (c) $181z + 13r - 2$
   (d) $x^2 + 3y^2 - 2y + 7x^2$
   (e) $4x^2 - 3x + 2x + 9$

2. Simplify

   (a) $5y + 8p - 17y + 9q$
   (b) $7x^2 - 11x^3 + 14x^2 + y^3$
   (c) $4xy + 3xy + y^2$
   (d) $xy + yx$
   (e) $xy - yx$

7.2 **Multiplication of algebraic expressions and removing brackets**

Recall that when multiplying two numbers together the order in which we write them is irrelevant. For example, both 5×4 and 4×5 equal 20.

When multiplying three or more numbers together the order in which we carry out the multiplication is also irrelevant. By this we mean, for example, that when asked to multiply $3 \times 4 \times 5$ we can think of this as either $(3 \times 4) \times 5$ or as $3 \times (4 \times 5)$. Check for yourself that the result is the same, 60, either way.

It is also important to appreciate that $3 \times 4 \times 5$ could have been written as $(3)(4)(5)$.

It is essential that you grasp these simple facts about numbers in order to understand the algebra which follows. This is because identical rules are applied. Rules for determining the sign of the answer when multiplying positive and negative alglebraic expressions are also the same as those used for multiplying numbers.

KEY POINT

When multiplying

$$\text{positive} \times \text{positive} = \text{positive}$$

$$\text{positive} \times \text{negative} = \text{negative}$$

$$\text{negative} \times \text{positive} = \text{negative}$$

$$\text{negative} \times \text{negative} = \text{positive}$$

We introduce the processes involved in removing brackets using some simple examples.

Worked examples

7.5   Simplify $3(4x)$.

Solution   Just as with numbers $3(4x)$ could be written as $3 \times (4 \times x)$, and then as $(3 \times 4) \times x$ which evaluates to $12x$.

So $3(4x) = 12x$.

7.6   Simplify $5(3y)$.

Solution $5(3y) = 5 \times 3 \times y = 15y$.

7.7   Simplify $(5a)(3a)$.

Solution   Here we can write $(5a)(3a) = (5 \times a) \times (3 \times a)$. Neither the order in which we carry out the multiplications nor the order in which we write down the terms matters and so we can write this as

$$(5a)(3a) = (5 \times 3)(a \times a)$$

As we have shown, it is usual to write numbers at the beginning of an expression. This simplifies to $15 \times a^2$, that is $15a^2$. Hence

$$(5a)(3a) = 15a^2$$

7.8   Simplify $4x^2 \times 7x^5$.

Solution   Recall that when multiplying, the order in which we write down the terms does not matter. Therefore we can write

$$4x^2 \times 7x^5 = 4 \times 7 \times x^2 \times x^5$$

which equals $28x^{2+5} = 28x^7$.

7.9   Simplify $7(2b^2)$.

Solution $7(2b^2) = 7 \times (2 \times b^2) = (7 \times 2) \times b^2 = 14b^2$.

7.10   Simplify $(a) \times (-b)$

Solution   Here we have the product of a positive and a negative quantity. The result will be negative. We write

$$(a) \times (-b) = -ab$$

7.11 Explain the distinction between ab^2 and $(ab)^2$.

Solution ab^2 means $a \times b \times b$ whereas $(ab)^2$ means $(ab) \times (ab)$ which equals $a \times b \times a \times b$. The latter could also be written as a^2b^2.

7.12 Simplify (a) $(6z)(8z)$ (b) $(6z) + (8z)$, noting the distinction between the two results.

Solution (a) $(6z)(8z) = 48z^2$.

(b) $(6z) + (8z)$ is the addition of like terms. This simplifies to $14z$.

7.13 Simplify (a) $(6x)(-2x)$ (b) $(-3y^2)(-2y)$.

Solution (a) $(6x)(-2x)$ means $(6x) \times (-2x)$ which equals $-12x^2$.

(b) $(-3y^2)(-2y) = (-3y^2) \times (-2y) = 6y^3$.

Exercise 7.2

1. Simplify each of the following:

   (a) $(4)(3)(7)$ (b) $(7)(4)(3)$ (c) $(3)(4)(7)$

2. Simplify

   (a) $5 \times (4 \times 2)$ (b) $(5 \times 4) \times 2$

3. Simplify each of the following:

   (a) $7(2z)$ (b) $15(2y)$ (c) $(2)(3)x$ (d) $9(3a)$

   (e) $(11)(5a)$ (f) $2(3x)$

4. Simplify each of the following:

   (a) $5(4x^2)$ (b) $3(2y^3)$ (c) $11(2u^2)$

   (d) $(2 \times 4) \times u^2$ (e) $(13)(2z^2)$

5. Simplify

   (a) $(7x)(3x)$ (b) $3a(7a)$ (c) $14a(a)$

6. Simplify

   (a) $5y(3y)$ (b) $5y + 3y$

   Explain why the two results are not the same.

7. Simplify the following:

   (a) $(abc)(a^2bc)$ (b) $x^2y(xy)$
   (c) $(xy^2)(xy^2)$

8. Explain the distinction, if any, between $(xy^2)(xy^2)$ and xy^2xy^2.

9. Explain the distinction, if any, between $(xy^2)(xy^2)$ and $(xy^2) + (xy^2)$.

   In both cases simplify the expressions.

10. Simplify

    (a) $(3z)(-7z)$ (b) $3z - 7z$

11. Simplify

    (a) $(-x)(3x)$ (b) $-x + 3x$

12. Simplify

    (a) $(-2x)(-x)$ (b) $-2x - x$

7.3   Removing brackets from $a(b+c)$, $a(b-c)$ and $(a+b)(c+d)$

Recall from your study of arithmetic that the expression $(5-4)+7$ is different from $5-(4+7)$ because of the position of the brackets. In order to simplify an expression it is often necessary to remove brackets.

Removing brackets from expressions of the form $a(b+c)$ and $a(b-c)$

In an expression such as $a(b+c)$, it is intended that the a multiplies all the bracketed terms:

KEY POINT

$a(b+c) = ab + ac$ Similarly: $a(b-c) = ab - ac$

Worked examples

7.14   Remove the brackets from

(a) $6(x+5)$ (b) $8(2x-4)$

Solution   (a) In the expression $6(x+5)$ it is intended that the 6 multiplies both terms in the brackets. Therefore
$$6(x+5) = 6x + 30$$

(b) In the expression $8(2x-4)$ the 8 multiplies both terms in the brackets so that
$$8(2x-4) = 16x - 32$$

7.15   Remove the brackets from the expression $7(5x+3y)$.

Solution   The 7 multiplies both the terms in the bracket. Therefore
$$7(5x+3y) = 7(5x) + 7(3y) = 35x + 21y$$

7.16   Remove the brackets from $-(x+y)$.

Solution   The expression $-(x+y)$ actually means $-1(x+y)$. It is intended that the -1 multiplies both terms in the brackets, therefore
$$-(x+y) = -1(x+y) = (-1) \times x + (-1) \times y = -x - y$$

7.17   Remove brackets from the expression

$$(x + y)z$$

Solution   Note that the order in which we write down the terms to be multiplied does not matter, so that we can write $(x + y)z$ as $z(x + y)$. Then

$$z(x + y) = zx + zy$$

Alternatively note that $(x + y)z = xz + yz$ which is an equivalent form of the answer.

7.18   Remove the brackets from the expressions

(a) $5(x - 2y)$ (b) $(x + 3)(-1)$

Solution   (a) $5(x - 2y) = 5x - 5(2y) = 5x - 10y$.

(b) $(x + 3)(-1) = (-1)(x + 3) = -1x - 3 = -x - 3$.

7.19   Simplify $x + 8(x - y)$.

Solution   An expression such as this is simplified by first removing the brackets and then collecting together like terms. Removing the brackets we find

$$x + 8(x - y) = x + 8x - 8y$$

Collecting like terms we obtain $9x - 8y$.

7.20   Remove the brackets from

(a) $\frac{1}{2}(x + 2)$ (b) $\frac{1}{2}(x - 2)$ (c) $-\frac{1}{3}(a + b)$

Solution   (a)   In the expression $\frac{1}{2}(x + 2)$ it is intended that the $\frac{1}{2}$ multiplies both the terms in the brackets. So

$$\frac{1}{2}(x + 2) = \frac{1}{2}x + \frac{1}{2}(2) = \frac{1}{2}x + 1$$

(b)   Similarly,

$$\frac{1}{2}(x - 2) = \frac{1}{2}x - \frac{1}{2}(2) = \frac{1}{2}x - 1$$

(c)   In the expression $-\frac{1}{3}(a + b)$, the term $-\frac{1}{3}$ multiplies both terms in the brackets. So

$$-\frac{1}{3}(a + b) = -\frac{1}{3}a - \frac{1}{3}b$$

Removing brackets from expressions of the form $(a + b)(c + d)$

In the expression $(a + b)(c + d)$ it is intended that the quantity $(c + d)$ is multiplied by both the a and the b in the first bracket. Therefore

$$(a + b)(c + d) = (a + b)c \quad + \quad (a + b)d$$

Each of these two terms can be expanded further to give

$$(a + b)c = ac + bc \quad \text{and} \quad (a + b)d = ad + bd$$

Therefore

KEY POINT

$$\boxed{(a + b)(c + d) = ac + bc + ad + bd}$$

Worked examples

7.21 Remove the brackets from $(3 + x)(2 + y)$.

Solution
$$\begin{aligned}(3 + x)(2 + y) &= (3 + x)(2) + (3 + x)y \\ &= 6 + 2x + 3y + xy\end{aligned}$$

7.22 Remove the brackets from $(x + 6)(x - 3)$.

Solution
$$\begin{aligned}(x + 6)(x - 3) &= (x + 6)x + (x + 6)(-3) \\ &= x^2 + 6x - 3x - 18 \\ &= x^2 + 3x - 18\end{aligned}$$

7.23 Remove the brackets from

(a) $(1 - x)(2 - x)$ (b) $(-x - 2)(2x - 1)$

Solution (a)
$$\begin{aligned}(1 - x)(2 - x) &= (1 - x)2 + (1 - x)(-x) \\ &= 2 - 2x - x + x^2 \\ &= 2 - 3x + x^2\end{aligned}$$

(b)
$$\begin{aligned}(-x - 2)(2x - 1) &= (-x - 2)(2x) + (-x - 2)(-1) \\ &= -2x^2 - 4x + x + 2 \\ &= -2x^2 - 3x + 2\end{aligned}$$

7.24   Remove the brackets from the expression $3(x+1)(x-1)$.

Solution   First consider the expression $(x+1)(x-1)$.

$$(x+1)(x-1) = (x+1)x + (x+1)(-1)$$
$$= x^2 + x - x - 1$$
$$= x^2 - 1$$

Then $3(x+1)(x-1) = 3(x^2 - 1) = 3x^2 - 3$.

Exercise 7.3

1. Remove the brackets from

   (a) $4(x+1)$ (b) $-4(x+1)$ (c) $4(x-1)$ (d) $-4(x-1)$

2. Remove the brackets from the following expressions:

   (a) $5(x-y)$ (b) $19(x+3y)$ (c) $8(a+b)$ (d) $(5+x)y$ (e) $12(x+4)$ (f) $17(x-9)$ (g) $-(a-2b)$ (h) $\frac{1}{2}(2x+1)$ (i) $-3m(-2+4m+3n)$

3. Remove the brackets and simplify the following:

   (a) $18 - 13(x+2)$ (b) $x(x+y)$

4. Remove the brackets and simplify the following expressions:

   (a) $(x+1)(x+6)$ (b) $(x+4)(x+5)$ (c) $(x-2)(x+3)$ (d) $(x+6)(x-1)$ (e) $(x+y)(m+n)$ (f) $(4+y)(3+x)$ (g) $(5-x)(5+x)$ (h) $(17x+2)(3x-5)$

5. Remove the brackets and simplify the following expressions:

   (a) $(x+3)(x-7)$ (b) $(2x-1)(3x+7)$ (c) $(4x+1)(4x-1)$ (d) $(x+3)(x-3)$ (e) $(2-x)(3+2x)$

6. Remove the brackets and simplify the following expressions:

   (a) $\frac{1}{2}(x+2y) + \frac{7}{2}(4x - y)$

   (b) $\frac{3}{4}(x-1) + \frac{1}{4}(2x+8)$

7. Remove the brackets from

   (a) $-(x-y)$ (b) $-(a+2b)$

   (c) $-\frac{1}{2}(3p+q)$

8. Remove the brackets from $(x+1)(x+2)$. Use your result to remove the brackets from $(x+1)(x+2)(x+3)$.

Test and assignment exercises 7

1. Simplify

   (a) $7x^2 + 4x^2 + 9x - 8x$
   (b) $y + 7 - 18y + 1$
   (c) $a^2 + b^2 + a^3 - 3b^2$

2. Simplify

   (a) $(3a^2b) \times (-a^3b^2c)$ (b) $\dfrac{x^3}{-x^2}$

3. Remove the brackets from

   (a) $(a + 3b)(7a - 2b)$ (b) $x^2(x + 2y)$
   (c) $x(x + y)(x - y)$

4. Remove the brackets from

   (a) $(7x + 2)(3x - 1)$ (b) $(1 - x)(x + 3)$ (c)
   $(5 + x)x$ (d) $(8x + 4)(7x - 2)$

5. Remove the brackets and simplify

   (a) $3x(x + 2) - 7x^2$
   (b) $-(2a + 3b)(a + b)$
   (c) $4(x + 7) + 13(x - 2)$
   (d) $5(2a + 5) - 3(5a - 2)$
   (e) $\dfrac{1}{2}(a + 4b) + \dfrac{3}{2}a$

8 Factorization

Objectives	This chapter
	• explains what is meant by the 'factors' of an algebraic expression
	• shows how an algebraic expression can be factorized

8.1 Factors and common factors

Recall from Chapter 1 that a number is **factorized** when it is written as a product. For example, 15 may be factorized into 3×5. We say that 3 and 5 are **factors** of 15. The number 16 can be written as 8×2 or 4×4, or even as 16×1, and so the factorization may not be unique.

Algebraic expressions can also be factorized. Consider the expression $5x + 20y$. Both $5x$ and $20y$ have the number 5 common to both terms. We say that 5 is a **common factor**. Any common factors can be written outside a bracketed term. Thus $5x + 20y = 5(x + 4y)$. Removal of the brackets will result in the original expression and can always be used to check your answer. We see that factorization can be thought of as reversing the process of removing brackets. Similarly, if we consider the expression $x^2 + 2x$, we note that both terms contain the factor x, and so $x^2 + 2x$ can be written as $x(x + 2)$. Hence x and $x + 2$ are both factors of $x^2 + 2x$.

Worked examples

8.1 Factorize $3x + 12$.

Solution The number 12 can be factorized as 3×4 so that 3 is a common factor of $3x$ and 12. We can write $3x + 12 = 3x + 3(4)$. Any common factors are written in front of a bracket and the contents of the bracket are adjusted accordingly. So

$$3x + 3(4) = 3(x + 4)$$

Note again that this answer can be checked by removing the brackets.

8.2   List the ways in which $15x^2$ can be written as a product of its factors.

Solution $15x^2$ can be written in many different ways. Some of these are $15x^2 \times 1$, $15x \times x$, $15 \times x^2$, $5x \times 3x$, $5 \times 3x^2$, and $3 \times 5x^2$.

8.3   Factorize $8x^2 - 12x$.

Solution   We can write $8x^2 - 12x = (4x)(2x) - (4x)3$ so that both terms contain the factor $4x$. This is placed at the front of a bracket to give

$$8x^2 - 12x = 4x(2x - 3)$$

8.4   What factors are common to the terms $5x^2$ and $15x^3$? Factorize $5x^2 + 15x^3$.

Solution   Both terms contain a factor of 5. Because x^3 can be written $x^2 \times x$, both $5x^2$ and $15x^3$ contain a factor x^2. Therefore

$$5x^2 + 15x^3 = 5x^2 + (5x^2)(3x) = 5x^2(1 + 3x)$$

8.5   Factorize $6x + 3x^2 + 9xy$.

Solution   By careful inspection of all of the terms we see that $3x$ is a factor of each term. Hence

$$6x + 3x^2 + 9xy = 3x(2 + x + 3y)$$

Hence the factors of $6x + 3x^2 + 9xy$ are $3x$ and $2 + x + 3y$.

Self-assessment questions 8.1

1. Explain what is meant by 'factorizing an expression'.

Exercise 8.1

1. Remove the brackets from

   (a) $9(x + 3)$ (b) $-5(x - 2)$ (c) $\dfrac{1}{2}(x + 1)$

   (d) $-(a - 3b)$ (e) $\dfrac{1}{2(x + y)}$ (f) $\dfrac{x}{y(x - y)}$

2. List all the factors of each of (a) $4x^2$, (b) $6x^3$.

3. Factorize

   (a) $3x + 18$ (b) $3y - 9$ (c) $-3y - 9$
   (d) $-3 - 9y$ (e) $20 + 5t$ (f) $20 - 5t$
   (g) $-5t - 20$ (h) $3x + 12$ (i) $17t + 34$
   (j) $-36 + 4t$

4. Factorize

   (a) $x^4 + 2x$ (b) $x^4 - 2x$ (c) $3x^4 - 2x$
   (d) $3x^4 + 2x$ (e) $3x^4 + 2x^2$ (f) $3x^4 + 2x^3$
   (g) $17z - z^2$ (h) $-xy + 3x$ (i) $-xy + 3y$
   (j) $x + 2xy + 3xyz$

5. Factorize

   (a) $10x + 20y$ (b) $12a + 3b$ (c) $4x - 6xy$
   (d) $7a + 14$ (e) $10m - 15$

   (f) $\dfrac{1}{5a + 35b}$ (g) $\dfrac{1}{5a^2 + 35ab}$

6. Factorize

   (a) $15x^2 + 3x$ (b) $4x^2 - 3x$ (c) $4x^2 - 8x$
   (d) $15 - 3x^2$ (e) $10x^3 + 5x^2 + 15x^2y$
   (f) $6a^2b - 12ab^2$ (g) $16abc - 8ab^2 + 24bc$

| 8.2 | **Factorizing quadratic expressions** |

Expressions of the form $ax^2 + bx + c$ where a, b, and c are numbers are called **quadratic expressions**. The numbers b or c may equal zero but a must not be zero. The number a is called the **coefficient** of x^2, b is the coefficient of x and c is called the **constant term**.

We see that

$$2x^2 + 3x - 1, \quad x^2 + 3x + 2, \quad x^2 + 7 \text{ and } 2x^2 - x$$

are all quadratic expressions.

KEY POINT

An expression of the form $ax^2 + bx + c$ where a, b and c are numbers is called a quadratic expression. The coefficient of x^2 is a, the coefficient of x is b and the constant term is c.

To factorize such an expression means to express it as a product of two terms. For example, removing the brackets from $(x + 3)(x + 2)$ gives $x^2 + 5x + 6$. Reversing the process, $x^2 + 5x + 6$ can be factorized to $(x + 3)(x + 2)$. Not all quadratic expressions can be factorized in this way. We shall now explore how such factorization is attempted.

Quadratic expressions where the coefficient of x^2 is 1

Consider the expression $(x + m)(x + n)$. Removing the brackets we find

$$(x + m)(x + n) = (x + m)x \quad + \quad (x + m)n$$
$$= x^2 + mx + nx + mn$$
$$= x^2 + (m + n)x + mn$$

Note that the coefficient of the x term is the sum $m + n$ and the constant term is the product mn. Using this information several quadratic expressions can be factorized by careful inspection. For example, suppose we wish to factorize $x^2 + 5x + 6$. We know that $x^2 + (m + n)x + mn$ can be factorized to $(x + m)(x + n)$. We seek values of m and n so that

$$x^2 + 5x + 6 = x^2 + (m + n)x + mn$$

Comparing the coefficients of x on both sides we require

$$5 = m + n$$

Comparing the constant terms on both sides we require

$$6 = mn$$

By inspection we see that $m = 3$ and $n = 2$ have this property and so

$$x^2 + 5x + 6 = (x + 3)(x + 2)$$

Note that the answer can be easily checked by removing the brackets again.

Worked examples

8.6   Factorize the quadratic expression $x^2 + 8x + 12$.

Solution   The factorization of $x^2 + 8x + 12$ will be of the form $(x + m)(x + n)$. This means that mn must equal 12 and $m + n$ must equal 8. The two numbers must therefore be 2 and 6. So

$$x^2 + 8x + 12 = (x + 2)(x + 6)$$

Note again that the answer can be checked by removing the brackets.

8.7   Factorize $x^2 + 10x + 25$.

Solution   We try to factorize in the form $(x + m)(x + n)$. We require $m + n$ to equal 10 and mn to equal 25. If $m = 5$ and $n = 5$ this requirement is met. Therefore $x^2 + 10x + 25 = (x + 5)(x + 5)$. It is usual practice to write this as $(x + 5)^2$.

8.8   Factorize $x^2 - 121$.

Solution   In this example the x term is missing. We still attempt to factorize as $(x + m)(x + n)$. We require $m + n$ to equal 0 and mn to equal -121. Some thought shows that if $m = 11$ and $n = -11$ this requirement is met. Therefore $x^2 - 121 = (x + 11)(x - 11)$.

8.9   Factorize $x^2 - 5x + 6$.

Solution   We try to factorize in the form $(x + m)(x + n)$. We require $m + n$ to equal -5 and mn to equal 6. By inspection we see that if $m = -3$ and $n = -2$ this requirement is met. Therefore $x^2 - 5x + 6 = (x - 3)(x - 2)$

Quadratic expressions where the coefficient of x^2 is not 1

These expressions are a little harder to factorize. All possible factors of the first and last terms must be found and various combinations of these should be attempted until the required answer is found. This involves trial and error along with educated guesswork and practice.

Worked examples

8.10   Factorize, if possible, the expression $2x^2 + 11x + 12$.

Solution   The factors of the first term, $2x^2$, are $2x$ and x. The factors of the last term, 12, are

$$12, 1 \quad -12, -1 \quad 6, 2 \quad -6, -2 \quad \text{and} \quad 4, 3 \quad -4, -3$$

We can try each of these combinations in turn to find which gives us a coefficient of x of 11. For example, removing the brackets from

$$(2x + 12)(x + 1)$$

gives

$$(2x + 12)(x + 1) = (2x + 12)x + (2x + 12)(1)$$
$$= 2x^2 + 12x + 2x + 12$$
$$= 2x^2 + 14x + 12$$

which has an incorrect middle term. By trying further combinations it turns out that the only one producing a middle term of $11x$ is $(2x + 3)(x + 4)$ because

$$(2x + 3)(x + 4) = (2x + 3)(x) + (2x + 3)(4)$$
$$= 2x^2 + 3x + 8x + 12$$
$$= 2x^2 + 11x + 12$$

so that $(2x + 3)(x + 4)$ is the correct factorization.

8.11   Factorize, if possible, $4x^2 + 6x + 2$.

Solution   Before we try to factorize this quadratic expression notice that there is a factor of 2 in each term so that we can write it as $2(2x^2 + 3x + 1)$. Now consider the quadratic expression $2x^2 + 3x + 1$. The factors of the first term, $2x^2$, are $2x$ and x. The factors of the last term, 1, are simply 1 and 1, or -1 and -1. We can try these combinations in turn to find which gives us a middle term of $3x$. Removing the brackets from $(2x + 1)(x + 1)$ gives $2x^2 + 3x + 1$ which has the correct middle term.

Finally, we can write

$$4x^2 + 6x + 2 = 2(2x^2 + 3x + 1) = 2(2x + 1)(x + 1)$$

8.12 Factorize $6x^2 + 7x - 3$.

Solution The first term may be factorized as $6x \times x$ and also as $3x \times 2x$. The factors of the last term are

$$3, -1 \quad \text{and} \quad -3, 1$$

We need to try each combination in turn to find which gives us a coefficient of x of 7. For example, removing the brackets from

$$(6x + 3)(x - 1)$$

gives

$$\begin{aligned}
(6x + 3)(x - 1) &= (6x + 3)x + (6x + 3)(-1) \\
&= 6x^2 + 3x - 6x - 3 \\
&= 6x^2 - 3x - 3
\end{aligned}$$

which has an incorrect middle term. By trying further combinations it turns out that the only one producing a middle term of $7x$ is $(3x - 1)(2x + 3)$ because

$$\begin{aligned}
(3x - 1)(2x + 3) &= (3x - 1)2x + (3x - 1)(3) \\
&= 6x^2 - 2x + 9x - 3 \\
&= 6x^2 + 7x - 3
\end{aligned}$$

The correct factorization is therefore $(3x - 1)(2x + 3)$.

Until you have sufficient experience at factorizing quadratic expressions you must be prepared to go through the process of trying all possible combinations until the correct answer is found.

Self-assessment questions 8.2

1. Not all quadratic expressions can be factorized. Try to find an example of one such expression.

Exercise 8.2

1. Factorize the following quadratic expressions:

   (a) $x^2 + 3x + 2$ (b) $x^2 + 13x + 42$
   (c) $x^2 + 2x - 15$ (d) $x^2 + 9x - 10$
   (e) $x^2 - 11x + 24$ (f) $x^2 - 100$
   (g) $x^2 + 4x + 4$ (h) $x^2 - 36$ (i) $x^2 - 25$
   (j) $x^2 + 10x + 9$ (k) $x^2 + 8x - 9$
   (l) $x^2 - 8x - 9$ (m) $x^2 - 10x + 9$
   (n) $x^2 - 5x$

2. Factorize the following quadratic expressions:

   (a) $2x^2 - 5x - 3$ (b) $3x^2 - 5x - 2$
   (c) $10x^2 + 11x + 3$ (d) $2x^2 + 12x + 16$
   (e) $2x^2 + 5x + 3$ (f) $3s^2 + 5s + 2$
   (g) $3z^2 + 17z + 10$ (h) $9x^2 - 36$
   (i) $4x^2 - 25$

3. (a) By removing the brackets show that

   $$(x + y)(x - y) = x^2 - y^2$$

This result is known as the **difference of two squares**.

   (b) Using the result in part (a) write down the factorization of

   (i) $16x^2 - 1$ (ii) $16x^2 - 9$
   (iii) $25t^2 - 16r^2$

4. Factorize the following quadratic expressions:

   (a) $x^2 + 3x - 10$ (b) $2x^2 - 3x - 20$
   (c) $9x^2 - 1$ (d) $10x^2 + 14x - 12$
   (e) $x^2 + 15x + 26$ (f) $-x^2 - 2x + 3$

5. Factorize

   (a) $100 - 49x^2$ (b) $36x^2 - 25y^2$

   (c) $\frac{1}{4} - 9v^2$ (d) $\frac{x^2}{y^2} - 4$

Test and assignment exercises 8

1. Factorize the following expressions:

   (a) $7x + 49$ (b) $121x + 22y$ (c) $a^2 + ab$
   (d) $ab + b^2$ (e) $ab^2 + ba^2$

2. Factorize the following quadratic expressions:

   (a) $3x^2 + x - 2$ (b) $x^2 - 144$
   (c) $s^2 - 5s + 6$ (d) $2y^2 - y - 15$

3. Factorize the following:

   (a) $1 - x^2$ (b) $x^2 - 1$ (c) $9 - x^2$
   (d) $x^2 - 81$ (e) $25 - y^2$

4. Factorize the denominators of the following expressions:

   (a) $\dfrac{1}{x^2 + 6x}$ (b) $\dfrac{3}{s^2 + 3s + 2}$

   (c) $\dfrac{3}{s^2 + s - 2}$ (d) $\dfrac{5}{x^2 + 11x + 28}$

   (e) $\dfrac{x}{2x^2 - 17x - 9}$

9 Algebraic fractions

Objectives	This chapter
	• explains how to simplify algebraic fractions by cancelling common factors
	• explains how algebraic fractions can be multiplied and divided
	• explains how algebraic fractions can be added and subtracted

9.1   Introduction

Just as one whole number divided by another is a numerical fraction, one algebraic expression divided by another is called an **algebraic fraction.**

$$\frac{x}{y} \qquad \frac{x^2+y}{x} \qquad \frac{3x+2}{7}$$

are all examples of algebraic fractions. The top line is known as the **numerator** of the fraction, and the bottom line is the **denominator**.

Rules for determining the sign of the answer when dividing positive and negative algebraic expressions are the same as those used for dividing numbers.

KEY POINT

When dividing

$$\frac{\text{positive}}{\text{positive}} = \text{positive}$$

$$\frac{\text{positive}}{\text{negative}} = \text{negative}$$

$$\frac{\text{negative}}{\text{positive}} = \text{negative}$$

$$\frac{\text{negative}}{\text{negative}} = \text{positive}$$

Using these rules we see that an algebraic expression can often be written in different but equivalent forms. For example note that

$$\frac{x}{-y} \text{ can be written as } -\frac{x}{y}$$

and that

$$\frac{-x}{y} \text{ can be written as } -\frac{x}{y}$$

and also that

$$\frac{-x}{-y} \text{ can be written as } \frac{x}{y}$$

9.2   Cancelling common factors

Cancellation of
common factors
was described in
detail in §2.2.

Consider the numerical fraction $\frac{3}{12}$. To simplify this we factorize both the numerator and the denominator. Any factors which appear in both the numerator and the denominator are called **common factors**. These can be cancelled. For example,

$$\frac{3}{12} = \frac{1 \times 3}{4 \times 3} = \frac{1 \times \cancel{3}}{4 \times \cancel{3}} = \frac{1}{4}$$

The same process is applied when dealing with algebraic fractions.

Worked examples

9.1   For each pair of expressions, state which factors are common to both.

(a) $3xy$ and $6xz$ (b) xy and $5y^2$ (c) $3(x+2)$ and $(x+2)^2$

(d) $3(x-1)$ and $(x-1)(x+4)$

Solution   (a) The expression $6xz$ can be written $(3)(2)xz$. We see that factors common to both this and $3xy$ are 3 and x.

(b) The expression $5y^2$ can be written $5(y)(y)$. We see that the only factor common to both this and xy is y.

(c) $(x+2)^2$ can be written $(x+2)(x+2)$. Thus $(x+2)$ is a factor common to both $(x+2)^2$ and $3(x+2)$.

(d) $3(x-1)$ and $(x-1)(x+4)$ have a common factor of $(x-1)$.

9.2 Simplify $\dfrac{18x^2}{6x}$

Solution First note that 18 can be factorized as 6×3. So there are factors of 6 and x in both the numerator and the denominator. Then common factors can be cancelled. That is

$$\frac{18x^2}{6x} = \frac{(6)(3)x^2}{6x} = \frac{3x}{1} = 3x$$

KEY POINT

When simplifying an algebraic fraction only factors common to both the numerator and denominator can be cancelled.

A fraction is expressed in its simplest form by factorizing the numerator and denominator and cancelling any common factors.

Worked examples

9.3 Simplify $\dfrac{5}{25 + 15x}$

Solution First of all note that the denominator can be factorized as $5(5 + 3x)$. There is therefore a factor of 5 in both the numerator and denominator. So 5 is a common factor. This can be cancelled. That is

$$\frac{5}{25 + 15x} = \frac{1 \times 5}{5(5 + 3x)} = \frac{1 \times \cancel{5}}{\cancel{5}(5 + 3x)} = \frac{1}{5 + 3x}$$

It is very important to note that the number 5 which has been cancelled is a common factor. It is incorrect to try to cancel terms which are not common factors.

9.4 Simplify $\dfrac{5x}{25x + 10y}$

Solution Factorizing the denominator we can write

$$\frac{5x}{25x + 10y} = \frac{5x}{5(5x + 2y)}$$

We see that there is a common factor of 5 in both numerator and denominator which can be cancelled. Thus

$$\frac{5x}{25x + 10y} = \frac{\cancel{5}x}{\cancel{5}(5x + 2y)} = \frac{x}{5x + 2y}$$

Note that no further cancellation is possible. x is not a common factor because it is not a factor of the denominator.

9.5   Simplify $\dfrac{4x}{3x^2 + x}$

Solution   Note that the denominator factorizes to $x(3x + 1)$. Once both numerator and denominator have been factorized, any common factors are cancelled. So

$$\frac{4x}{3x^2 + x} = \frac{4x}{x(3x + 1)} = \frac{4\cancel{x}}{\cancel{x}(3x + 1)} = \frac{4}{3x + 1}$$

Note that the factor x is common to both numerator and denominator and so has been cancelled.

9.6   Simplify $\dfrac{x}{x^2 + 2x}$

Solution   Note that the denominator factorizes to $x(x + 2)$. Also note that the numerator can be written as $1 \times x$. So

$$\frac{x}{x^2 + 2x} = \frac{1 \times x}{x(x + 2)} = \frac{1 \times \cancel{x}}{\cancel{x}(x + 2)} = \frac{1}{x + 2}$$

9.7   Simplify

(a) $\dfrac{2(x - 1)}{(x + 3)(x - 1)}$ (b) $\dfrac{x - 4}{(x - 4)^2}$

Solution   (a) There is a factor of $(x - 1)$ common to both the numerator and denominator. This is cancelled to give

$$\frac{2(x - 1)}{(x + 3)(x - 1)} = \frac{2}{x + 3}$$

(b) There is a factor of $x - 4$ in both numerator and denominator. This is cancelled as follows:

$$\frac{x - 4}{(x - 4)^2} = \frac{1(x - 4)}{(x - 4)(x - 4)} = \frac{1}{x - 4}$$

9.8   Simplify $\dfrac{x + 2}{x^2 + 3x + 2}$

Solution   The denominator is factorized and then any common factors are cancelled:

$$\frac{x + 2}{x^2 + 3x + 2} = \frac{1(x + 2)}{(x + 2)(x + 1)} = \frac{1}{x + 1}$$

9.9 Simplify

(a) $\dfrac{3x + xy}{x^2 + 5x}$ (b) $\dfrac{x^2 - 1}{x^2 + 3x + 2}$

Solution The numerator and denominator are both factorized and any common factors are cancelled.

(a) $\dfrac{3x + xy}{x^2 + 5x} = \dfrac{x(3 + y)}{x^2 + 5x} = \dfrac{\cancel{x}(3 + y)}{\cancel{x}(x + 5)} = \dfrac{3 + y}{5 + x}$

(b) $\dfrac{x^2 - 1}{x^2 + 3x + 2} = \dfrac{(x + 1)(x - 1)}{(x + 1)(x + 2)} = \dfrac{x - 1}{x + 2}$

Self-assessment questions 9.2

1. Explain why no cancellation is possible in the expression $\dfrac{3x}{3x + y}$.

2. Explain why no cancellation is possible in the expression $\dfrac{x + 1}{x + 3}$.

3. Explain why it is possible to perform a cancellation in the expression $\dfrac{x + 1}{2x + 2}$, and perform it.

Exercise 9.2

1. Simplify

   (a) $\dfrac{9x}{3y}$ (b) $\dfrac{9x}{x^2}$ (c) $\dfrac{9xy}{3x}$ (d) $\dfrac{9xy}{3y}$

   (e) $\dfrac{9xy}{xy}$ (f) $\dfrac{9xy}{3xy}$

2. Simplify

   (a) $\dfrac{15x}{3y}$ (b) $\dfrac{15x}{5y}$ (c) $\dfrac{15xy}{x}$ (d) $\dfrac{15xy}{xy}$

   (e) $\dfrac{x^5}{-x^3}$ (f) $\dfrac{-y^3}{y^7}$ (g) $\dfrac{-y}{-y^2}$ (h) $\dfrac{-y^{-3}}{-y^4}$

3. Simplify the following algebraic fractions:

   (a) $\dfrac{4}{12 + 8x}$ (b) $\dfrac{5 + 10x}{5}$ (c) $\dfrac{2}{4 + 14x}$

   (d) $\dfrac{2x}{4 + 14x}$ (e) $\dfrac{2x}{2 + 14x}$ (f) $\dfrac{7}{49x + 7y}$

   (g) $\dfrac{7y}{49x + 7y}$ (h) $\dfrac{7x}{49x + 7y}$

4. Simplify

   (a) $\dfrac{15x + 3}{3}$ (b) $\dfrac{15x + 3}{3x + 6y}$ (c) $\dfrac{12}{4x + 8}$

   (d) $\dfrac{12x}{4xy + 8x}$ (e) $\dfrac{13x}{x^2 + 5x}$ (f) $\dfrac{17y}{9y^2 + 4y}$

5. Simplify the following:

   (a) $\dfrac{5}{15 + 10x}$ (b) $\dfrac{2x}{x^2 + 7x}$ (c) $\dfrac{2x + 8}{x^2 + 2x - 8}$

   (d) $\dfrac{7ab}{a^2b^2 + 9ab}$ (e) $\dfrac{xy}{xy + x}$

6. Simplify

   (a) $\dfrac{x - 4}{(x - 4)(x - 2)}$ (b) $\dfrac{2x - 4}{x^2 + x - 6}$

   (c) $\dfrac{3x}{3x^2 + 6x}$ (d) $\dfrac{x^2 + 2x + 1}{x^2 - 2x - 3}$ (e) $\dfrac{2(x - 3)}{(x - 3)^2}$

   (f) $\dfrac{x - 3}{(x - 3)^2}$ (g) $\dfrac{x - 3}{2(x - 3)^2}$ (h) $\dfrac{4(x - 3)}{2(x - 3)^2}$

   (i) $\dfrac{x + 4}{2(x + 4)^2}$ (j) $\dfrac{x + 4}{2(x + 4)}$ (k) $\dfrac{2(x + 4)}{(x + 4)}$

   (l) $\dfrac{(x + 4)(x - 3)}{x - 3}$ (m) $\dfrac{x + 4}{(x - 3)(x + 4)}$

   (n) $\dfrac{x + 3}{x^2 + 7x + 12}$ (o) $\dfrac{x + 4}{2x + 8}$ (p) $\dfrac{x + 4}{2x + 9}$

9.3   Multiplication and division of algebraic fractions

To multiply two algebraic fractions together we multiply their numerators together and multiply their denominators together:

KEY POINT

$$\frac{a}{b} \times \frac{c}{d} = \frac{a \times c}{b \times d}$$

Any common factors in the result should be cancelled.

Worked examples

9.10   Simplify

$$\frac{4}{5} \times \frac{x}{y}$$

Solution   We multiply the numerators together and multiply the denominators together. That is,

$$\frac{4}{5} \times \frac{x}{y} = \frac{4x}{5y}$$

9.11   Simplify

$$\frac{4}{x} \times \frac{3y}{16}$$

Solution   The numerators are multiplied together and the denominators are multiplied together. Therefore,

$$\frac{4}{x} \times \frac{3y}{16} = \frac{4 \times 3y}{16x}$$

Because $16x = 4 \times 4x$, the common factor 4 can be cancelled. So

$$\frac{4 \times 3y}{16x} = \frac{\cancel{4} \times 3y}{\cancel{4} \times 4x} = \frac{3y}{4x}$$

9.12   Simplify

(a) $\frac{1}{2} \times x$ (b) $\frac{1}{2} \times (a + b)$

Solution   (a) Writing x as $\frac{x}{1}$ we can state

$$\frac{1}{2} \times x = \frac{1}{2} \times \frac{x}{1} = \frac{1 \times x}{2 \times 1} = \frac{x}{2}$$

(b) Writing $a + b$ as $\frac{a+b}{1}$ we can state

$$\frac{1}{2} \times (a + b) = \frac{1}{2} \times \frac{(a + b)}{1} = \frac{1 \times (a + b)}{2 \times 1} = \frac{a + b}{2}$$

9.13   Simplify

$$\frac{4x^2}{y} \times \frac{3x^3}{yz}$$

Solution   We multiply the numerators together and multiply the denominators together.

$$\frac{4x^2}{y} \times \frac{3x^3}{yz} = \frac{4x^2 \times 3x^3}{y \times yz} = \frac{12x^5}{y^2z}$$

9.14   Simplify

$$5 \times \left(\frac{x-3}{25} \right)$$

Solution   This means

$$\frac{5}{1} \times \frac{x-3}{25}$$

which equals

$$\frac{5 \times (x-3)}{1 \times 25}$$

A common factor of 5 can be cancelled from the numerator and denominator to give

$$\frac{(x-3)}{5}$$

9.15   Simplify

$$-\frac{1}{5} \times \frac{3x-4}{8}$$

Solution   We can write

$$-\frac{1}{5} \times \frac{3x-4}{8} = -\frac{1 \times (3x-4)}{5 \times 8} = -\frac{3x-4}{40}$$

Note that the answer can also be expressed as $\frac{(4-3x)}{40}$ because

$$-\frac{3x-4}{40} = \frac{-1}{1} \times \frac{3x-4}{40} = \frac{-3x+4}{40} = \frac{4-3x}{40}$$

You should be aware from the last example that a solution can often be expressed in a number of equivalent ways.

Worked examples

9.16   Simplify

$$\frac{a}{a+b} \times \frac{b}{5a^2}$$

Solution

$$\frac{a}{a+b} \times \frac{b}{5a^2} = \frac{ab}{5a^2(a+b)}$$

Cancelling the common factor of a in numerator and denominator gives

$$\frac{b}{5a(a+b)}$$

9.17 Simplify

$$\frac{x^2 + 4x + 3}{2x + 8} \times \frac{x + 4}{x + 1}$$

Solution Before multiplying the two fractions together we should try to factorize if possible so that common factors can be identified. By factorizing, we can write the given expressions as

$$\frac{(x+1)(x+3)}{2(x+4)} \times \frac{x+4}{x+1} = \frac{(x+1)(x+3)(x+4)}{2(x+4)(x+1)}$$

Cancelling common factors this simplifies to just

$$\frac{x + 3}{2}$$

Division is performed by inverting the second fraction and multiplying.

KEY POINT

$$\frac{a}{b} \div \frac{c}{d} = \frac{a}{b} \times \frac{d}{c}$$

Worked examples

9.18 Simplify

$$\frac{10a}{b} \div \frac{a^2}{3b}$$

Solution The second fraction is inverted and then multiplied by the first, that is,

$$\frac{10a}{b} \div \frac{a^2}{3b} = \frac{10a}{b} \times \frac{3b}{a^2} = \frac{30ab}{a^2 b} = \frac{30}{a}$$

9.19 Simplify

$$\frac{x^2 y^3}{z} \div \frac{y}{x}$$

Solution

$$\frac{x^2 y^3}{z} \div \frac{y}{x} = \frac{x^2 y^3}{z} \times \frac{x}{y} = \frac{x^3 y^3}{zy}$$

Any common factors in the result can be cancelled. So

$$\frac{x^3 y^3}{zy} = \frac{x^3 y^2}{z}$$

Self-assessment questions 9.3

1. The technique of multiplying and dividing algebraic fractions is identical to that used for numbers – true or false?

Exercise 9.3

1. Simplify

(a) $\dfrac{1}{2} \times \dfrac{y}{3}$ (b) $\dfrac{1}{3} \times \dfrac{z}{2}$ (c) $\dfrac{2}{5}$ of $\dfrac{1}{y}$ (d) $\dfrac{2}{5}$ of $\dfrac{1}{x}$

(e) $\dfrac{3}{4}$ of $\dfrac{x}{y}$ (f) $\dfrac{3}{5} \times \dfrac{x^2}{y}$ (g) $\dfrac{x}{y} \times \dfrac{3}{5}$ (h) $\dfrac{7}{8} \times \dfrac{x}{2y}$

(i) $\dfrac{1}{2} \times \dfrac{1}{2x}$ (j) $\dfrac{1}{2} \times \dfrac{x}{2}$ (k) $\dfrac{1}{2} \times \dfrac{2}{x}$ (l) $\dfrac{1}{3} \times \dfrac{x}{3}$

(m) $\dfrac{1}{3} \times \dfrac{3}{x}$ (n) $\dfrac{1}{3} \times \dfrac{1}{3x}$ (o) $\dfrac{1}{3} \times \dfrac{3x}{2}$

2. Simplify

(a) $\dfrac{1}{2} \div \dfrac{x}{2}$ (b) $\dfrac{1}{2} \div \dfrac{2}{x}$ (c) $\dfrac{2}{x} \div \dfrac{2}{x}$ (d) $\dfrac{x}{2} \div \dfrac{1}{2}$

(e) $\dfrac{2}{x} \div 2$ (f) $\dfrac{2}{x} \div \dfrac{1}{2}$ (g) $\dfrac{3}{x} \div \dfrac{1}{2}$

3. Simplify the following:

(a) $\dfrac{5}{4} \times \dfrac{a}{25}$ (b) $\dfrac{5}{4} \times \dfrac{a}{b}$ (c) $\dfrac{8a}{b^2} \times \dfrac{b}{16a^2}$

(d) $\dfrac{9x}{3y} \times \dfrac{2x}{y^2}$ (e) $\dfrac{3}{5a} \times \dfrac{b}{a}$

(f) $\dfrac{1}{4} \times \dfrac{x}{y}$ (g) $\dfrac{1}{3} \times \dfrac{x}{x+y}$

(h) $\dfrac{x-3}{x+4} \times \dfrac{1}{3x-9}$

4. Simplify the following:

(a) $\dfrac{3}{x} \times \dfrac{xy}{z^3}$ (b) $\dfrac{(3+x)}{x} \div \dfrac{y}{x}$ (c) $\dfrac{4}{3} \div \dfrac{16}{x}$

(d) $\dfrac{a}{bc^2} \times \dfrac{b^2c}{a}$

5. Simplify

(a) $\dfrac{x+2}{(x+5)(x+4)} \times \dfrac{x+5}{x+2}$

(b) $\dfrac{x-2}{4} \div \dfrac{x}{16}$ (c) $\dfrac{12ab}{5ef} \div \dfrac{4ab^2}{f}$

(d) $\dfrac{x+3y}{2x} \div \dfrac{y}{4x^2}$ (e) $\dfrac{3}{x} \times \dfrac{3}{y} \times \dfrac{1}{z}$

6. Simplify

$\dfrac{1}{x+1} \times \dfrac{2x+2}{x+3}$

7. Simplify

$\dfrac{x+1}{x+2} \times \dfrac{x^2+6x+8}{x^2+4x+3}$

9.4   Addition and subtraction of algebraic fractions

For revision of adding and subtracting fractions see §2.3.

The method is the same as that for adding or subtracting numerical fractions. Note that it is not correct to simply add or subtract the numerator and denominator. The lowest common denominator must first be found. This is the simplest expression which contains all original denominators as its factors. Each fraction is then written with this common denominator. The fractions can then be added or subtracted by adding or subtracting just the numerators, and dividing the result by the common denominator.

Worked examples

9.20 Add the fractions $\frac{3}{4}$ and $\frac{1}{x}$.

Solution We must find $\frac{3}{4} + \frac{1}{x}$. To do this we must first rewrite the fractions to ensure they have a common denominator. The common denominator is the simplest expression that has the given denominators as its factors. The simplest such expression is $4x$. We write

$$\frac{3}{4} \text{ as } \frac{3x}{4x}, \quad \text{and} \quad \frac{1}{x} \text{ as } \frac{4}{4x}$$

Then

$$\frac{3}{4} + \frac{1}{x} = \frac{3x}{4x} + \frac{4}{4x}$$

$$= \frac{3x + 4}{4x}$$

No further simplification is possible.

9.21 Simplify

$$\frac{3}{x} + \frac{4}{x^2}$$

Solution The expression $\frac{3}{x}$ is rewritten as $\frac{3x}{x^2}$ which makes the denominators of both terms x^2 but leaves the value of the expression unaltered. Note that both the original denominators, x and x^2, are factors of the new denominator. We call this denominator the **lowest common denominator.** The fractions are then added by adding just the numerators. That is,

$$\frac{3x}{x^2} + \frac{4}{x^2} = \frac{3x + 4}{x^2}$$

9.22 Express $\frac{5}{a} - \frac{4}{b}$ as a single fraction.

Solution Both fractions are rewritten to have the same denominator. The simplest expression containing both a and b as its factors is ab. Therefore ab is the lowest common denominator. Then

$$\frac{5}{a} - \frac{4}{b} = \frac{5b}{ab} - \frac{4a}{ab} = \frac{5b - 4a}{ab}$$

9.23 Write $\frac{4}{x+y} - \frac{3}{y}$ as a single fraction.

Solution   The simplest expression which contains both denominators as its factors is $(x + y)y$. We must rewrite each term so that it has this denominator.

$$\frac{4}{x + y} = \frac{4}{x + y} \times \frac{y}{y} = \frac{4y}{(x + y)y}$$

Similarly,

$$\frac{3}{y} = \frac{3}{y} \times \frac{x + y}{x + y} = \frac{3(x + y)}{(x + y)y}$$

The fractions are then subtracted by subtracting just the numerators:

$$\frac{4}{x + y} - \frac{3}{y} = \frac{4y}{(x + y)y} - \frac{3(x + y)}{(x + y)y} = \frac{4y - 3(x + y)}{(x + y)y}$$

which simplifies to

$$\frac{y - 3x}{(x + y)y}$$

9.24   Express as a single fraction

$$\frac{2}{x + 3} + \frac{5}{x - 1}$$

Solution   The simplest expression having both $x + 3$ and $x - 1$ as its factors is

$$(x + 3)(x - 1)$$

This is the lowest common denominator. Each term is rewritten so that it has this denominator. Thus

$$\frac{2}{x + 3} = \frac{2(x - 1)}{(x + 3)(x - 1)} \qquad \text{and} \qquad \frac{5}{x - 1} = \frac{5(x + 3)}{(x + 3)(x - 1)}$$

Then

$$\frac{2}{x + 3} + \frac{5}{x - 1} = \frac{2(x - 1)}{(x + 3)(x - 1)} + \frac{5(x + 3)}{(x + 3)(x - 1)}$$

$$= \frac{2(x - 1) + 5(x + 3)}{(x + 3)(x - 1)}$$

which simplifies to

$$\frac{7x + 13}{(x + 3)(x - 1)}$$

9.25   Express as a single fraction

$$\frac{1}{x - 4} + \frac{1}{(x - 4)^2}$$

Solution   The simplest expression having $x - 4$ and $(x - 4)^2$ as its factors is $(x - 4)^2$.

Both fractions are rewritten with this denominator

$$\frac{1}{x - 4} + \frac{1}{(x - 4)^2} = \frac{(x - 4)}{(x - 4)^2} + \frac{1}{(x - 4)^2}$$

$$= \frac{x - 4 + 1}{(x - 4)^2}$$

$$= \frac{x - 3}{(x - 4)^2}$$

Self-assessment questions 9.4

1. Explain what is meant by the 'lowest common denominator' and how it is found.

Exercise 9.4

1. Express each of the following as a single fraction:

   (a) $\dfrac{z}{2} + \dfrac{z}{3}$ (b) $\dfrac{x}{3} + \dfrac{x}{4}$ (c) $\dfrac{y}{5} + \dfrac{y}{25}$

2. Express each of the following as a single fraction:

   (a) $\dfrac{1}{2} + \dfrac{1}{x}$ (b) $\dfrac{1}{2} + x$ (c) $\dfrac{1}{3} + y$

   (d) $\dfrac{1}{3} + \dfrac{1}{y}$ (e) $8 + \dfrac{1}{y}$

3. Express each of the following as a single fraction:

   (a) $\dfrac{5}{x} - \dfrac{1}{2}$ (b) $\dfrac{5}{x} + 2$ (c) $\dfrac{3}{x} - \dfrac{1}{3}$

   (d) $\dfrac{x}{3} - \dfrac{1}{2}$ (e) $\dfrac{3}{x} + \dfrac{1}{3}$

4. Express each of the following as a single fraction:

   (a) $\dfrac{3}{x} + \dfrac{4}{y}$ (b) $\dfrac{3}{x^2} + \dfrac{4y}{x}$ (c) $\dfrac{4ab}{x} + \dfrac{3ab}{2y}$

   (d) $\dfrac{4xy}{a} + \dfrac{3xy}{2b}$ (e) $\dfrac{3}{x} - \dfrac{6}{2x}$ (f) $\dfrac{3x}{2y} - \dfrac{7y}{4x}$

   (g) $\dfrac{3}{x + y} - \dfrac{2}{y}$ (h) $\dfrac{1}{a + b} - \dfrac{1}{a - b}$

   (i) $2x + \dfrac{1}{2x}$ (j) $2x - \dfrac{1}{2x}$

5. Express each of the following as a single fraction:

   (a) $\dfrac{x}{y} + \dfrac{3x^2}{z}$ (b) $\dfrac{4}{a} + \dfrac{5}{b}$ (c) $\dfrac{6x}{y} - \dfrac{2y}{x}$

   (d) $3x - \dfrac{3x + 1}{4}$ (e) $\dfrac{5a}{12} + \dfrac{9a}{18}$ (f) $\dfrac{x - 3}{4} + \dfrac{3}{5}$

6. Express each of the following as a single fraction:

   (a) $\dfrac{1}{x + 1} + \dfrac{1}{x + 2}$ (b) $\dfrac{1}{x - 1} + \dfrac{2}{x + 3}$

   (c) $\dfrac{3}{x + 5} + \dfrac{1}{x + 4}$ (d) $\dfrac{1}{x - 2} + \dfrac{3}{x - 4}$

   (e) $\dfrac{3}{2x + 1} + \dfrac{1}{x + 1}$ (f) $\dfrac{3}{1 - 2x} + \dfrac{1}{x}$

   (g) $\dfrac{3}{x + 1} + \dfrac{4}{(x + 1)^2}$ (h) $\dfrac{1}{x - 1} + \dfrac{1}{(x - 1)^2}$

Test and assignment exercises 9

1. Simplify

   (a) $\dfrac{5a}{4a + 3ab}$ (b) $\dfrac{5a}{30a + 15b}$ (c) $\dfrac{5ab}{30a + 15b}$

   (d) $\dfrac{5ab}{ab + 7ab}$ (e) $\dfrac{y}{13y + y^2}$ (f) $\dfrac{13y + y^2}{y}$

2. Simplify the following:

   (a) $\dfrac{5a}{7} \times \dfrac{14b}{2}$ (b) $\dfrac{3}{x} + \dfrac{7}{3x}$ (c) $t - \dfrac{4 - t}{2}$

   (d) $4(x + 3) - \dfrac{(4x - 5)}{3}$ (e) $\dfrac{7}{x} + \dfrac{3}{2x} + \dfrac{5}{3x}$

   (f) $x + \dfrac{3x}{y}$ (g) $xy + \dfrac{1}{xy}$

   (h) $\dfrac{1}{x + y} + \dfrac{2}{x - y}$

3. Simplify the following:

   (a) $\dfrac{y}{9} + \dfrac{2y}{7}$ (b) $\dfrac{3}{x} - \dfrac{5}{3x} + \dfrac{4}{5x}$ (c) $\dfrac{3x}{2y} + \dfrac{5y}{6x}$

   (d) $m + \dfrac{m + n}{2}$ (e) $m - \dfrac{m + n}{2}$

   (f) $m - \dfrac{m - n}{2}$ (g) $\dfrac{3s - 5}{10} - \dfrac{2s - 3}{15}$

4. Simplify

   $\dfrac{x^2 - x}{x - 1}$

10 Transposing formulae

Objectives	This chapter
	• explains how formulae can be rearranged or transposed

10.1 Rearranging a formula

In the formula for the area of a circle, $A = \pi r^2$, we say that A is the **subject** of the formula. The subject appears by itself on one side, usually the left, of the formula, and nowhere else. If we are asked to **transpose** the formula for r, then we must rearrange the formula so that r becomes the subject. The rules for transposing formulae are quite simple. Essentially whatever you do to one side you must also do to the whole of the other side.

KEY POINT

To transpose a formula you may:

- add the same quantity to both sides
- subtract the same quantity from both sides
- multiply or divide both sides by the same quantity
- perform operations on both sides, such as 'square both sides', 'square root both sides' etc.

Worked examples

10.1 The circumference C of a circle is given by the formula $C = 2\pi r$. Transpose this to make r the subject.

Solution The intention is to obtain r by itself on the left-hand side. Starting with $C = 2\pi r$ we must try to isolate r. Dividing both sides by 2π we find

$$C = 2\pi r$$

$$\frac{C}{2\pi} = \frac{2\pi r}{2\pi}$$

$$\frac{C}{2\pi} = r \quad \text{by cancelling the common factor } 2\pi \text{ on the right}$$

Finally we can write $r = \frac{C}{2\pi}$ and the formula has been transposed for r.

10.2 Transpose the formula $y = 3(x + 7)$ for x.

Solution We must try to obtain x on its own on the left-hand side. This can be done in stages. Dividing both sides by 3 we find

$$\frac{y}{3} = \frac{3(x+7)}{3} = x + 7 \qquad \text{by cancelling the common factor 3}$$

Subtracting 7 from both sides then gives

$$\frac{y}{3} - 7 = x + 7 \ - 7$$

so that

$$\frac{y}{3} - 7 = x$$

Equivalently we can write $x = \dfrac{y}{3} - 7$, and the formula has been transposed for x.

Alternatively, again starting from $y = 3(x + 7)$ we could proceed as follows. Removing the brackets we obtain $y = 3x + 21$. Then, subtracting 21 from both sides,

$$y - 21 = 3x + 21 - 21 = 3x$$

Dividing both sides by 3 will give x on its own:

$$\frac{y - 21}{3} = x \qquad \text{so that} \qquad x = \frac{y - 21}{3}$$

Noting that

$$\frac{y - 21}{3} = \frac{y}{3} - \frac{21}{3} = \frac{y}{3} - 7$$

we see that this answer is equivalent to the expression obtained previously.

10.3   Transpose the formula $y - z = 3(x + 2)$ for x.

Solution   Dividing both sides by 3 we find

$$\frac{y - z}{3} = x + 2$$

Subtracting 2 from both sides gives

$$\frac{y - z}{3} - 2 = x + 2 - 2$$

$$= x$$

so that

$$x = \frac{y - z}{3} - 2$$

10.4   Transpose $x + xy = 7$, (a) for x, (b) for y.

Solution   (a) In the expression $x + xy$, note that x is a common factor. We can then write the given formula as $x(1 + y) = 7$. Dividing both sides by $1 + y$ will isolate x. That is,

$$\frac{x(1 + y)}{1 + y} = \frac{7}{1 + y}$$

Cancelling the common factor $1 + y$ on the left-hand side gives

$$x = \frac{7}{1 + y}$$

and the formula has been transposed for x.

(b) To transpose $x + xy = 7$ for y we first subtract x from both sides to give $xy = 7 - x$. Finally dividing both sides by x gives

$$\frac{xy}{x} = \frac{7 - x}{x}$$

Cancelling the common factor of x on the left-hand side gives

$$y = \frac{7 - x}{x}$$ which is the required transposition.

10.5   Transpose the formula $I = \frac{V}{R}$ for R.

Solution   Starting with $I = \frac{V}{R}$ we first multiply both sides by R. This gives

$$IR = \frac{V}{R} \times R = V$$

Then, dividing both sides by I gives

$$\frac{IR}{I} = \frac{V}{I}$$

so that $R = \frac{V}{I}$.

10.6   Make x the subject of the formula $y = \dfrac{4}{x-7}$

Solution   We must try to obtain x on its own on the left-hand side. It is often useful to multiply both sides of the formula by the same quantity in order to remove fractions. Multiplying both sides by $(x-7)$ we find

$$(x-7)y = (x-7) \times \frac{4}{x-7} = 4$$

Removing the brackets on the left-hand side we find

$$xy - 7y = 4 \qquad \text{so that} \qquad xy = 7y + 4$$

Finally, dividing both sides by y we obtain

$$\frac{xy}{y} = \frac{7y+4}{y}$$

that is,

$$x = \frac{7y+4}{y}$$

10.7   Transpose the formula

$$T = 2\pi\sqrt{\frac{l}{g}}$$

to make l the subject.

Solution   We must attempt to isolate l. We can do this in stages as follows. Dividing both sides by 2π we find

$$\frac{T}{2\pi} = \frac{2\pi\sqrt{\frac{l}{g}}}{2\pi} = \sqrt{\frac{l}{g}} \qquad \text{by cancelling the common factor } 2\pi$$

Recall from §6.3 that $(\sqrt{a})^2 = a$.

The square root sign over the l/g can be removed by squaring both sides. Recall that if a term containing a square root is squared then the square root will disappear.

$$\left(\frac{T}{2\pi}\right)^2 = \frac{l}{g}$$

Then multiplying both sides by g we find

$$g\left(\frac{T}{2\pi}\right)^2 = l$$

Equivalently we have

$$l = g\left(\frac{T}{2\pi}\right)^2$$

and the formula has been transposed for l.

Self-assessment questions 10.1

1. Explain what is meant by the 'subject' of a formula.
2. In what circumstances might you want to transpose a formula?
3. Explain what processes are allowed when transposing a formula.

Exercise 10.1

1. Transpose each of the following formulae to make x the subject:

    (a) $y = 3x$ (b) $y = \dfrac{1}{x}$ (c) $y = 7x - 5$

    (d) $y = \dfrac{1}{2}x - 7$ (e) $y = \dfrac{1}{2x}$

    (f) $y = \dfrac{1}{2x + 1}$ (g) $y = \dfrac{1}{2x} + 1$

    (h) $y = 18x - 21$ (i) $y = 19 - 8x$

2. Transpose $y = mx + c$, (a) for m, (b) for x, (c) for c.

3. Transpose the following formulae for x:

    (a) $y = 13(x - 2)$ (b) $y = x\left(1 + \dfrac{1}{x}\right)$

    (c) $y = a + t(x - 3)$

4. Transpose each of the following formulae to make the given variable the subject:

    (a) $y = 7x + 11$, for x (b) $V = IR$, for I

    (c) $V = \dfrac{4}{3}\pi r^3$, for r (d) $F = ma$, for m

5. Make n the subject of the formula $l = a + (n - 1)d$.

6. If $m = n + t\sqrt{x}$, find an expression for x.

7. Make x the subject of the following formulae:

    (a) $y = 1 - x^2$ (b) $y = \dfrac{1}{1 - x^2}$

    (c) $y = \dfrac{1 - x^2}{1 + x^2}$

Test and assignment exercises 10

1. Make x the subject of the following formulae:

    (a) $y = 13x - 18$ (b) $y = \dfrac{13}{x + 18}$

    (c) $y = \dfrac{13}{x} + 18$

2. Transpose the following formulae:

    (a) $y = \dfrac{7 + x}{14}$ for x (b) $E = \dfrac{1}{2}mv^2$ for v

    (c) $8x - 13y = 12$ for y (d) $y = \dfrac{x^2}{2g}$ for x

3. Make x the subject of the formula $v = k/\sqrt{x}$.

4. Transpose the formula $V = \pi r^2 h$ for h.

5. Make r the subject of the formula $y = H + Cr$.

6. Transpose the formula $s = ut + \dfrac{1}{2}at^2$ for a.

7. Transpose the formula $v^2 = u^2 + 2as$ for u.

8. Transpose the formula $k = pv^3$ for v.

11 Solving equations

Objectives

This chapter

- explains what is meant by an equation and its solution
- shows how to solve linear, simultaneous and quadratic equations

An **equation** states that two quantities are equal, and will always contain an **unknown quantity** which we wish to find. For example, in the equation $5x + 10 = 20$ the unknown quantity is x. To **solve** an equation means to find all values of the unknown quantity which can be substituted into the equation so that the left side equals the right side. Each such value is called a **solution** or alternatively a **root** of the equation. In the example above, the solution is $x = 2$ because when $x = 2$ is substituted both the left side and the right side equal 20. The value $x = 2$ is said to **satisfy** the equation.

11.1 Solving linear equations

A **linear** equation is one of the form $ax + b = 0$ where a and b are numbers and the unknown quantity is x. The number a is called the **coefficient** of x. The number b is called the **constant term**. For example, $3x + 7 = 0$ is a linear equation. The coefficient of x is 3 and the constant term is 7. Similarly, $-2x + 17.5 = 0$ is a linear equation. The coefficient of x is -2 and the constant term is 17.5. Note that the unknown quantity only occurs to the first power, that is, as x, and not as x^2, x^3, $x^{1/2}$ etc. Linear equations may appear in other forms which may seem to be different but are nevertheless equivalent. Thus

$$4x + 13 = -7, \qquad 3 - 14x = 0 \qquad \text{and} \qquad 3x + 7 = 2x - 4$$

are all linear equations which could be written in the form $ax + b = 0$ if necessary.

An equation such as $3x^2 + 2 = 0$ is not linear because the unknown quantity occurs to the power 2. Linear equations are solved by trying to obtain the unknown quantity on its own on the left-hand side, that is, by making the unknown quantity the subject of the equation. This is done using the same rules given for transposing formulae in Chapter 10. Consider the following examples.

Worked example

11.1 Solve the equation $x + 10 = 0$.

Solution We make x the subject by subtracting 10 from both sides to give $x = -10$. Therefore $x = -10$ is the solution. It can be easily checked by substituting into the original equation:

$$(-10) + 10 = 0 \text{ as required}$$

Note from the last example that the solution should be checked by substitution to ensure it satisfies the given equation. If it does not then a mistake has been made.

Worked examples

11.2 Solve the equation $4x + 8 = 0$.

Solution In order to find the unknown quantity x we attempt to make it the subject of the equation. Subtracting 8 from both sides we find

$$4x + 8 - 8 = 0 - 8 = -8$$

that is,

$$4x = -8$$

Then, dividing both sides by 4 gives

$$\frac{4x}{4} = \frac{-8}{4}$$

so that $x = -2$. The solution of the equation $4x + 8 = 0$ is $x = -2$.

11.3 Solve the equation $5x + 17 = 4x - 3$.

Solution First we collect all terms involving x together. This is done by subtracting $4x$ from both sides to remove this term from the right. This gives

$$5x - 4x + 17 = -3$$

that is, $x + 17 = -3$. To make x the subject of this equation we now subtract 17 from both sides to give $x = -3 - 17 = -20$. The solution of the equation $5x + 17 = 4x - 3$ is $x = -20$. Note that the answer can be easily checked by substituting $x = -20$ into the original equation and verifying that the left side equals the right side.

11.4   Solve the equation $\dfrac{x-3}{4} = 1$.

Solution   We attempt to obtain x on its own. First note that if we multiply both sides of the equation by 4 this will remove the 4 in the denominator. That is,

$$4 \times \left(\frac{x-3}{4}\right) = 4 \times 1$$

so that

$$x - 3 = 4$$

Finally adding 3 to both sides gives $x = 7$.

Self-assessment questions 11.1

1. Explain what is meant by a root of an equation.
2. Explain what is meant by a linear equation.
3. State the rules which can be used to solve a linear equation.
4. You may think that a formula and an equation look very similar. Try to explain the distinction between a formula and an equation.

Exercise 11.1

1. Verify that the given values of x satisfy the given equations:

   (a) $x = 7$ satisfies $3x + 4 = 25$
   (b) $x = -5$ satisfies $2x - 11 = -21$
   (c) $x = -4$ satisfies $-x - 8 = -4$
   (d) $x = \frac{1}{2}$ satisfies $8x + 4 = 8$
   (e) $x = -\frac{1}{3}$ satisfies $27x + 8 = -1$
   (f) $x = 4$ satisfies $3x + 2 = 7x - 14$

2. Solve the following linear equations:

   (a) $3x = 9$ (b) $\dfrac{x}{3} = 9$ (c) $3t + 6 = 0$

   (d) $3x - 13 = 2x + 9$ (e) $3x + 17 = 21$
   (f) $4x - 20 = 3x + 16$
   (g) $5 - 2x = 2 + 3x$
   (h) $\dfrac{x+3}{2} = 3$ (i) $\dfrac{3x+2}{2} + 3x = 1$

3. Solve the following equations:

   (a) $5(x + 2) = 13$
   (b) $3(x - 7) = 2(x + 1)$
   (c) $5(1 - 2x) = 2(4 - 2x)$

4. Solve the following equations:

   (a) $3t + 7 = 4t - 2$ (b) $3v = 17 - 4v$
   (c) $3s + 2 = 14(s - 1)$

5. Solve the following linear equations:

   (a) $5t + 7 = 22$ (b) $7 - 4t = -13$
   (c) $7 - 4t = 27$ (d) $5 = 14 - 3t$

   (e) $4x + 13 = -x + 25$ (f) $\dfrac{x+3}{2} = \dfrac{x-3}{4}$

   (g) $\dfrac{1}{3}x + 6 = \dfrac{1}{2}x + 2$ (h) $\dfrac{1}{5}x + 7 = \dfrac{1}{3}x + 5$

   (i) $\dfrac{2x+4}{5} = \dfrac{x-3}{2}$ (j) $\dfrac{x-7}{8} = \dfrac{3x+1}{5}$

6. The following equations may not appear to be linear at first sight but they can all be rewritten in the standard form of a linear equation. Find the solution of each equation.

   (a) $\dfrac{1}{x} = 5$ (b) $\dfrac{1}{x} = \dfrac{5}{2}$

   (c) $\dfrac{1}{x+1} = \dfrac{5}{2}$ (d) $\dfrac{1}{x-1} = \dfrac{5}{2}$

   (e) $\dfrac{1}{x} = \dfrac{1}{2x+1}$ (f) $\dfrac{3}{x} = \dfrac{1}{2x+1}$

   (g) $\dfrac{1}{x+1} = \dfrac{1}{3x+2}$ (h) $\dfrac{3}{x+1} = \dfrac{2}{4x+1}$

| 11.2 | **Solving simultaneous equations** |

Sometimes equations contain more than one unknown quantity. When this happens there are usually two or more equations. For example in the two equations

$$x + 2y = 14 \qquad 3x + y = 17$$

the unknowns are x and y. Such equations are called **simultaneous equations** and to solve them we must find values of x and y which satisfy both equations at the same time. If we substitute $x = 4$ and $y = 5$ into either of the two equations above we see that the equation is satisfied. We shall demonstrate how simultaneous equations can be solved by removing, or **eliminating**, one of the unknowns.

Worked examples

11.5   Solve the simultaneous equations

$$x + 3y = 14 \tag{11.1}$$

$$2x - 3y = -8 \tag{11.2}$$

Solution   Note that if these two equations are added the unknown y is removed or eliminated:

$$\begin{array}{r} x + 3y = 14 \\ 2x - 3y = -8 \\ \hline 3x = 6 \end{array} \; +$$

so that $x = 2$. To find y we substitute $x = 2$ into either equation. Substituting into Equation 11.1 gives

$$2 + 3y = 14$$

Solving this linear equation will give y. We have

$$2 + 3y = 14$$
$$3y = 14 - 2 = 12$$
$$y = \frac{12}{3} = 4$$

Therefore the solution of the simultaneous equations is $x = 2$ and $y = 4$. Note that these solutions should be checked by substituting back into both given equations to check that the left-hand side equals the right-hand side.

11.6   Solve the simultaneous equations

$$5x + 4y = 7 \tag{11.3}$$
$$3x - y = 11 \tag{11.4}$$

Solution   Note that in this example, if we multiply the first equation by 3 and the second by 5 we shall have the same coefficient of x in both equations. This gives

$$15x + 12y = 21 \tag{11.5}$$
$$15x - 5y = 55 \tag{11.6}$$

We can now eliminate x by subtracting Equation 11.6 from Equation 11.5, giving

$$
\begin{aligned}
15x + 12y &= 21 \\
15x - 5y &= 55 \\
\hline
17y &= -34
\end{aligned}
$$

from which $y = -2$. In order to find x we substitute our solution for y into either of the given equations. Substituting into Equation 11.3 gives

$$5x + 4(-2) = 7 \quad \text{so that} \quad 5x = 15 \quad \text{or} \quad x = 3$$

The solution of the simultaneous equations is therefore $x = 3$, $y = -2$.

Exercise 11.2

1. Verify that the given values of x and y satisfy the given simultaneous equations.

   (a) $x = 7$, $y = 1$ satisfy $2x - 3y = 11$,
       $3x + y = 22$

   (b) $x = -7$, $y = 2$ satisfy $2x + y = -12$,
       $x - 5y = -17$

   (c) $x = -1$, $y = -1$ satisfy $7x - y = -6$,
       $x - y = 0$

2. Solve the following pairs of simultaneous equations:

   (a) $3x + y = 1$, $2x - y = 2$
   (b) $4x + 5y = 21$, $3x + 5y = 17$

   (c) $2x - y = 17$, $x + 3y = 12$
   (d) $-2x + y = -21$, $x + 3y = -14$
   (e) $-x + y = -10$, $3x + 7y = 20$
   (f) $4x - 2y = 2$, $3x - y = 4$

3. Solve the following simultaneous equations.

   (a) $5x + y = 36$, $3x - y = 20$
   (b) $x - 3y = -13$, $4x + 2y = -24$
   (c) $3x + y = 30$, $-5x + 3y = -50$
   (d) $3x - y = -5$, $-7x + 3y = 15$
   (e) $11x + 13y = -24$, $x + y = -2$

11.3   ## Solving quadratic equations

A **quadratic equation** is an equation of the form $ax^2 + bx + c = 0$ where a, b and c are numbers and x is the unknown quantity we wish to find. The number a is the **coefficient** of x^2, b is the coefficient of x and c is the **constant term**. Sometimes b or c may be zero, although a can never be zero. For example

$$x^2 + 7x + 2 = 0 \qquad 3x^2 - 2 = 0 \qquad -2x^2 + 3x = 0 \qquad 8x^2 = 0$$

are all quadratic equations.

KEY POINT

A quadratic equation has the form $ax^2 + bx + c = 0$ where a, b and c are numbers, and x represents the unknown we wish to find.

Worked examples

11.7   State the coefficient of x^2 and the coefficient of x in the following quadratic equations:

(a) $4x^2 + 3x - 2 = 0$ (b) $x^2 - 23x + 17 = 0$ (c) $-x^2 + 19 = 0$

Solution   (a) In the equation $4x^2 + 3x - 2 = 0$ the coefficient of x^2 is 4 and the coefficient of x is 3.

(b) In the equation $x^2 - 23x + 17 = 0$ the coefficient of x^2 is 1 and the coefficient of x is -23.

(c) In the equation $-x^2 + 19 = 0$ the coefficient of x^2 is -1 and the coefficient of x is zero, since there is no term involving just x.

11.8   Verify that both $x = -7$ and $x = 5$ satisfy the quadratic equation $x^2 + 2x - 35 = 0$.

Solution   We substitute $x = -7$ into the left-hand side of the equation. This yields

$$(-7)^2 + 2(-7) - 35$$

that is

$$49 - 14 - 35$$

This simplifies to zero, and so the left-hand side equals the right-hand side of the given equation. Therefore $x = -7$ is a solution.

Similarly, if $x = 5$ we find

$$(5^2) + 2(5) - 35 = 25 + 10 - 35$$

which also simplifies to zero. We conclude that $x = 5$ is also a solution.

Solution by factorization

If the quadratic expression on the left-hand side of the equation can be factorized solutions can be found using the method in the following examples.

Worked examples

11.9 Solve the quadratic equation $x^2 + 3x - 10 = 0$.

Solution The left-hand side of the equation can be factorized to give

$$x^2 + 3x - 10 = (x + 5)(x - 2) = 0$$

Whenever the product of two quantities equals zero, then one or both of these quantities must be zero. It follows that either $x + 5 = 0$ or $x - 2 = 0$ from which $x = -5$ and $x = 2$ are the required solutions.

11.10 Solve the quadratic equation $6x^2 + 5x - 4 = 0$.

Solution The left-hand side of the equation can be factorized to give

$$6x^2 + 5x - 4 = (2x - 1)(3x + 4) = 0$$

It follows that either $2x - 1 = 0$ or $3x + 4 = 0$ from which $x = \frac{1}{2}$ and $x = -\frac{4}{3}$ are the required solutions.

11.11 Solve the quadratic equation $x^2 - 8x = 0$.

Solution Factorizing the left-hand side gives

$$x^2 - 8x = x(x - 8) = 0$$

from which either $x = 0$ or $x - 8 = 0$. The solutions are therefore $x = 0$ and $x = 8$.

11.12 Solve the quadratic equation $x^2 - 36 = 0$.

Solution The left-hand side factorizes to give

$$x^2 - 36 = (x + 6)(x - 6) = 0$$

so that $x + 6 = 0$ or $x - 6 = 0$. The solutions are therefore $x = -6$ and $x = 6$.

11.13 Solve the equation $x^2 = 81$.

Solution Writing this in the standard form of a quadratic equation we obtain $x^2 - 81 = 0$. The left-hand side can be factorized to give

$$x^2 - 81 = (x - 9)(x + 9) = 0$$

from which $x - 9 = 0$ or $x + 9 = 0$. The solutions are then $x = 9$ and $x = -9$.

Solution of quadratic equations using the formula

It should be noted that not all quadratic equations have solutions. However, when it is difficult or impossible to factorize the quadratic expression $ax^2 + bx + c$, solutions of a quadratic equation can be sought using the following formula:

KEY POINT

> If $ax^2 + bx + c = 0$, then $x = \dfrac{-b \pm \sqrt{b^2 - 4ac}}{2a}$

This formula gives possibly two solutions: one solution is obtained by taking the positive square root and the second solution by taking the negative square root.

Worked examples

11.14   Solve the equation $x^2 + 9x + 20 = 0$ using the formula.

Solution   Comparing the given equation with the standard form $ax^2 + bx + c = 0$ we see that $a = 1$, $b = 9$ and $c = 20$. These values are substituted into the formula

$$x = \frac{-b \pm \sqrt{b^2 - 4ac}}{2a}$$

$$= \frac{-9 \pm \sqrt{81 - 4(1)(20)}}{(2)(1)}$$

$$= \frac{-9 \pm \sqrt{81 - 80}}{2}$$

$$= \frac{-9 \pm \sqrt{1}}{2}$$

$$= \frac{-9 \pm 1}{2}$$

$$= \begin{cases} -4 & \text{by taking the positive square root} \\ -5 & \text{by taking the negative square root} \end{cases}$$

The two solutions are therefore $x = -4$ and $x = -5$.

11.15   Solve the equation $2x^2 - 3x - 7 = 0$ using the formula.

Solution   In this example $a = 2$, $b = -3$ and $c = -7$. Care should be taken with the negative signs. Substituting these into the formula we find

$$x = \frac{-(-3) \pm \sqrt{(-3)^2 - 4(2)(-7)}}{2(2)}$$

$$= \frac{3 \pm \sqrt{9 + 56}}{4}$$

$$= \frac{3 \pm \sqrt{65}}{4}$$

$$= \frac{3 \pm 8.062}{4}$$

$$= \begin{cases} 2.766 & \text{by taking the } +\text{ve square root} \\ -1.266 & \text{by taking the } -\text{ve square root} \end{cases}$$

The two solutions are therefore $x = 2.766$ and $x = -1.266$.

If the values of a, b and c are such that $b^2 - 4ac$ is positive the formula will produce two solutions known as **distinct real roots** of the equation. If $b^2 - 4ac = 0$ there will be a single root known as a **repeated root**. Some books refer to the equation having **equal roots**. If the equation is such that $b^2 - 4ac$ is negative the formula requires us to find the square root of a negative number. In ordinary arithmetic this is impossible and we say that in such a case the quadratic equation does not possess any real roots.

Worked examples

11.16   Can the equation $x^2 + 49 = 0$ be solved using the formula?

Solution   In this example $a = 1$, $b = 0$ and $c = 49$. Therefore $b^2 - 4ac = 0 - 4(1)(49)$ which is negative. Therefore this equation does not possess any real roots.

11.17   Use the formula to solve the equation $4x^2 + 4x + 1 = 0$.

Solution   In this example $a = 4$, $b = 4$ and $c = 1$. Applying the formula gives

$$x = \frac{-4 \pm \sqrt{4^2 - 4(4)(1)}}{(2)(4)}$$

$$= \frac{-4 \pm \sqrt{16 - 16}}{8}$$

$$= \frac{-4 \pm 0}{8} = -\frac{1}{2}$$

There is a single, repeated root $x = -\frac{1}{2}$. Note that $b^2 - 4ac = 0$.

Self-assessment questions 11.3

1. Under what conditions will a quadratic equation possess distinct real roots?
2. Under what conditions will a quadratic equation possess a repeated root?

Exercise 11.3

1. Solve the following quadratic equations by factorization:

   (a) $x^2 + x - 2 = 0$ (b) $x^2 - 8x + 15 = 0$
   (c) $4x^2 + 6x + 2 = 0$ (d) $x^2 - 6x + 9 = 0$
   (e) $x^2 - 81 = 0$ (f) $x^2 + 4x + 3 = 0$
   (g) $x^2 + 2x - 3 = 0$ (h) $x^2 + 3x - 4 = 0$
   (i) $x^2 + 6x + 5 = 0$ (j) $x^2 - 12x + 35 = 0$
   (k) $x^2 + 12x + 35 = 0$ (l) $2x^2 + x - 3 = 0$
   (m) $2x^2 - x - 6 = 0$ (n) $2x^2 - 7x - 15 = 0$
   (o) $3x^2 - 2x - 1 = 0$ (p) $9x^2 - 12x - 5 = 0$
   (q) $7x^2 + x = 0$ (r) $4x^2 + 12x + 9 = 0$

2. Solve the following quadratic equations using the formula:

   (a) $3x^2 - 6x - 5 = 0$
   (b) $x^2 + 3x - 77 = 0$ (c) $2x^2 - 9x + 2 = 0$
   (d) $x^2 + 3x - 4 = 0$ (e) $3x^2 - 3x - 4 = 0$
   (f) $4x^2 + x - 1 = 0$ (g) $x^2 - 7x - 3 = 0$
   (h) $x^2 + 7x - 3 = 0$ (i) $11x^2 + x + 1 = 0$
   (j) $2x^2 - 3x - 7 = 0$

3. Solve the following quadratic equations:

   (a) $6x^2 + 13x + 6 = 0$
   (b) $3t^2 + 13t + 12 = 0$ (c) $t^2 - 7t + 3 = 0$

Test and assignment exercises 11

1. Solve the following equations:

   (a) $13t - 7 = 2t + 5$
   (b) $-3t + 9 = 13 - 7t$ (c) $-5t = 0$

2. Solve the following equations:

   (a) $4x = 16$ (b) $\dfrac{x}{12} = 9$ (c) $4x - 13 = 3$

   (d) $4x - 14 = 2x + 8$ (e) $3x - 17 = 4$

   (f) $7 - x = 9 + 3x$ (g) $4(2 - x) = 8$

3. Solve the following equations:

   (a) $2y = 8$ (b) $5u = 14u + 3$ (c) $5 = 4I$

   (d) $13i + 7 = 2i - 9$ (e) $\dfrac{1}{4}x + 9 = 3 - \dfrac{1}{2}x$

   (f) $\dfrac{4x + 2}{2} + 8x = 0$

4. Solve the following equations:

   (a) $3x^2 - 27 = 0$ (b) $2x^2 + 18x + 14 = 0$
   (c) $x^2 = 16$ (d) $x^2 - x - 72 = 0$
   (e) $2x^2 - 3x - 44 = 0$
   (f) $x^2 - 4x - 21 = 0$

5. The solutions of the equation $x^2 + 4x = 0$ are identical to the solutions of $3x^2 + 12x = 0$. True or false?

6. Solve the following simultaneous equations:

   (a) $x - 2y = -11, \quad 7x + y = -32$
   (b) $2x - y = 2, \quad x + 3y = 29$
   (c) $x + y = 19, \quad -x + y = 1$

12 Functions

Objectives

This chapter

- explains what is meant by a function
- describes the notation used to write functions
- explains the terms 'independent variable' and 'dependent variable'
- explains what is meant by a composite function
- explains what is meant by the inverse of a function

12.1   Definition of a function

A **function** is a rule which receives an input and produces an output. It is shown schematically in Figure 12.1. For example, the rule may be 'add 2 to the input'. If 6 is the input, then $6 + 2 = 8$ will be the output. If -5 is

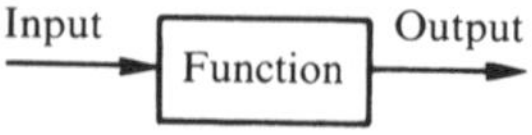

Figure 12.1
A function produces an output from an input

the input then $-5 + 2 = -3$ will be the output. In general, if x is the input then $x + 2$ will be the output. Figure 12.2 illustrates this function schematically.

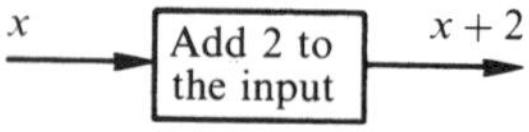

Figure 12.2
The function adds 2 to the input

For a rule to be a function then it is crucial that only **one** output is produced for any given input.

KEY POINT

> A function is a rule which produces a *single* output for any given input.

The input to a function can usually take many values and so is called a **variable**. The output, too, varies depending upon the value of the input, and so is also a variable. The input is referred to as the **independent variable** because we are free to choose its value. The output is called the **dependent variable** because its value depends upon the value of the input.

12.2   Notation used for functions

We usually denote the input, the output and the function by letters or symbols. Commonly we use x to represent the input, y the output and f the function, although other letters will be used as well.

Consider again the example from §12.1. We let f be the function 'add 2 to the input' and we let x be the input. In mathematical notation we write

$$f : x \rightarrow x + 2$$

This means that the function f takes an input x and produces an output $x + 2$.

An alternative, but commonly used, notation is

$$f(x) = x + 2$$

The quantity $f(x)$ does not mean f times x but rather indicates that the function f acts on the quantity in the brackets. Because we also call the output y we can write $y = f(x) = x + 2$, or simply $y = x + 2$.

We could represent the same function using different letters. If h represents the function and t the input then we could write

$$h(t) = t + 2$$

Worked examples

12.1   A function multiplies the input by 4. Write down the function in mathematical notation.

Solution   Let us call the function f and the input x. Then we have

$$f : x \rightarrow 4x \quad \text{or alternatively} \quad f(x) = 4x$$

If we call the output y, we can write $y = f(x) = 4x$, or simply $y = 4x$.

12.2   A function divides the input by 6 and then adds 3 to the result. Write the function in mathematical notation.

Solution   Let us call the function z and the input t. Then we have

$$z(t) = \frac{t}{6} + 3$$

12.3   A function f is given by the rule $f: x \rightarrow 9$, or alternatively as $f(x) = 9$. Describe in words what this function does.

Solution   Whatever the value of the input to this function, the output is always 9.

12.4   A function squares the input and then multiplies the result by 6. Write down the function using mathematical notation.

Solution   Let us call the function f, the input x and the output y. Then

$$y = f(x) = 6x^2$$

12.5   Describe in words what the following functions do:

(a) $h(x) = \frac{1}{x}$ (b) $g(t) = t + t^2$

Solution   (a) The function $h(x) = 1/x$ divides 1 by the input.
(b) The function $g(t) = t + t^2$ adds the input to the square of the input.

Often we are given a function and need to calculate the output from a given input.

Worked examples

12.6   A function f is defined by $f(x) = 3x + 1$. Calculate the output when the input is (a) 4, (b) −1, (c) 0.

Solution   The function f multiplies the input by 3 and then adds 1 to the result.

(a) When the input is 4, the output is $3 \times 4 + 1 = 12 + 1 = 13$. We write

$$f(x = 4) = 3(4) + 1 = 12 + 1 = 13$$

or more simply

$$f(4) = 13$$

Note that 4 has been substituted for x in the formula for f.

(b) We require the output when the input is −1, that is, $f(-1)$.

$$f(x = -1) = f(-1) = 3(-1) + 1 = -3 + 1 = -2$$

The output is -2 when the input is -1.

(c) We require the output when the input is 0, that is, $f(0)$.

$$f(x = 0) = f(0) = 3(0) + 1 = 0 + 1 = 1$$

12.7 A function g is defined by $g(t) = 2t^2 - 1$. Find

(a) $g(3)$ (b) $g(0.5)$ (c) $g(-2)$

Solution (a) We obtain $g(3)$ by substituting 3 for t.

$$g(3) = 2(3)^2 - 1 = 2(9) - 1 = 17$$

(b) $g(0.5) = 2(0.5)^2 - 1 = 0.5 - 1 = -0.5$.
(c) $g(-2) = 2(-2)^2 - 1 = 8 - 1 = 7$.

12.8 A function h is defined by $h(x) = \dfrac{x}{3} + 1$. Find

(a) $h(3)$ (b) $h(t)$ (c) $h(\alpha)$ (d) $h(2\alpha)$ (e) $h(2x)$

Solution (a) If $h(x) = \dfrac{x}{3} + 1$ then $h(3) = \dfrac{3}{3} + 1 = 1 + 1 = 2$.

(b) The function h divides the input by 3 and then adds 1. We require the output when the input is t, that is, we require $h(t)$. Now

$$h(t) = \frac{t}{3} + 1$$

since the input t has been divided by 3 and then 1 has been added to the result. Note that $h(t)$ is obtained by substituting t in place of x in $h(x)$.

(c) We require the output when the input is α. This is obtained by substituting α in place of x. We find

$$h(\alpha) = \frac{\alpha}{3} + 1$$

(d) We require the output when the input is 2α. We substitute 2α in place of x. This gives

$$h(2\alpha) = \frac{2\alpha}{3} + 1$$

(e) We require the output when the input is $2x$. We substitute $2x$ in place of x. That is,

$$h(2x) = \frac{2x}{3} + 1$$

12.9 Given $f(x) = x^2 + x - 1$ write expressions for

(a) $f(\alpha)$ (b) $f(x+1)$ (c) $f(2t)$

Solution   (a) Substituting α in place of x we obtain

$$f(\alpha) = \alpha^2 + \alpha - 1$$

(b) Substituting $x + 1$ for x we obtain

$$f(x + 1) = (x + 1)^2 + (x + 1) - 1$$
$$= x^2 + 2x + 1 + x + 1 - 1$$
$$= x^2 + 3x + 1$$

(c) Substituting $2t$ in place of x we obtain

$$f(2t) = (2t)^2 + 2t - 1 = 4t^2 + 2t - 1$$

Sometimes a function uses different rules on different intervals. For example we could define a function as

$$f(x) = \begin{cases} 3x & \text{when} & 0 \le x \le 4 \\ 2x + 6 & \text{when} & 4 < x < 5 \\ 9 & \text{when} & x \ge 5 \end{cases}$$

Here the function is defined in three 'pieces'. The value of x determines which part of the definition is used to evaluate the function.

Worked examples

12.10   A function is defined by

$$y(x) = \begin{cases} x^2 + 1 & \text{when} & -1 \le x \le 2 \\ 3x & \text{when} & 2 < x \le 6 \\ 2x + 1 & \text{when} & x > 6 \end{cases}$$

Evaluate   (a) $y(0)$ (b) $y(4)$ (c) $y(2)$ (d) $y(7)$

Solution   (a) We require the value of y when $x = 0$. Since 0 lies between -1 and 2 we use the first part of the definition, that is, $y = x^2 + 1$. Hence

$$y(0) = 0^2 + 1 = 1$$

(b) We require y when $x = 4$. The second part of the definition must be used because x lies between 2 and 6. Therefore

$$y(4) = 3(4) = 12$$

(c) We require y when $x = 2$. The value $x = 2$ occurs in the first part of the definition. Therefore

$$y(2) = 2^2 + 1 = 5$$

(d) We require y when $x = 7$. The final part of the function must be used. Therefore

$$y(7) = 2(7) + 1 = 15$$

Self-assessment questions 12.2

1. Explain what is meant by a function.
2. Explain the meaning of the terms 'dependent variable' and 'independent variable'.
3. Given $f(x)$, is the statement '$f(1/x)$ means $1/f(x)$' true or false?
4. Give an example of a function $f(x)$ such that $f(2) = f(3)$, that is the outputs for the inputs 2 and 3 are identical.

Exercise 12.2

1. Describe in words each of the following functions:

   (a) $h(t) = 10t$ (b) $g(x) = -x + 2$

   (c) $h(t) = 3t^4$ (d) $f(x) = \dfrac{4}{x^2}$

   (e) $f(x) = 3x^2 - 2x + 9$ (f) $f(x) = 5$
   (g) $f(x) = 0$

2. Describe in words each of the following functions:

   (a) $f(t) = 3t^2 + 2t$ (b) $g(x) = 3x^2 + 2x$

   Comment upon your answers.

3. Write the following functions using mathematical notation.

   (a) The input is cubed and the result is divided by 12.

   (b) The input is added to 3 and the result is squared.

   (c) The input is squared and added to 4 times the input. Finally, 10 is subtracted from the result.

   (d) The input is squared and added to 5. Then the input is divided by this result.

   (e) The input is cubed and then 1 is subtracted from the result.

   (f) 1 is subtracted from the input and the result is squared.

   (g) Twice the input is subtracted from 7 and the result is divided by 4.

   (h) The output is always -13 whatever the value of the input.

4. Given the function $A(n) = n^2 - n + 1$ evaluate

   (a) $A(2)$ (b) $A(3)$ (c) $A(0)$ (d) $A(-1)$

5. Given $y(x) = (2x - 1)^2$ evaluate

   (a) $y(1)$ (b) $y(-1)$ (c) $y(-3)$ (d) $y(0.5)$
   (e) $y(-0.5)$

6. The function f is given by $f(t) = 4t + 6$. Write expressions for

   (a) $f(t + 1)$ (b) $f(t + 2)$
   (c) $f(t + 1) - f(t)$ (d) $f(t + 2) - f(t)$

7. The function $f(x)$ is defined by $f(x) = 2x^2 - 3$. Write expressions for

   (a) $f(n)$ (b) $f(z)$ (c) $f(t)$ (d) $f(2t)$

   (e) $f\left(\dfrac{1}{z}\right)$ (f) $f\left(\dfrac{3}{n}\right)$ (g) $f(-x)$

   (h) $f(-4x)$ (i) $f(x + 1)$ (j) $f(2x - 1)$

continued

8. Given the function $a(p) = p^2 + 3p + 1$ write an expression for $a(p+1)$. Verify that $a(p+1) - a(p) = 2p + 4$.

9. Sometimes the output from one function forms the input to another function. Suppose we have two functions f given by $f(t) = 2t$, and h given by $h(t) = t + 1$. $f(h(t))$ means that t is input to h, and the output from h is input to f. Evaluate

   (a) $f(3)$ (b) $h(2)$ (c) $f(h(2))$ (d) $h(f(3))$

10. The functions f and h are defined as in Question 9. Write down expressions for

   (a) $f(h(t))$ (b) $h(f(t))$

11. A function is defined by

$$f(x) = \begin{cases} x & 0 \le x < 1 \\ 2 & x = 1 \\ 1 & x > 1 \end{cases}$$

   Evaluate

   (a) $f(0.5)$ (b) $f(1.1)$ (c) $f(1)$

12.3   Composite functions

Sometimes we wish to apply two or more functions, one after the other. The output of one function becomes the input of the next function.

Suppose $f(x) = 2x$ and $g(x) = x + 3$. We note that the function $f(x)$ doubles the input while the function $g(x)$ adds 3 to the input. Now, we let the output of $g(x)$ become the input to $f(x)$. Figure 12.3 illustrates the position.

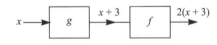

Figure 12.3
The output of g is the input of f

We have

$$g(x) = x + 3$$
$$f(x + 3) = 2(x + 3) = 2x + 6$$

Note that $f(x + 3)$ may be written as $f(g(x))$. Referring to Figure 12.3 we see that the initial input is x and that the final output is $2x + 6$. The functions $g(x)$ and $f(x)$ have been combined. We call $f(g(x))$ a **composite function**. It is composed of the individual functions $f(x)$ and $g(x)$. In this example we have

$$f(g(x)) = 2x + 6$$

Worked examples

12.11   Given $f(x) = 2x$ and $g(x) = x + 3$ find the composite function $g(f(x))$.

Solution   The output of $f(x)$ becomes the input to $g(x)$. Figure 12.4 illustrates this.

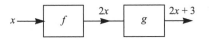

Figure 12.4
The composite function
$g(f(x))$

We see that

$$g(f(x)) = g(2x)$$
$$= 2x + 3$$

Note that in general $f(g(x))$ and $g(f(x))$ are different functions.

12.12   Given $f(t) = t^2 + 1$, $g(t) = \frac{3}{t}$ and $h(t) = 2t$ determine each of the following composite functions.

(a) $f(g(t))$ (b) $g(h(t))$ (c) $f(h(t))$ (d) $f(g(h(t)))$ (e) $g(f(h(t)))$

Solution   (a) $f(g(t)) = f\left(\dfrac{3}{t}\right) = \left(\dfrac{3}{t}\right)^2 + 1 = \dfrac{9}{t^2} + 1$

(b) $g(h(t)) = g(2t) = \dfrac{3}{2t}$

(c) $f(h(t)) = f(2t) = (2t)^2 + 1 = 4t^2 + 1$

(d) $f(g(h(t))) = f\left(\dfrac{3}{2t}\right)$ using (b)

$$= \left(\dfrac{3}{2t}\right)^2 + 1$$

$$= \dfrac{9}{4t^2} + 1$$

$$\text{(e) } g(f(h(t))) = g(4t^2 + 1) \qquad \text{using (c)}$$

$$= \frac{3}{4t^2 + 1}$$

Self-assessment questions 12.3

1. Explain the term 'composite function'.
2. Give examples of functions $f(x)$ and $g(x)$ such that $f(g(x))$ and $g(f(x))$ are equal.

Exercise 12.3

1. Given $f(x) = 4x$ and $g(x) = 3x - 2$ find

   (a) $f(g(x))$ (b) $g(f(x))$

2. If $x(t) = t^3$ and $y(t) = 2t$ find

   (a) $y(x(t))$ (b) $x(y(t))$

3. Given $r(x) = \dfrac{1}{2x}$, $s(x) = 3x$ and $t(x) = x - 2$ find

   (a) $r(s(x))$ (b) $t(s(x))$ (c) $t(r(s(x)))$
   (d) $r(t(s(x)))$ (e) $r(s(t(x)))$

4. A function can be combined with itself. This is known as **self-composition**. Given $v(t) = 2t + 1$ find

   (a) $v(v(t))$ (b) $v(v(v(t)))$

5. Given $m(t) = (t + 1)^3$, $n(t) = t^2 - 1$ and $p(t) = t^2$ find

   (a) $m(n(t))$ (b) $n(m(t))$ (c) $m(p(t))$
   (d) $p(m(t))$ (e) $n(p(t))$ (f) $p(n(t))$
   (g) $m(n(p(t)))$ (h) $p(p(t))$ (i) $n(n(t))$
   (j) $m(m(t))$

12.4 The inverse of a function

Note that the symbol f^{-1} does not mean $\dfrac{1}{f}$.

We have described a function f as a rule which receives an input, say x, and generates an output, say y. We now consider the reversal of that process, namely finding a function which receives y as input and generates x as the output. If such a function exists it is called the inverse function of f. Figure 12.5 illustrates this schematically. The inverse of $f(x)$ is denoted by $f^{-1}(x)$.

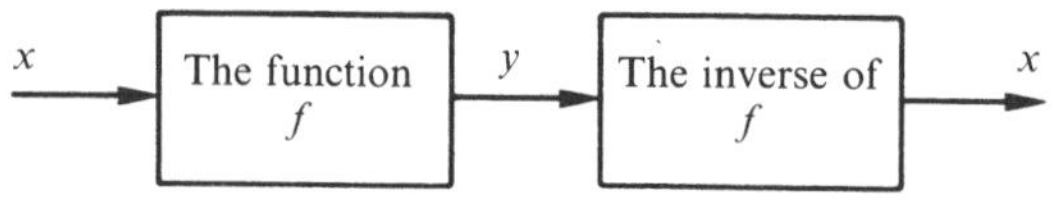

Figure 12.5
The inverse of f reverses the effect of f

Worked examples

12.13 The functions f and g are defined by

$$f(x) = 2x \qquad g(x) = \frac{x}{2}$$

(a) Verify that f is the inverse of g.
(b) Verify that g is the inverse of f.

Solution (a) The function g receives an input of x and generates an output of $x/2$, that is, it halves the input. In order to reverse the process, the inverse of g should receive $x/2$ as input and generate x as output. Now consider the function $f(x) = 2x$. This function doubles the input. Hence

$$f\left(\frac{x}{2}\right) = 2\left(\frac{x}{2}\right) = x$$

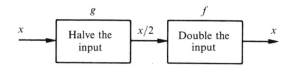

Figure 12.6
The function f is the inverse of g

The function f has received $x/2$ as input and generated x as output. Hence f is the inverse of g. This is shown schematically in Figure 12.6.

(b) The function f receives x as input and generates $2x$ as output. In order to reverse the process, the inverse of f should receive $2x$ as input and generate x as output. Now $g(x) = x/2$, that is, the input is halved, and so

$$g(2x) = \frac{2x}{2} = x$$

Hence g is the inverse of f. This is shown schematically in Figure 12.7.

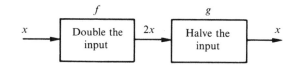

Figure 12.7
The function g is the inverse of f

12.14   Find the inverse of the function $f(x) = 3x - 4$.

Solution   The function f multiplies the input by 3 and subtracts 4 from the result. To reverse the process, the inverse function, g say, must add 4 to the input and then divide the result by 3. Hence

$$g(x) = \frac{x + 4}{3}$$

12.15   Find the inverse of $h(t) = -\frac{1}{2}t + 5$.

Solution   The function h multiplies the input by $-\frac{1}{2}$ and then adds 5 to the result. Therefore the inverse function, g say, must subtract 5 from the input and then divide the result by $-\frac{1}{2}$. Hence

$$g(t) = \frac{(t - 5)}{-1/2} = -2(t - 5) = -2t + 10$$

There is an algebraic method of finding an inverse function which is often easier to apply. Suppose we wish to find the inverse of the function $f(x) = 6 - 2x$. We let

$$y = 6 - 2x$$

and then transpose this for x. This gives

$$x = \frac{6 - y}{2}$$

Finally, we interchange x and y to give $y = (6 - x)/2$. This is the required inverse function. To summarize these stages:

KEY POINT

To find the inverse of $y = f(x)$,

- transpose the formula to make x the subject
- interchange x and y

The result is the required inverse function.

We shall meet some functions which do not have an inverse function. For example, consider the function $f(x) = x^2$. If 3 is the input, the output is 9. Now if -3 is the input, the output will also be 9 since $(-3)^2 = 9$. In order to reverse this process an inverse function would have to take an input of 9 and produce outputs of both 3 and -3. However, this contradicts the definition of a function which states that a function must have only *one* output for a given input. We say that $f(x) = x^2$ does not have an inverse function.

Self-assessment questions 12.4

1. Explain what is meant by the inverse of a function.
2. Explain why the function $f(x) = 4x^4$ does not possess an inverse function.

Exercise 12.4

1. Find the inverse of each of the following functions:

   (a) $f(x) = 3x$ (b) $f(x) = \dfrac{x}{4}$

   (c) $f(x) = x + 1$ (d) $f(x) = x - 3$
   (e) $f(x) = 3 - x$ (f) $f(x) = 2x + 6$

   (g) $f(x) = 7 - 3x$ (h) $f(x) = \dfrac{1}{x}$

   (i) $f(x) = \dfrac{3}{x}$ (j) $f(x) = -\dfrac{3}{4x}$

2. Find the inverse, $f^{-1}(x)$, when $f(x)$ is given by:

   (a) $6x$ (b) $6x + 1$ (c) $x + 6$ (d) $\dfrac{x}{6}$ (e) $\dfrac{6}{x}$

3. Find the inverse, $g^{-1}(t)$, when $g(t)$ is given by:

   (a) $3t + 1$ (b) $\dfrac{1}{3t + 1}$ (c) t^3 (d) $3t^3$

   (e) $3t^3 + 1$ (f) $\dfrac{3}{t^3 + 1}$

4. The functions $g(t)$ and $h(t)$ are defined by:

   $$g(t) = 2t - 1, \; h(t) = 4t + 3$$

   Find
   (a) the inverse of $h(t)$, that is, $h^{-1}(t)$
   (b) the inverse of $g(t)$, that is, $g^{-1}(t)$
   (c) $g^{-1}(h^{-1}(t))$ (d) $h(g(t))$
   (e) the inverse of $h(g(t))$
   What observations do you make from (d) and (e)?

Test and assignment exercises 12

1. Given $r(t) = t^2 - t/2 + 4$ evaluate

   (a) $r(0)$ (b) $r(-1)$ (c) $r(2)$ (d) $r(3.6)$
   (e) $r(-4.6)$

2. A function is defined as $h(t) = t^2 - 7$:
   (a) state the dependent variable;
   (b) state the independent variable.

3. Given the functions $a(x) = x^2 + 1$ and $b(x) = 2x + 1$ write expressions for

   (a) $a(\alpha)$ (b) $b(t)$ (c) $a(2x)$ (d) $b\left(\dfrac{x}{3}\right)$

   (e) $a(x + 1)$ (f) $b(x + h)$ (g) $b(x - h)$
   (h) $a(b(x))$ (i) $b(a(x))$

4. Find the inverse of each of the following functions:

   (a) $f(x) = \pi - x$ (b) $h(t) = \dfrac{t}{3} + 2$

   (c) $r(n) = \dfrac{1}{n}$ (d) $r(n) = \dfrac{1}{n - 1}$

   (e) $r(n) = \dfrac{2}{n - 1}$ (f) $r(n) = \dfrac{a}{n - b}$

   where a and b are constants.

5. Given $A(n) = n^2 + n - 6$ find expressions for

   (a) $A(n + 1)$ (b) $A(n - 1)$
   (c) $2A(n + 1) - A(n) + A(n - 1)$

continued

6. Find $h^{-1}(x)$ when $h(x)$ is given by:

   (a) $\dfrac{x+1}{3}$ (b) $\dfrac{3}{x+1}$ (c) $\dfrac{x+1}{x}$ (d) $\dfrac{x}{x+1}$

7. Given $v(t) = 4t - 2$ find

   (a) $v^{-1}(t)$ (b) $v(v(t))$

8. Given $h(x) = 9x - 6$ and $g(x) = \dfrac{1}{3x}$ find

   (a) $h(g(x))$ (b) $g(h(x))$

9. Given $f(x) = (x+1)^2$, $g(x) = 4x$ and $h(x) = x - 1$ find

   (a) $f(g(x))$ (b) $g(f(x))$ (c) $f(h(x))$
   (d) $h(f(x))$ (e) $g(h(x))$ (f) $h(g(x))$
   (g) $f(g(h(x)))$ (h) $g(h(f(x)))$

10. Given $x(t) = t^3$, $y(t) = \dfrac{1}{t+1}$ and $z(t) = 3t - 1$ find

    (a) $y(x(t))$ (b) $y(z(t))$
    (c) $x(y(z(t)))$ (d) $z(y(x(t)))$

11. Write each of the following functions using mathematical notation.

    (a) The input is multiplied by 7.

    (b) Five times the square of the input is subtracted from twice the cube of the input.

    (c) The output is 6.

    (d) The input is added to the reciprocal of the input.

    (e) The input is multiplied by 11 and then 6 is subtracted from this. Finally, 9 is divided by this result.

12. Find the inverse of

    $$f(x) = \frac{x+1}{x-1}$$

13 Graphs of functions

Objectives

This chapter

- shows how coordinates are plotted on the $x - y$ plane
- introduces notation to denote intervals on the x axis
- introduces symbols to denote 'greater than' and 'less than'
- shows how to draw a graph of a function
- explains what is meant by the domain and range of a function
- explains how to use the graph of a function to solve equations
- explains how to solve simultaneous equations using graphs

In Chapter 12 we introduced functions and represented them by algebraic equations. Another useful way of representing functions is pictorially, by means of **graphs**.

13.1 The x–y plane

We introduce horizontal and vertical **axes** as shown in Figure 13.1. These axes intersect at a point O called the **origin**. The horizontal axis is used to represent the independent variable, commonly x, and the vertical axis is used to represent the dependent variable, commonly y. The region shown is then referred to as the x–y plane.

A **scale** is drawn on both axes in such a way that at the origin $x = 0$ and $y = 0$. Positive x values lie to the right and negative x values to the left. Positive y values are above the origin, negative y values are below the origin. It is not essential that the scales on both axes are the same. Note that anywhere on the x axis, the value of y must be zero. Anywhere on the y axis the value of x must be zero. Each point in the plane corresponds to a specific value of x and y. We usually call the x value the x **coordinate** and call the y value the y **coordinate**. To refer to a specific point we give both its coordinates in brackets in the form (x, y), always giving the x coordinate first.

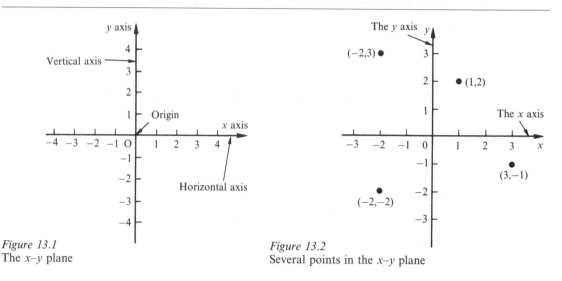

Figure 13.1
The *x–y* plane

Figure 13.2
Several points in the *x–y* plane

Worked example

13.1 Draw the *x–y* plane and on it mark the points whose coordinates are
$(1, 2), (3, -1), (-2, 3), (-2, -2)$.

Solution The plane is drawn in Figure 13.2, and the given points are indicated. Note that the first coordinate in each bracket is the *x* coordinate.

Self-assessment questions 13.1

1. Explain the terms 'horizontal axis', 'vertical axis' and 'origin'.
2. When giving the coordinates of a point in the *x–y* plane, which coordinate is always given first?
3. State the coordinates of the origin.

Exercise 13.1

1. Plot and label the following points in the *x–y* plane:
$(-2, 0), (2, 0), (0, -2), (0, 2), (0, 0)$.

2. A set of points is plotted and then these points are joined by a straight line which is parallel to the *x* axis. What can you deduce about the coordinates of these points?

3. A set of points is plotted and then these points are joined by a straight line which is parallel to the *y* axis. What can you deduce about the coordinates of these points?

| 13.2 | **Inequalities and intervals** |

We often need only part of the x axis when plotting graphs, for example, we may be interested only in that part of the x axis running from $x = 1$ to $x = 3$ or from $x = -2$ to $x = 7.5$. Such parts of the x axis are called **intervals**. In order to describe concisely intervals on the x axis, we introduce some new notation.

Greater than and less than

We use the symbol $>$ to mean 'greater than'. For example, $6 > 5$ and $-4 > -5$ are both true statements. If $x_1 > x_2$ then x_1 is to the right of x_2 on the x axis. The symbol $\geq$ means 'greater than or equal to' so, for example, $10 \geq 8$ and $8 \geq 8$ are both true.

The symbol $<$ means 'less than'. We may write $4 < 5$ for example. If $x_1 < x_2$ then x_1 is to the left of x_2 on the x axis. Finally, $\leq$ means 'less than or equal to'. Hence $6 \leq 9$ and $6 \leq 6$ are both true statements.

Intervals

$\mathbb{R}$ is called the set of real numbers.

We often need to represent intervals on the number line. To help us do this we introduce the set, $\mathbb{R}$. $\mathbb{R}$ is the symbol we use to denote all numbers from minus infinity to plus infinity. All numbers, including integers, fractions and decimals belong to $\mathbb{R}$. To show that a number, x, is in this set, we write $x \in \mathbb{R}$ where $\in$ means 'belongs to'.

There are three different kinds of intervals.

(a) *The closed interval* An interval which includes its end-points is called a closed interval. All the numbers from 1 to 3, including both 1 and 3, comprise a closed interval and this is denoted using square brackets, [1,3]. Any number in this closed interval must be greater than or equal to 1, and also less than or equal to 3. Thus if x is any number in the interval, then $x \geq 1$ and also $x \leq 3$. We write this compactly as $1 \leq x \leq 3$. Finally we need to show explicitly that x is any number on the x axis, and not just an integer for example. So we write $x \in \mathbb{R}$. Hence the interval [1,3] can be expressed as:

$$\{x : x \in \mathbb{R}, 1 \leq x \leq 3\}$$

This means the set contains all the numbers x with $x \geq 1$ and $x \leq 3$.

(b) *The open interval*   Any interval which does not include its end-points is called an open interval. For example, all the numbers from 1 to 3, but excluding 1 and 3, comprise an open interval. Such an interval is denoted using round brackets, (1,3). The interval may be written using set notation as

$$\{x : x \in \mathbb{R}, \ 1 < x < 3\}$$

We say that x is **strictly greater** than 1, and **strictly less** than 3, so that the values of 1 and 3 are excluded from the interval.

(c) *The semi-open or semi-closed interval*   An interval may be open at one end and closed at the other. Such an interval is called semi-open or, as some authors say, semi-closed. The interval (1,3] is a semi-open interval. The square bracket next to the 3 shows that 3 is included in the interval; the round bracket next to the 1 shows that 1 is not included in the interval. Using set notation we would write

$$\{x : x \in \mathbb{R}, \ 1 < x \le 3\}$$

When marking intervals on the x-axis there is a notation to show whether or not the end-point is included. We use $\bullet$ to show that an end-point is included (i.e. closed) while $\circ$ is used to denote an end-point that is not included (i.e. open).

Worked examples

13.2   Describe the interval $[-3, 4]$ using set notation and illustrate it on the x axis.

Solution   The interval $[-3, 4]$ is given in set notation by

$$\{x : x \in \mathbb{R}, \ -3 \le x \le 4\}$$

Figure 13.3 illustrates the interval.

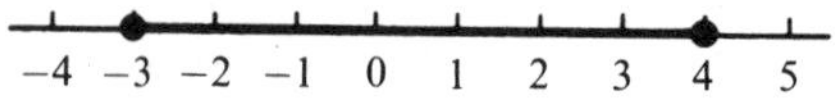

Figure 13.3
The interval $[-3,4]$

13.3   Describe the interval (1,4) using set notation and illustrate it on the x axis.

Solution   The interval (1,4) is expressed as

$$\{x : x \in \mathbb{R}, \ 1 < x < 4\}$$

Figure 13.4 illustrates the interval on the x axis.

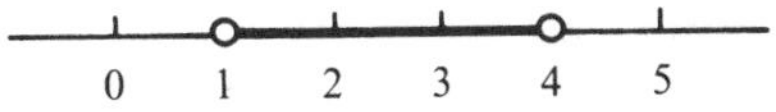

Figure 13.4
The interval (1,4)

Self-assessment questions 13.2

1. Explain what is meant by (a) a closed interval (b) an open interval (c) a semi-closed interval.
2. Describe the graphical notation used to denote that an end-point is (a) included (b) not included.

Exercise 13.2

1. Describe the following intervals using set notation. Draw the intervals on the x axis

   (a) [2,6] (b) (6,8] (c) (−2,0)
   (d) [−3, −1.5)

2. Which of the following are true?

   (a) $8 > 3$ (b) $-3 < 8$ (c) $-3 \leq -3$
   (d) $0.5 \geq 0.25$ (e) $0 > 9$ (f) $0 \leq 9$
   (g) $-7 \geq 0$ (h) $-7 < 0$

13.3	**Plotting the graph of a function**

The method of plotting a graph is best illustrated by example.

Worked examples

13.4   Plot a graph of $y = 2x - 1$ for $-3 \leq x \leq 3$.

Solution   We first calculate the value of y for several values of x. A table of x values and corresponding y values is drawn up as shown in Table 13.1.

The independent variable x varies from -3 to 3; the dependent variable y varies from -7 to 5. In order to accommodate all these values the scale on the x axis must vary from -3 to 3, and that on the y axis from -7 to 5. Each pair of values of x and y is represented by a unique point in the x–y plane, with coordinates (x, y). Each pair of values in the table is plotted as a point and then the points are joined

Table 13.1
Values of x and y when
$y = 2x - 1$

x	-3	-2	-1	0	1	2	3
y	-7	-5	-3	-1	1	3	5

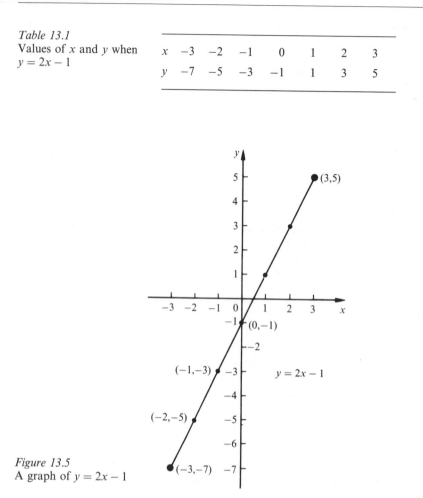

Figure 13.5
A graph of $y = 2x - 1$

to form the graph (Figure 13.5). By joining the points we see that, in this example, the graph is a straight line. We were asked to plot the graph for $-3 \leq x \leq 3$. Therefore, we have indicated each end point of the graph by a • to show that it is included. This follows the convention used for labelling closed intervals given in Section 13.2.

In the previous example, although we have used x and y as independent and dependent variables, clearly other letters could be used. The points to remember are that the axis for the independent variable is horizontal and the axis for the dependent variable is vertical. When co-ordinates are given the value of the independent variable is always given first.

Worked example

13.5 (a) Plot a graph of $y = x^2 - 2x + 2$ for $-2 \le x \le 3$.

(b) Use your graph to determine which of the following points lie on the graph: $(1,2)$, $(0,2)$, and $(1,1)$.

Solution (a) Table 13.2 gives values of x and the corresponding values of y. The independent variable is x and so the x axis is horizontal. The calculated points are plotted and joined. Figure 13.6 shows the graph. In this example the graph is not a straight line but rather a curve.

(b) The point $(1,2)$ has an x coordinate of 1 and a y coordinate of 2. The point is labelled by A on Figure 13.6. Points $(0,2)$ and $(1,1)$ are plotted and labelled by B and C. From the figure we can see that $(0,2)$ and $(1,1)$ lie on the graph.

x	-2	-1	0	1	2	3
y	10	5	2	1	2	5

Table 13.2
Values of x and y when
$y = x^2 - 2x + 2$

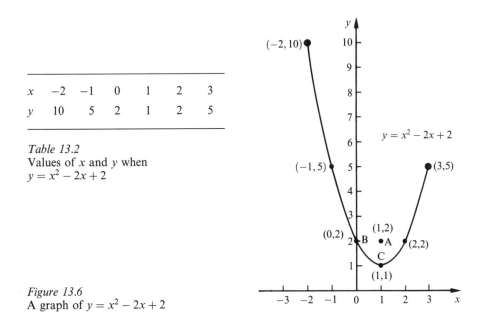

Figure 13.6
A graph of $y = x^2 - 2x + 2$

Self-assessment questions 13.3

1. Which variable is plotted vertically on a graph – the independent or dependent?
2. Suppose we wished to plot a graph of the function $g = 4t^2 + 3t - 2$. Upon which axis would the variable t be plotted?

Exercise 13.3

1. Plot graphs of the following functions:

   (a) $y = f(x) = 2x + 7$ for $-4 \leq x \leq 4$.
   (b) $y = f(x) = -2x$ for values of x between -3 and 3 inclusive.
   (c) $y = f(x) = x$ for $-5 \leq x \leq 5$.
   (d) $y = f(x) = -x$ for $-5 \leq x \leq 5$.

2. Plot a graph of the function
   $y = f(x) = 7$, for $-2 \leq x \leq 2$.

3. A function is given by $y = f(x) = x^3 + x$ for $-3 \leq x \leq 3$.

   (a) Plot a graph of the function.
   (b) Determine which of the following points lie on the curve: $(0,1)$, $(1,0)$, $(1,2)$, and $(-1,-2)$.

4. Plot, on the same axes, graphs of the functions $f(x) = x + 2$ for $-3 \leq x \leq 3$, and $g(x) = -\frac{1}{2}x + \frac{1}{2}$ for $-3 \leq x \leq 3$. State the coordinates of the point where the graphs intersect.

5. Plot a graph of the function $f(x) = x^2 - x - 1$ for $-2 \leq x \leq 3$. State the coordinates of the points where the curve cuts (a) the horizontal axis, (b) the vertical axis.

6. (a) Plot a graph of the function $y = 1/x$ for $0 < x \leq 3$.
   (b) Deduce the graph of the function $y = -1/x$ for $0 < x \leq 3$ from your graph in part (a).

7. Plot a graph of the function defined by

   $$r(t) = \begin{cases} t^2 & 0 \leq t \leq 3 \\ -2t + 15 & 3 < t \leq 7.5 \end{cases}$$

8. On the same axes plot a graph of $y = 2x$, $y = 2x + 1$ and $y = 2x + 3$, for $-3 \leq x \leq 3$. What observations can you make?

13.4   The domain and range of a function

The set of values which we allow the independent variable to take is called the **domain** of the function. If the domain is not actually specified in any particular example, it is taken to be the largest set possible. The set of values taken by the output is called the **range** of the function.

Worked examples

13.6 The function f is given by $y = f(x) = 2x$, for $1 \le x \le 3$.
(a) State the independent variable.
(b) State the dependent variable.
(c) State the domain of the function.
(d) Plot a graph of the function.
(e) State the range of the function.

Solution (a) The independent variable is x.
(b) The dependent variable is y.
(c) Since we are given $1 \le x \le 3$ the domain of the function is the interval $[1, 3]$, that is, all values from 1 to 3 inclusive.
(d) The graph of $y = f(x) = 2x$ is shown in Figure 13.7.
(e) The range is the set of values taken by the output, y. From the graph we see that as x varies from 1 to 3 then y varies from 2 to 6. Hence the range of the function is $[2,6]$.

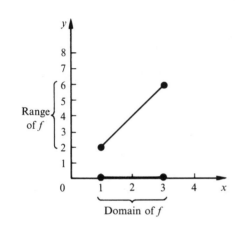

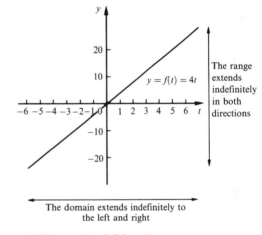

Figure 13.7 Graph of $y = f(x) = 2x$ *Figure 13.8* A graph of $f(t) = 4t$

13.7 Consider the function $y = f(t) = 4t$.

(a) State which is the independent variable and which is the dependent variable.

(b) Plot a graph of the function and give its domain and range.

Solution (a) The independent variable is t and the dependent variable is y.

(b) No domain is specified so it is taken to be the largest set possible. The domain is $\mathbb{R}$, the set of all (real) numbers. However, it would be impractical to draw a graph whose domain was the whole extent of the t axis and so a selected portion is shown in Figure 13.8.

Similarly only a restricted portion of the range can be drawn although in this example the range is also $\mathbb{R}$.

13.8   Consider the function $y = f(x) = x^2 - 1$: (a) state the independent variable; (b) state the dependent variable; (c) state the domain; (d) plot a graph of f and determine the range.

Solution   (a) The independent variable is x.
(b) The dependent variable is y.
(c) No domain is specified so it is taken to be the largest set possible. The domain is $\mathbb{R}$.
(d) A table of values and graph of $f(x) = x^2 - 1$ is shown in Figure 13.9. From the graph we see that the smallest value of y is -1 which occurs when $x = 0$. Whether x increases or decreases from 0, the value of y increases. The range is thus all values greater than or equal to -1. We can write the range using the set notation introduced in §13.2 as

$$\text{range} = \{y : y \in \mathbb{R}, y \geq -1\}$$

13.9   Given the function $y = 1/x$, state the domain and range.

Solution   The domain is the largest set possible. All values of x are permissible except $x = 0$ since $\frac{1}{0}$ is not defined. It is never possible to divide by 0. Thus the domain comprises all real numbers except 0. A table of values and graph is shown in Figure 13.10. As x varies, y can take on any value except 0. In this example we see that the range is thus the same as the domain. We note that the graph is split at $x = 0$. For small,

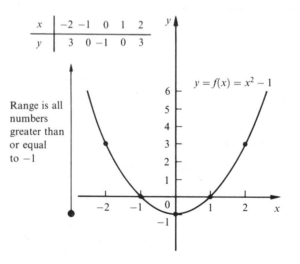

x	-2	-1	0	1	2
y	3	0	-1	0	3

$y = f(x) = x^2 - 1$

Range is all numbers greater than or equal to -1

Figure 13.9
A graph of $f(x) = x^2 - 1$

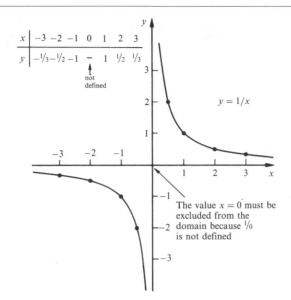

x	−3	−2	−1	0	1	2	3
y	−1/3	−1/2	−1	—	1	1/2	1/3

not defined

$y = 1/x$

The value $x = 0$ must be excluded from the domain because $1/0$ is not defined

Figure 13.10
A graph of $y = 1/x$

positive values of x, y is large and positive. The graph approaches the y axis for very small positive values of x. For small, negative values of x, y is large and negative. Again, the graph approaches the y axis for very small negative values of x. We say that the y axis is an **asymptote**. In general, if the graph of a function approaches a straight line we call that line an asymptote.

Self-assessment questions 13.4

1. Explain the terms 'domain of a function' and 'range of a function'.
2. Explain why the value $x = 3$ must be excluded from any domain of the function $f(x) = 7/(x − 3)$.
3. Give an example of a function for which we must exclude the value $x = −2$ from the domain.

Exercise 13.4

1. Plot graphs of the following functions. In each case state (i) the independent variable, (ii) the dependent variable, (iii) the domain, (iv) the range.

(a) $y = x − 1$, $6 \le x \le 10$
(b) $h = t^2 + 3$, $4 \le t < 5$
(c) $m = 3n − 2$, $−1 < n < 1$
(d) $y = 2x + 4$
(e) $X = y^3$
(f) $k = 4r^2 + 5$, $5 \le r \le 10$

2. A function is defined by

$$q(t) = \begin{cases} t^2 & 0 < t < 3 \\ 6 + t & 3 \le t < 5 \\ −2t + 21 & 5 \le t \le 10.5 \end{cases}$$

(a) Plot a graph of this function.
(b) State the domain of the function.
(c) State the range of the function.
(d) Evaluate $q(1)$, $q(3)$, $q(5)$ and $q(7)$.

13.5   Solving equations using graphs

In Chapter 11 we used algebraic methods to solve equations. However, many equations cannot be solved exactly using such methods. In such cases it is often useful to use a graphical approach. The following examples illustrate the method of using graphs to solve equations.

Worked examples

13.10   Find graphically all solutions in the interval $[-3, 3]$ of the equation $x^2 + x - 3 = 0$.

Solution   We draw a graph of $y(x) = x^2 + x - 3$. Table 13.3 gives x and y values and Figure 13.11 shows a graph of the function. We have plotted $y = x^2 + x - 3$ and wish to solve $0 = x^2 + x - 3$. Thus we read from the graph the coordinates of the points at which $y = 0$. Such points must be on the x axis. From the graph the points are (1.34,0) and (−2.30,0). So the solutions of $x^2 + x - 3 = 0$ are $x = 1.34$ and $x = -2.30$. Note that these are approximate solutions, being dependent upon the accuracy of the graph and the x and y scales used. Increased accuracy can be achieved by drawing an enlargement of the graph

Table 13.3
Values of x and y when
$y = x^2 + x - 3$

x	−3	−2	−1	0	1	2	3
y	3	−1	−3	−3	−1	3	9

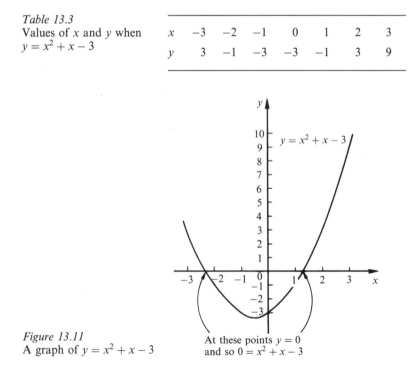

Figure 13.11
A graph of $y = x^2 + x - 3$

At these points $y = 0$
and so $0 = x^2 + x - 3$

around the values $x = 1.34$ and $x = -2.30$. So, for example, we could draw $y = x^2 + x - 3$ for $1.2 \leq x \leq 1.5$ and $y = x^2 + x - 3$ for $-2.4 \leq x \leq -2.2$.

13.11 Find solutions in the interval $[-2,2]$ of the equation $x^3 = 2x + \frac{1}{2}$ using a graphical method.

Solution The problem of solving $x^3 = 2x + \frac{1}{2}$ is identical to that of solving $x^3 - 2x - \frac{1}{2} = 0$. So we plot a graph of $y = x^3 - 2x - \frac{1}{2}$ for $-2 \leq x \leq 2$ and then locate the points where the curve cuts the x axis, that is, where the y coordinate is 0. Table 13.4 gives x and y values and Figure 13.12 shows a graph of the function. We now consider points on the graph where the y coordinate is zero. These points are marked A, B and C. Their x coordinates are -1.27, -0.26 and 1.53. Hence the solutions of $x^3 = 2x + \frac{1}{2}$ are approximately $x = -1.27$, $x = -0.26$ and $x = 1.53$.

Table 13.4
Values of x and y when
$y = x^3 - 2x - \frac{1}{2}$

x	-2	-1.5	-1	-0.5	0	0.5	1	1.5	2
y	-4.5	-0.875	0.5	0.375	-0.5	-1.375	-1.5	-0.125	3.5

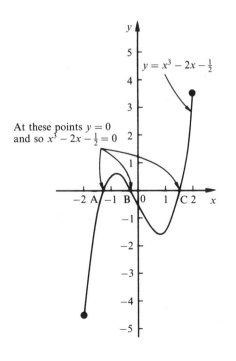

At these points $y = 0$
and so $x^3 - 2x - \frac{1}{2} = 0$

Figure 13.12
A graph of
$y = x^3 - 2x - \frac{1}{2}$

Self-assessment questions 13.5

1. Explain how the graph of the function f can be used to solve the equation
   (a) $f(x) = 0$ (b) $f(x) = 1$

Exercise 13.5

1. Plot a graph of
   $$y = x^2 + \frac{3x}{2} - 2$$
   for $-3 \leq x \leq 2$. Hence solve the equation
   $$x^2 + \frac{3x}{2} - 2 = 0$$

2. Plot $y = x^2 - x - 1$ for $-3 \leq x \leq 3$.
   Hence solve $x^2 - x - 1 = 0$.

3. Plot $y = x^2 - 0.4x - 4$ for $-3 \leq x \leq 3$.
   Hence solve $x^2 - 0.4x - 4 = 0$.

4. (a) Plot $y = 3 + \frac{x}{2} - x^2$ for $-3 \leq x \leq 3$.
   (b) Use your graph to solve
   $$x^2 - \frac{x}{2} - 3 = 0.$$

13.6   Solving simultaneous equations graphically

In the previous section we saw how an approximate solution to an equation could be found by drawing an appropriate graph. We now extend that technique to find an approximate solution of simultaneous equations. Recall that simultaneous equations were introduced algebraically in Chapter 11.

Given two simultaneous equations in x and y, then a solution is a pair of x and y values which satisfy both equations. For example, $x = 2$, $y = -3$ is a solution of

$$3x - y = 9$$
$$x + 2y = -4$$

because both equations are satisfied by these particular values. Example 13.12 illustrates the graphical method.

Worked examples

13.12 Solve graphically

$$3x - y = 9 \tag{13.1}$$
$$x + 2y = -4 \tag{13.2}$$

Solution   Both equations are rearranged so that y is the subject. This gives

$$y = 3x - 9 \tag{13.3}$$

$$y = -\frac{x}{2} - 2 \tag{13.4}$$

Since equations (13.3) and (13.4) are simple rearrangements of equations (13.1) and (13.2) then they have the same solution. Equations (13.3) and (13.4) are drawn; Figure 13.13 illustrates this.

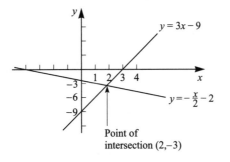

Figure 13.13
Points of intersection give the solution to simultaneous equations

We seek values of x and y which fit both equations. For a point to lie on both graphs, the point must be at the intersection of the graphs. In other words, solutions to simultaneous equations are given by the points of intersection. Reading from the graph the point of intersection is at $x = 2$, $y = -3$. Hence $x = 2$, $y = -3$ is a solution of the given simultaneous equations.

13.13   Solve graphically

$$4x - y = 0$$
$$3x + y = 7$$

Solution   We write the equations with y as subject.

$$y = 4x$$
$$y = -3x + 7$$

These are now plotted as shown in Figure 13.14.

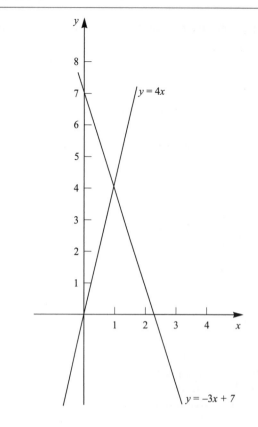

Figure 13.14
The graphs intersect
at $x = 1$, $y = 4$

The point of intersection is $x = 1$, $y = 4$ and so this is the solution of the given simultaneous equations.

13.14   Solve graphically

$$x^2 - y = -1$$
$$2x - y = -3$$

Solution   Writing the equations with y as subject gives

$$y = x^2 + 1$$
$$y = 2x + 3$$

These are plotted as shown in Figure 13.15.

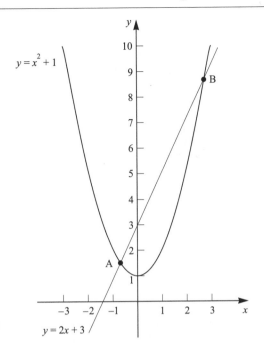

$y = x^2 + 1$

B

A

$y = 2x + 3$

Figure 13.15
The graphs have two
points of intersection
and hence there are two
solutions

There are two points of intersection, A and B. From the graph it is difficult to extract an accurate estimate. However, by using a graphics calculator or package, greater accuracy can be achieved. The coordinates of A are $x = -0.73$, $y = 1.54$; the coordinates of B are $x = 2.73$, $y = 8.46$. Hence there are two solutions to the given equations; $x = -0.73$, $y = 1.54$ and $x = 2.73$, $y = 8.46$.

13.15 (a) On the same axes draw graphs of $y = x^3$ and $y = 2x + \frac{1}{2}$ for $-2 \le x \le 2$.

(b) Note the x coordinates of the points where the two graphs intersect. By referring to Worked Example 13.11 what do you conclude? Can you explain your findings?

Solution (a) Table 13.5 gives x values and the corresponding values of x^3 and $2x + \frac{1}{2}$. Figure 13.16 shows a graph of $y = x^3$ together with a graph of $y = 2x + \frac{1}{2}$.

Table 13.5
Values of x and y when
$y = x^3$ and $y = 2x + \frac{1}{2}$

x	−2	−1.5	−1	−0.5	0	0.5	1	1.5	2
x^3	−8	−3.375	−1	−0.125	0	0.125	1	3.375	8
$2x + \frac{1}{2}$	−3.5	−2.5	−1.5	−0.5	0.5	1.5	2.5	3.5	4.5

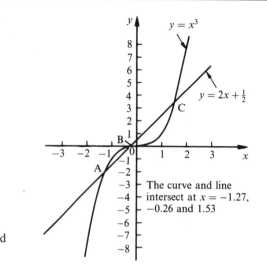

Figure 13.16
Graphs of $y = x^3$ and
$y = 2x + \frac{1}{2}$

The curve and line
intersect at $x = -1.27$,
-0.26 and 1.53

(b) The two graphs intersect at A, B and C. The x coordinates of
these points are -1.27, -0.26 and 1.53. We note from Worked
Example 13.11 that these values are the solutions of $x^3 = 2x + \frac{1}{2}$.
We explain this as follows. Where $y = x^3$ and $y = 2x + \frac{1}{2}$ intersect,
their y values are identical and so at these points

$$x^3 = 2x + \frac{1}{2}$$

By reading the x coordinates at these points of intersection we are
finding those x values for which $x^3 = 2x + \frac{1}{2}$.

Self-assessment questions 13.6

1. Explain the significance of points of intersection when solving simultaneous equations
   graphically.
2. The graphs of a pair of simultaneous equations are drawn and they intersect at three
   points. How many solutions do the simultaneous equations have?
3. The graphs of a pair of simultaneous equations are drawn. The graphs do not intersect.
   What can you conclude, if anything, about the solution of the simultaneous equations?

Exercise 13.6

1. Solve the following pairs of simultaneous equations graphically. In each case use x values from -4 to 3.

   (a) $3x + 2y = 4$
   $x - y = 3$

   (b) $2x + y = -2$
   $-\dfrac{x}{2} + 2y = 5$

   (c) $2x + y = 4$
   $4x - 3y = -7$

   (d) $-x + 4y = 7$
   $2x - y = -7$

   (e) $x + 2y = 1$
   $\dfrac{x}{2} + 5y = -\dfrac{1}{2}$

2. Solve graphically the simultaneous equations

   $$y = x^2$$
   $$2x + y = 1$$

   Take x values from -3 to 2.

3. Solve graphically

   $$x^2 + y = 3$$
   $$4x - 3y = -3$$

   Take x values from -3 to 2.

4. Solve graphically

   $$y = \dfrac{x^3}{2}$$
   $$x + y = 3$$

5. Solve graphically, taking x values from -2 to 2.

   $$x^3 - y = 0$$
   $$-1.5x^2 + 5x - y = -2$$

Test and assignment exercises 13

1. Plot a graph of each of the following functions. In each case state the independent variable, the dependent variable, the domain and the range.

   (a) $y = f(x) = 3x - 7$
   (b) $y = f(x) = 3x - 7$, $-2 \leq x \leq 2$
   (c) $y = 4x^2$, $x \geq 0$
   (d) $y = \dfrac{1}{x^2}$, $x \neq 0$

2. A function is defined by

   $$f(x) = \begin{cases} 2x + 4 & 0 \leq x \leq 4 \\ 6 & 4 < x < 6 \\ x & 6 \leq x \leq 8 \end{cases}$$

   (a) Plot a graph of this function, stating its domain and range.
   (b) Evaluate $f(0)$, $f(4)$ and $f(6)$.

3. Plot graphs of the function $f(x) = x + 4$ and $g(x) = 2 - x$ for values of x between -4 and 4. Hence solve the simultaneous equations $y - x = 4$, $y + x = 2$.

4. Plot a graph of the function $f(x) = 3x^2 + 10x - 8$ for values of x between -5 and 3. Hence solve the equation $3x^2 + 10x - 8 = 0$.

5. Plot a graph of the function $f(t) = t^3 + 2t^2 - t - 2$ for $-3 \leq t \leq 3$. Hence solve the equation $t^3 + 2t^2 - t - 2 = 0$.

6. Plot a graph of $h = 2 + 2x - x^2$ for $-3 \leq x \leq 3$. Hence solve the equation $2 + 2x - x^2 = 0$.

7. (a) On the same axes plot graphs of $y = x^3$ and $y = 5x^2 - 3$ for $-2 \leq x \leq 2$.
   (b) Use your graphs to solve the equation $x^3 - 5x^2 + 3 = 0$.

8. (a) Plot a graph of $y = x^3 - 2x^2 - x + 1$ for $-2 \leq x \leq 3$.
   (b) Hence find solutions of $x^3 - 2x^2 - x + 1 = 0$.
   (c) Use your graph to solve $x^3 - 2x^2 - x + 2 = 0$.

9. Solve each of the following pairs of simultaneous equations graphically. In each case take x values from -3 to 3.

   (a) $2x - y = 0$
        $x + y = -3$

   (b) $3x + 4y = 6$
        $x - 2y = -8$

   (c) $3x + y = 8$
        $x - 3y = -1.5$

   (d) $2x - 3y = 4.7$
        $\dfrac{x}{2} + y = -0.4$

   (e) $x + 3y = 1.2$
        $-2x + 5y = 10.8$

10. Solve graphically

    $$x^2 - y = -2$$
    $$x^2 + 2x + y = 3$$

    Take x values from -3 to 3.

11. Solve graphically

    $$x^3 - y = 0$$
    $$1.5x^2 + 6x - y = 9$$

    Take x values from -3 to 3.

14 The straight line

This chapter

- describes some special properties of straight line graphs
- explains the equation $y = mx + c$
- explains the terms 'vertical intercept' and 'gradient'
- shows how the equation of a line can be calculated
- explains what is meant by a tangent to a curve
- explains what is meant by the gradient of a curve
- explains how the gradient of a curve can be estimated by drawing a tangent

14.1 Straight line graphs

In the previous chapter we explained how to draw graphs of functions. Several of the resulting graphs were straight lines. In this chapter we focus specifically on the straight line and some of its properties.

Any equation of the form $y = mx + c$, where m and c are constants, will have a straight line graph. For example,

$$y = 3x + 7 \qquad y = -2x + \frac{1}{2} \qquad y = 3x - 1.5$$

will all result in straight line graphs. It is important to note that the variable x only occurs to the power 1. The values of m and c may be zero, so that $y = -3x$ is a straight line for which the value of c is zero, and $y = 17$ is a straight line for which the value of m is zero.

KEY POINT

Any straight line has an equation of the form $y = mx + c$ where m and c are constants.

Worked examples

14.1 Which of the following equations have straight line graphs? For those that do, identify the values of m and c.

(a) $y = 7x + 5$ (b) $y = \dfrac{13x - 5}{2}$ (c) $y = 3x^2 + x$ (d) $y = -19$

Solution (a) $y = 7x + 5$ is an equation of the form $y = mx + c$ where $m = 7$ and $c = 5$. This equation has a straight line graph.

(b) The equation

$$y = \frac{13x - 5}{2}$$

can be written as $y = \frac{13}{2}x - \frac{5}{2}$ which is in the form $y = mx + c$ with $m = \frac{13}{2}$ and $c = -\frac{5}{2}$. This has a straight line graph.

(c) $y = 3x^2 + x$ contains the term x^2. Such a term is not allowed in the equation of a straight line. The graph will not be a straight line.

(d) $y = -19$ is in the form of the equation of a straight line for which $m = 0$ and $c = -19$.

14.2 Plot each of the following graphs: $y = 2x + 3$, $y = 2x + 1$, and $y = 2x - 2$. Comment upon the resulting lines.

Solution The three graphs are shown in Figure 14.1. Note that all three graphs have the same steepness or **slope**. However, each one cuts the vertical axis at a different point. This point can be obtained directly from the equation $y = 2x + c$ by looking at the value of c. For example,

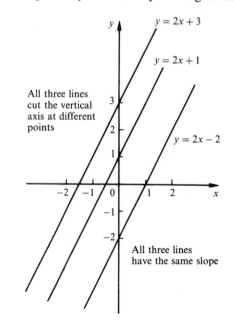

Figure 14.1
Graphs of $y = 2x + 3$, $y = 2x + 1$ and $y = 2x - 2$

$y = 2x + 1$ intersects the vertical axis at $y = 1$. The graph of $y = 2x + 3$ cuts the vertical axis at $y = 3$, and the graph of $y = 2x - 2$ cuts this axis at $y = -2$.

The point where a graph cuts the vertical axis is called the **vertical intercept**. The vertical intercept can be obtained from $y = mx + c$ by looking at the value of c.

KEY POINT

> In the equation $y = mx + c$ the value of c gives the y coordinate of the point where the line cuts the vertical axis.

Worked example

14.3   On the same graph plot the following: $y = x + 2$, $y = 2x + 2$, $y = 3x + 2$. Comment upon the graphs.

Solution   The three straight lines are shown in Figure 14.2. Note that the steepness of the line is determined by the coefficient of x, with $y = 3x + 2$ being steeper than $y = 2x + 2$, and this in turn being steeper than $y = x + 2$.

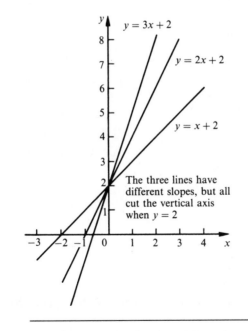

Figure 14.2
Graphs of $y = x + 2$,
$y = 2x + 2$, and
$y = 3x + 2$

The value of m in the equation $y = mx + c$ determines the steepness of the straight line. The larger the value of m, the steeper is the line. The value m is known as the slope or **gradient** of the straight line.

> **KEY POINT**
>
> In the equation $y = mx + c$ the value m is known as the gradient and is a measure of the steepness of the line.

If m is positive, the line will rise as we move from left to right. If m is negative, the line will fall. If $m = 0$ the line will be horizontal. We say it has zero gradient. These points are summarized in Figure 14.3.

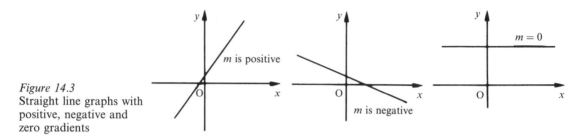

Figure 14.3
Straight line graphs with positive, negative and zero gradients

---

Self-assessment questions 14.1

1. Give the standard form of the equation of a straight line.
2. Explain the meaning of m and c in the equation $y = mx + c$.

Exercise 14.1

1. Identify, without plotting, which of the following functions will give straight line graphs:

   (a) $y = 3x + 9$ (b) $y = -9x + 2$
   (c) $y = -6x$ (d) $y = x^2 + 3x$
   (e) $y = 17$ (f) $y = x^{-1} + 7$

2. Identify the gradient and vertical intercept of each of the following lines:

   (a) $y = 9x - 11$ (b) $y = 8x + 1.4$

   (c) $y = \dfrac{1}{2}x - 11$ (d) $y = 17 - 2x$

   (e) $y = \dfrac{2x + 1}{3}$ (f) $y = \dfrac{4 - 2x}{5}$

   (g) $y = 3(x - 1)$ (h) $y = 4$

3. Identify (i) the gradient and (ii) the vertical intercept of the following lines:

   (a) $y + x = 6$ (b) $y - 2x + 1 = 0$

   (c) $2y - 4x + 3 = 0$ (d) $3x - 4y + 12 = 0$

   (e) $3x + \dfrac{y}{2} - 9 = 0$

14.2   Finding the equation of a straight line from its graph

If we are given the graph of a straight line it is often necessary to find its equation, $y = mx + c$. This amounts to finding the values of m and c. Finding the vertical intercept is straightforward because we can look directly for the point where the line cuts the y axis. The y coordinate of this point gives the value of c. The gradient m can be determined from knowledge of any two points on the line using the formula

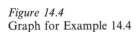

KEY POINT

$$\text{gradient} = \frac{\text{difference between the } y \text{ coordinates}}{\text{difference between the } x \text{ coordinates}}$$

Worked examples

14.4   A straight line graph is shown in Figure 14.4. Determine its equation.

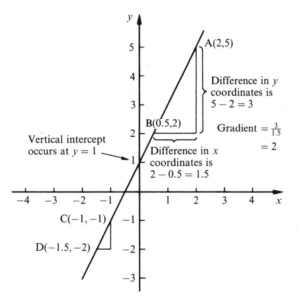

Figure 14.4
Graph for Example 14.4

Solution   We require the equation of the line in the form $y = mx + c$. From the graph it is easy to see that the vertical intercept occurs at $y = 1$. Therefore the value of c is 1. To find the gradient m we choose any two points on the line. We have chosen the point A with coordinates $(2,5)$ and the point B with coordinates $(0.5,2)$. The difference between their y coordinates is then $5 - 2 = 3$. The difference between their x coordinates is $2 - 0.5 = 1.5$. Then

$$\text{gradient} = \frac{\text{difference between their } y \text{ coordinates}}{\text{difference between their } x \text{ coordinates}}$$

$$= \frac{3}{1.5} = 2$$

The gradient m is equal to 2. Note that as we move from left to right the line is rising and so the value of m is positive. The equation of the line is then $y = 2x + 1$. There is nothing special about the points A and B. Any two points are sufficient to find m. For example, using the points C with coordinates $(-1, -1)$ and D with coordinates $(-1.5, -2)$ we would find

$$\text{gradient} = \frac{\text{difference between their } y \text{ coordinates}}{\text{difference between their } x \text{ coordinates}}$$

$$= \frac{-1 - (-2)}{-1 - (-1.5)}$$

$$= \frac{1}{0.5} = 2$$

as before.

14.5   A straight line graph is shown in Figure 14.5. Find its equation.

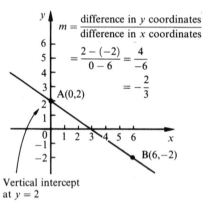

$$m = \frac{\text{difference in } y \text{ coordinates}}{\text{difference in } x \text{ coordinates}}$$

$$= \frac{2 - (-2)}{0 - 6} = \frac{4}{-6}$$

$$= -\frac{2}{3}$$

A(0,2)

B(6,−2)

Figure 14.5
Graph for Example 14.5

Vertical intercept
at $y = 2$

Solution   We need to find the equation in the form $y = mx + c$. From the graph we see immediately that the value of c is 2. To find the gradient we have selected any two points, $A(0, 2)$ and $B(6, -2)$. The difference between their y coordinates is $2 - (-2) = 4$. The difference between their x coordinates is $0 - 6 = -6$. Then

$$\text{gradient} = \frac{\text{difference between their } y \text{ coordinates}}{\text{difference between their } x \text{ coordinates}}$$

$$= \frac{4}{-6}$$

$$= -\frac{2}{3}$$

The equation of the line is therefore $y = -\frac{2}{3}x + 2$. Note in particular that because the line is sloping downwards as we move from left to right, the gradient is negative. Note also that the coordinates of A and B both satisfy the equation of the line, that is, for $A(0, 2)$,

$$2 = -\frac{2}{3}(0) + 2$$

and for $B(6, -2)$,

$$-2 = -\frac{2}{3}(6) + 2$$

The coordinates of any other point on the line must also satisfy the equation.

The point noted at the end of example 14.5 is important:

KEY POINT

> If the point (a, b) lies on the line $y = mx + c$ then this equation is satisfied by letting $x = a$ and $y = b$.

Worked example

14.6   Find the equation of the line shown in Figure 14.6.

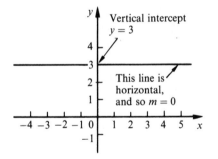

Figure 14.6
Graph for Example 14.6

Solution   We are required to express the equation in the form $y = mx + c$. From the graph we notice that the line is horizontal. This means that its gradient is 0, that is, $m = 0$. Furthermore the line cuts the vertical axis at $y = 3$ and so the equation of the line is $y = 0x + 3$ or simply $y = 3$.

It is not necessary to sketch a graph in order to find the equation. Consider the following example which illustrates an algebraic method.

Worked examples

14.7   A straight line passes through $A(7, 1)$ and $B(-3, 2)$. Find its equation.

Solution   The equation must be of the form $y = mx + c$. The gradient of the line can be found from

$$\text{gradient} = m = \frac{\text{difference between their } y \text{ coordinates}}{\text{difference between their } x \text{ coordinates}}$$

$$= \frac{1 - 2}{7 - (-3)}$$

$$= \frac{-1}{10}$$

$$= -0.1$$

Hence $y = -0.1x + c$. We can find c by noting that the line passes through (7,1), that is, the point where $x = 7$ and $y = 1$. Substituting these values into the equation $y = -0.1x + c$ gives

$$1 = -0.1(7) + c$$

so that $c = 1 + 0.7 = 1.7$. Therefore the equation of the line is $y = -0.1x + 1.7$.

14.8   Determine the equation of the line which passes through $(4, -1)$ and has gradient -2.

Solution   Let the equation of the line be $y = mx + c$. We are told that the gradient of the line is -2, that is $m = -2$ and so we have

$$y = -2x + c$$

The point $(4, -1)$ lies on this line; hence when $x = 4, y = -1$. These values are substituted into the equation of the line:

$$-1 = -2(4) + c$$
$$c = 7$$

The equation of the line is thus $y = -2x + 7$.

Self-assessment questions 14.2

1. State the formula for finding the gradient of a straight line when two points upon it are known. If the two points are (x_1, y_1) and (x_2, y_2) write down an expression for the gradient.
2. Explain how the value of c in the equation $y = mx + c$ can be found by inspecting the straight line graph.

Exercise 14.2

1. A straight line passes through the two points $(1,7)$ and $(2,9)$. Sketch a graph of the line and find its equation.

2. Find the equation of the line which passes through the two points $(2,2)$ and $(3,8)$.

3. Find the equation of the line which passes through $(8,2)$ and $(-2,2)$.

4. Find the equation of the straight line which has gradient 1 and passes through the origin.

5. Find the equation of the straight line which has gradient -1 and passes through the origin.

6. Find the equation of the straight line passing through $(-1,6)$ with gradient 2.

7. Which of the following points lie on the line $y = 4x - 3$?

   (a) $(1,2)$ (b) $(2,5)$ (c) $(5,17)$
   (d) $(-1,-7)$ (e) $(0,2)$

8. Find the equation of the straight line passing through $(-3,7)$ with gradient -1.

9. Determine the equation of the line passing through $(-1,-6)$ which is parallel to the line $y = 3x + 17$.

10. Find the equation of the line with vertical intercept -2 passing through $(3,10)$.

14.3   Gradients of tangents to curves

Figure 14.7 shows a graph of $y = x^2$. If you study the graph you will notice that as we move from left to right, at some points the y values are decreasing, whilst at others the y values are increasing. It is intuitively obvious that the slope of the curve changes from point to point. At some points, such as A, the curve appears quite steep and falling. At points such as B the curve appears quite steep and rising. Unlike a straight line, the slope of a curve is not fixed but changes as we move from one point to another. A useful way of measuring the slope at any point is to draw a **tangent** to the curve at that point. The tangent is a straight line which just touches the curve at the point of interest. In Figure 14.7 a tangent to the curve $y = x^2$ has been drawn at the point $(2,4)$. If we calculate the gradient of this tangent, this gives the gradient of the curve at the point $(2,4)$.

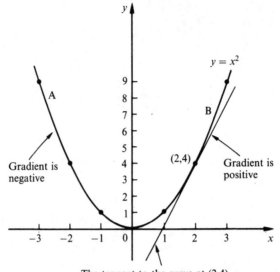

Figure 14.7
A graph of $y = x^2$

The tangent to the curve at (2,4)

    The gradient of a curve at any point is equal to the gradient of the tangent at that point.

Worked example

14.9 (a) Plot a graph of $y = x^2 - x$ for values of x between -2 and 4.

(b) Draw in tangents at the points $A(-1, 2)$ and $B(3, 6)$.

(c) By calculating the gradients of these tangents find the gradient of the curve at A and at B.

Solution (a) A table of values and the graph is shown in Figure 14.8.

(b) We now draw tangents at A and B. At present, the best we can do is estimate these by eye.

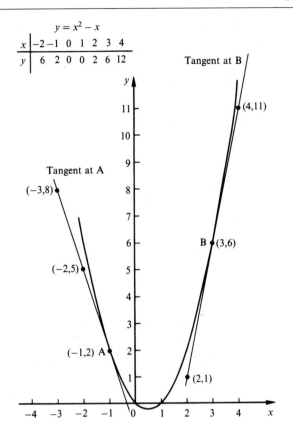

$y = x^2 - x$

x	-2	-1	0	1	2	3	4
y	6	2	0	0	2	6	12

Figure 14.8
A graph of $y = x^2 - x$

(c) We now calculate the gradient of the tangent at A. We select any two points on the tangent and calculate the difference between their y coordinates and the difference between their x coordinates. We have chosen the points $(-3, 8)$ and $(-2, 5)$. Referring to Figure 14.8 we see that

$$\text{gradient of tangent at } A = \frac{8 - 5}{-3 - (-2)} = \frac{3}{-1}$$
$$= -3$$

Hence the gradient of the curve at A is -3. Similarly, to find the gradient of the tangent at B we have selected two points on this tangent, namely $(4, 11)$ and $(2, 1)$. We find

$$\text{gradient of tangent at } B = \frac{11 - 1}{4 - 2} = \frac{10}{2}$$
$$= 5$$

Hence the gradient of the tangent at B is 5.

Clearly, the accuracy of our answer depends to a great extent upon how well we can draw and measure the gradient of the tangent.

Worked example

14.10   (a) Sketch a graph of the curve $y = x^3$ for $-2 \leq x \leq 2$.

(b) Draw the tangent to the graph at the point where $x = 1$.

(c) Estimate the gradient of this tangent and find its equation.

Solution   (a) A graph is shown in Figure 14.9.

(b) The tangent has been drawn at $x = 1$.

(c) Let us write the equation of the tangent as $y = mx + c$. Two points on the tangent have been selected in order to estimate the gradient. These are $(2, 4)$ and $(-1, -5)$. From these we find

$$\text{gradient of tangent is approximately } \frac{4 - (-5)}{2 - (-1)} = \frac{9}{3} = 3$$

Therefore $m = 3$. The value of c is found by noting that the vertical intercept of the tangent is -2. The equation of the tangent is then $y = 3x - 2$.

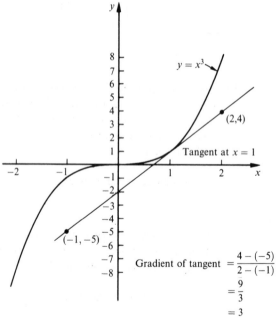

Figure 14.9
Graph of $y = x^3$

Of course, this method will usually result in an approximation based upon how well we have drawn the graph and its tangent. A much more precise method for calculating gradients of curves is given in Chapter 24.

Self-assessment questions 14.3

1. Explain what is meant by the 'tangent' to a curve at a point.
2. Explain how a tangent is used to determine the gradient of a curve.

Exercise 14.3

1. Draw the graph of $y = 2x^2 - 1$ for values of x between -3 and 3. By drawing a tangent, estimate the gradient of the curve at $A(2, 7)$ and $B(-1, 1)$.

2. Draw the graph of $y = -2x^2 + 2$ for values of x between -3 and 3. Draw the tangent at the point where $x = 1$ and calculate its equation.

Test and assignment exercises 14

1. Which of the following will have straight line graphs?

   (a) $y = 2x - 11$ (b) $y = 5x + 10$
   (c) $y = x^2 - 1$ (d) $y = -3 + 3x$
   (e) $y = \dfrac{2x + 3}{2}$

   For each straight line, identify the gradient and vertical intercept.

2. Find the equation of the straight line which passes through the points $(1, 11)$ and $(2, 18)$. Show that the line also passes through $(-1, -3)$.

3. Find the equation of the line which has gradient -2 and passes through the point $(1, 1)$.

4. Find the equation of the line which passes through $(-1, 5)$ and $(1, 5)$. Does the line also pass through $(2, 6)$?

5. Draw a graph of $y = -x^2 + 3x$ for values of x between -3 and 3. By drawing in tangents estimate the gradient of the curve at the points $(-2, -10)$ and $(1, 2)$.

6. Find the equations of the lines passing through the origin with gradients (a) -2, (b) -4, (c) 4.

7. Find the equation of the line passing through $(4, 10)$ and parallel to $y = 6x - 3$.

8. Find the equation of the line with vertical intercept 3 and passing through $(-1, 9)$.

9. Find where the line joining $(-2, 4)$ and $(3, 10)$ cuts (a) the x axis (b) the y axis.

10. A line cuts the x axis at $x = -2$ and the y axis at $y = 3$. Determine the equation of the line.

11. Determine where the line $y = 4x - 1$ cuts

    (a) the y axis
    (b) the x axis
    (c) the line $y = 2$.

15 The exponential function

Objectives

This chapter

- shows how to simplify exponential expressions
- describes the form of the exponential function
- illustrates graphs of exponential functions
- lists the properties of the exponential function
- shows how to solve equations with exponential terms using a graphical technique

15.1   Exponential expressions

An expression of the form a^x is called an **exponential expression**. The number a is called the **base** and x is the **power** or **index**. For example, 4^x and 0.5^x are both exponential expressions. The first has base 4 and the second has base 0.5. An exponential expression can be evaluated using the power button on your calculator. For example, the exponential expression with base 4 and index 0.3 is

$$4^{0.3} = 1.516$$

The letter e always denotes the constant 2.71828 . . . .

One of the most commonly used values for the base in an exponential expression is 2.71828 . . . . This number is denoted by the letter e.

KEY POINT

The most common exponential expression is

e^x

where e is the exponential constant, 2.71828 . . . .

The expression e^x is found to occur in many natural phenomena, for example, population growth, spread of bacteria and radioactive decay.

Scientific calculators are pre-programmed to evaluate e^x for any value of x. Usually the button is marked e^x or exp x. Use the next example to check you can use this facility on your calculator.

Worked example

15.1 Use a scientific calculator to evaluate

(a) e^2 (b) e^3 (c) $e^{1.3}$ (d) $e^{-1.3}$

Solution We use the e^x button giving:

(a) $e^2 = 7.3891$

(b) $e^3 = 20.0855$

(c) $e^{1.3} = 3.6693$

(d) $e^{-1.3} = 0.2725$

Exponential expressions can be simplified using the normal rules of algebra. The laws of indices apply to exponential expressions. We note

$$e^a e^b = e^{a+b}$$

$$\frac{e^a}{e^b} = e^{a-b}$$

$$e^0 = 1$$

$$(e^a)^b = e^{ab}$$

Worked examples

15.2 Simplify the following exponential expressions:

(a) $e^2 e^4$ (b) $\dfrac{e^4}{e^3}$ (c) $(e^2)^{2.5}$

Solution (a) $e^2 e^4 = e^{2+4} = e^6$

(b) $\dfrac{e^4}{e^3} = e^{4-3} = e^1 = e$

(c) $(e^2)^{2.5} = e^{2\times2.5} = e^5$

15.3 Simplify the following exponential expressions:

(a) $e^x e^{3x}$ (b) $\dfrac{e^{4t}}{e^{3t}}$ (c) $(e^t)^4$

Solution   (a) $e^x e^{3x} = e^{x+3x} = e^{4x}$

(b) $\dfrac{e^{4t}}{e^{3t}} = e^{4t-3t} = e^t$

(c) $(e^t)^4 = e^{4t}$

15.4   Simplify

(a) $\sqrt{e^{6t}}$ (b) $e^{3y}(1 + e^y) - e^{4y}$ (c) $(2e^t)^2(3e^{-t})$

$\sqrt{a} = a^{\frac{1}{2}}$, that is, a square root sign is equivalent to the power $\frac{1}{2}$.

Solution   (a) $\sqrt{e^{6t}} = (e^{6t})^{1/2} = e^{6t/2} = e^{3t}$

(b) $e^{3y}(1 + e^y) - e^{4y} = e^{3y} + e^{3y}e^y - e^{4y}$

$= e^{3y} + e^{4y} - e^{4y}$

$= e^{3y}$

(c) $(2e^t)^2(3e^{-t}) = (2e^t)(2e^t)(3e^{-t})$

$= 2.2.3.e^t e^t e^{-t}$

$= 12e^{t+t-t}$

$= 12e^t$

15.5   (a) Verify that

$$(e^t + 1)^2 = e^{2t} + 2e^t + 1$$

(b) Hence simplify $\sqrt{e^{2t} + 2e^t + 1}$

Solution   (a) $(e^t + 1)^2 = (e^t + 1)(e^t + 1)$

$= e^t e^t + e^t + e^t + 1$

$= e^{2t} + 2e^t + 1$

(b) $\sqrt{e^{2t} + 2e^t + 1} = \sqrt{(e^t + 1)^2} = e^t + 1$

Exercise 15.1

1. Use a scientific calculator to evaluate
   (a) $e^{2.3}$ (b) $e^{1.9}$ (c) $e^{-0.6}$ (d) $\dfrac{1}{e^{-2}+1}$

   (e) $\dfrac{3e^2}{e^2-10}$

2. Simplify

   (a) $e^2.e^7$ (b) $\dfrac{e^7}{e^4}$ (c) $\dfrac{e^{-2}}{e^3}$ (d) $\dfrac{e^3}{e^{-1}}$

   (e) $\dfrac{(4e)^2}{(2e)^3}$ (f) $e^2\left(e+\dfrac{1}{e}\right)-e$

   (g) $e^{1.5}e^{2.7}$ (h) $\sqrt{e}\,\sqrt{4e}$

3. Simplify each of the following expressions as far as possible.
   (a) $e^x e^{-3x}$ (b) $e^{3x}e^{-x}$ (c) $e^{-3x}e^{-x}$

   (d) $(e^{-x})^2 e^{2x}$ (e) $\dfrac{e^{3x}}{e^x}$ (f) $\dfrac{e^{-3x}}{e^{-x}}$

4. Simplify as far as possible

   (a) $\dfrac{e^t}{e^3}$ (b) $\dfrac{2e^{3x}}{4e^x}$ (c) $\dfrac{(e^x)^3}{3e^x}$

   (d) $e^x + 2e^{-x} + e^x$ (e) $\dfrac{e^x e^y e^z}{e^{x/2-y}}$ (f) $\dfrac{(e^{t/3})^6}{e^t + e^t}$

   (g) $\dfrac{e^t}{e^{-t}}$ (h) $\dfrac{1+e^{-t}}{e^{-t}}$ (i) $(e^z e^{-z/2})^2$

   (j) $\dfrac{e^{-3+t}}{2e^t}$ (k) $\left(\dfrac{e^{-1}}{e^{-x}}\right)^{-1}$

15.2   **The exponential function and its graph**

The **exponential function** has the form

$$y = e^x$$

The function $y = e^x$ is found to occur in many natural phenomena. Table 15.1 gives values of x and e^x. Using Table 15.1, a graph of $y = e^x$ can be drawn. This is shown in Figure 15.1. From the table and graph we note the following important properties of $y = e^x$.

(a) The exponential function is never negative.
(b) When $x = 0$, the function value is 1.
(c) As x increases, then e^x increases. This is known as **exponential growth**.
(d) By choosing x large enough, the value of e^x can be made larger than any given number. We say e^x increases without bound as x increases.
(e) As x becomes large and negative, e^x gets nearer and nearer to 0.

Table 15.1

x	-3	-2	-1	0	1	2	3
e^x	0.0498	0.1353	0.3679	1	2.7183	7.3891	20.086

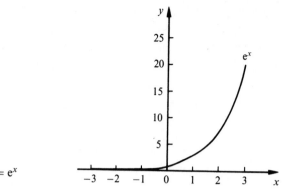

Figure 15.1
A graph of $y = e^x$

Worked example

15.6   Sketch a graph of $y = 2e^x$ and $y = e^{2x}$ for $-3 \leq x \leq 3$.

Solution   Table 15.2 gives values of $2e^x$ and e^{2x} and the graphs are shown in Figure 15.2.

Table 15.2

x	-3	-2	-1	0	1	2	3
$2e^x$	0.0996	0.2707	0.7358	2	5.4366	14.7781	40.1711
e^{2x}	0.0025	0.0183	0.1353	1	7.3891	54.5982	403.429

Figure 15.2
Graphs of $y = e^{2x}$ and $y = 2e^x$

We now turn attention to an associated function: $y = e^{-x}$. Values are listed in Table 15.3 and the function is illustrated in Figure 15.3. From the table and the graph we note the following properties of $y = e^{-x}$.

(a) The function is never negative.

(b) When $x = 0$, the function has a value of 1.

(c) As x increases, then e^{-x} decreases, getting nearer to 0. This is known as **exponential decay**.

Table 15.3

x	-3	-2	-1	0	1	2	3
$-x$	3	2	1	0	-1	-2	-3
e^{-x}	20.086	7.3891	2.7183	1	0.3679	0.1353	0.0498

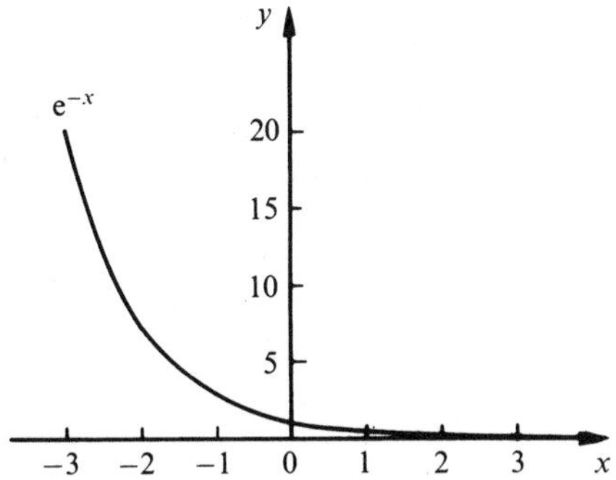

Figure 15.3
Graph of $y = e^{-x}$

Self-assessment questions 15.2

1. List properties which are common to $y = e^x$ and $y = e^{-x}$.
2. By choosing x large enough, then e^{-x} can be made to be negative. True or false?

Exercise 15.2

1. Draw up a table of values of $y = e^{3x}$ for x between -1 and 1 at intervals of 0.2. Sketch the graph of $y = e^{3x}$. Comment upon its shape and properties.

2. A species of animal has population $P(t)$ at time t given by

   $P = 10 - 5e^{-t}$

   (a) Sketch a graph of P against t for $0 \leq t \leq 5$.

   (b) What is the size of the population of the species as t becomes very large?

3. The concentration, $c(t)$, of a chemical in a reaction is modelled by the equation

   $c(t) = 6 + 3e^{-2t}$

   (a) Sketch a graph of c against t for $0 \leq t \leq 2$.

   (b) What value does the concentration approach as t becomes large?

4. The hyperbolic functions $\sinh x$ and $\cosh x$ are defined by

   $$\sinh x = \frac{e^x - e^{-x}}{2} \quad \cosh x = \frac{e^x + e^{-x}}{2}$$

   (a) Show $e^x = \cosh x + \sinh x$.

   (b) Show $e^{-x} = \cosh x - \sinh x$.

5. Express $3 \sinh x + 7 \cosh x$ in terms of exponential functions.

6. Express $6e^x - 9e^{-x}$ in terms of $\sinh x$ and $\cosh x$.

7. Show $(\cosh x)^2 - (\sinh x)^2 = 1$.

15.3   Solving equations involving exponential terms using a graphical method

Many equations involving exponential terms can be solved using graphs. A graphical method will yield an approximate solution which can then be refined by choosing a smaller interval for the domain. The following examples illustrate the technique.

Worked example

15.7   (a) Plot $y = e^{x/2}$ and $y = 2e^{-x}$ for $-1 \leq x \leq 1$.

(b) Hence solve the equation

$$e^{x/2} - 2e^{-x} = 0$$

Solution   (a) A table of values is drawn up for $e^{x/2}$ and $2e^{-x}$ so that the graphs can be drawn. Table 15.4 lists the values and Figure 15.4 shows the graphs of the functions.

Table 15.4

x	-1	-0.75	-0.5	-0.25	0	0.25	0.5	0.75	1
$e^{x/2}$	0.607	0.687	0.779	0.882	1	1.133	1.284	1.455	1.649
$2e^{-x}$	5.437	4.234	3.297	2.568	2	1.558	1.213	0.945	0.736

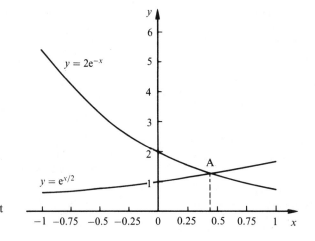

Figure 15.4
The graphs of $y = e^{x/2}$ and $y = 2e^{-x}$ intersect at A

(b) We note that the graphs intersect; the point of intersection is A. The equation

$$e^{x/2} - 2e^{-x} = 0$$

is equivalent to

$$e^{x/2} = 2e^{-x}$$

The graphs of $y = e^{x/2}$ and $y = 2e^{-x}$ intersect at A, where $x = 0.44$. Hence $x = 0.44$ is an approximate solution of $e^{x/2} - 2e^{-x} = 0$. The exact answer can be shown to be 0.46 (2 d.p.).

Exercise 15.3

1. (a) Plot $y = 15 - x^2$ and $y = e^x$ for $0 \le x \le 3$.

   (b) Hence find an approximate solution to
   $$e^x + x^2 = 15$$

2. (a) Plot $y = 12x^2 - 1$ and $y = 3e^{-x}$ for $-1 \le x \le 1$.

   (b) Hence state two approximate solutions of $3e^{-x} = 12x^2 - 1$.

3. By drawing $y = e^x$ for $-3 \le x \le 3$ and additional appropriate lines solve the following equations:

   (a) $e^x + x = 0$

   (b) $e^x - 1.5 = 0$

   (c) $e^x - x - 5 = 0$

   (d) $\dfrac{e^x}{2} + \dfrac{x}{2} - 5 = 0$

4. (a) Draw $y = 2 + 6e^{-t}$ for $0 \le t \le 3$

   (b) Use your graph to find an approximate solution to
   $$2 + 6e^{-t} = 5$$

   (c) Add the line $y = t + 4$ to your graph and hence find an approximate solution to
   $$6e^{-t} = t + 2$$

Test and assignment exercises 15

1. Use a calculator to evaluate

   (a) $e^{-3.1}$ (b) $e^{0.2}$ (c) $\dfrac{1}{e}$

2. (a) Sketch the graphs of

   $$y = e^{-x} \text{ and } y = \frac{x}{2}$$

   for $0 \le x \le 2$.

   (b) Hence solve the equation

   $$2e^{-x} - x = 0$$

3. Simplify where possible:

   (a) $e^x e^{2x} e^{-3x}$ (b) $\dfrac{e^{2x} e^{3x}}{e^{4x}}$

   (c) $(e^{2x})^3 e^{-3x}$ (d) $e^{2x} + e^{3x} - e^{5x}$

   (e) $\dfrac{e^{3x} + e^{5x}}{e^{2x}}$

4. Expand the brackets of:

   (a) $(e^x - 1)^2$ (b) $(e^{-x} - 1)^2$

   (c) $e^{2x}(e^x + 1)$ (d) $2e^x(e^{-x} + 3e^x)$

5. Simplify

   (a) $e^t e^{t/2}$ (b) $e^2 e^3 e$

   (c) $e^{3x}(e^x - e^{2x}) - (e^{2x})^2$

   (d) $e^{3+t} e^{5-2t}$ (e) $\dfrac{e^{2z+3} e^{5-z}}{e^{z+2}}$

   (f) $\sqrt{e^{4z}}$ (g) $(e^{x/2} e^x)^2$

   (h) $\sqrt{e^{2t} + 2e^{2t+1} + e^{2t+2}}$

   (i) $\dfrac{e^t + e^{2t}}{e^{2t} + e^{3t}}$

6. Use a graphical method to solve

   (a) $e^{2x} = 2x + 2$

   (b) $e^{-x} = 2 - x^2$

   (c) $e^x = 2 + \dfrac{x^3}{2}$

7. (a) Sketch the graphs of

   $$y = 2 + e^{-x} \text{ and } y = e^x$$

   for $0 \le x \le 2$.

   (b) Use your graphs to solve

   $$e^x - e^{-x} = 2$$

8. Remove the brackets from the following expressions:

   (a) $(e^{2t})^4$ (b) $e^t(e^t - e^{-t})$

   (c) $e^{2y}(e^{3y} + e^{-2y} + e^y)$ (d) $(1 + e^t)^2$

9. The height of mercury in an experiment, $H(t)$, varies according to

   $$H(t) = 10 + 3e^{-t} - 4e^{-2t}$$

   What value does H approach as t gets very large?

10. Given

    $$X(t) = \frac{3 + 2e^{2t}}{4 + e^{2t}}$$

    find the value that X approaches as t gets very large.

16 The logarithm function

<table>
<tr><td>

Objectives

</td><td>

This chapter

- explains the term 'base of a logarithm'
- shows how to calculate the logarithm of a number to any base
- states the laws of logarithms and uses them to simplify expressions
- shows how to solve exponential and logarithmic equations
- defines the logarithm function
- illustrates graphically the logarithm function

</td></tr>
</table>

16.1 Introducing logarithms

Given an equation such as $125 = 5^3$, we call 5 the base and 3 the power or index. We can use **logarithms** to write the equation in another form. The logarithm form is

$$\log_5 125 = 3$$

This is read as 'logarithm to the base 5 of 125 is 3'. In general if

$$y = a^x$$

then

$$\log_a y = x$$

KEY POINT

$y = a^x$ and $\log_a y = x$ are equivalent.

'The logarithm to the base a of y is x' is equivalent to saying that 'y is a to the power x'. The word logarithm is usually shortened to just 'log'.

Worked example

16.1   Write down the logarithm form of the following:

(a) $16 = 4^2$ (b) $8 = 2^3$ (c) $25 = 5^2$

Solution   (a) $16 = 4^2$ may be written as

$$2 = \log_4 16$$

that is, 2 is the logarithm to the base 4 of 16.

(b) $8 = 2^3$ may be expressed as

$$3 = \log_2 8$$

which is read as '3 is the logarithm to the base 2 of 8'.

(c) $25 = 5^2$ may be expressed as

$$2 = \log_5 25$$

that is, 2 is the logarithm to the base 5 of 25.

Given an equation such as $16 = 4^2$ then 'taking logs' to base 4 will result in the logarithmic form $\log_4 16 = 2$.

Worked example

16.2   Write the exponential form of the following:

(a) $\log_2 16 = 4$ (b) $\log_3 27 = 3$ (c) $\log_5 125 = 3$ (d) $\log_{10} 100 = 2$

Solution   (a) Here the base is 2 and so we may write $16 = 2^4$.

(b) The base is 3 and so $27 = 3^3$.

(c) The base is 5 and so $125 = 5^3$.

(d) The base is 10 and so $100 = 10^2$.

Recall that e is the constant 2.71828..., which occurs frequently in natural phenomena.

Although the base of a logarithm can be any positive number, the commonly used bases are 10 and e. Logarithms to base 10 are often denoted by 'log' or 'log$_{10}$'; logarithms to base e are denoted by 'ln' or 'log$_e$' and referred to as **natural logarithms**. Most scientific calculators possess 'log' and 'ln' buttons which are used to evaluate logarithms to base 10 and base e.

Worked examples

16.3   Use a scientific calculator to evaluate the following:

(a) $\log 71$ (b) $\ln 3.7$ (c) $\log 0.4615$ (d) $\ln 0.5$ (e) $\ln 1000$

Solution   Using a scientific calculator we obtain

(a) $\log 71 = 1.8513$ (b) $\ln 3.7 = 1.3083$ (c) $\log 0.4615 = -0.3358$

(d) $\ln 0.5 = -0.6931$ (e) $\ln 1000 = 6.9078$

You should ensure that you know how to use your calculator to verify these results.

16.4   Given $10^{0.6990} = 5$ evaluate the following:

(a) $10^{1.6990}$ (b) $\log 5$ (c) $\log 500$

Solution   (a) $10^{1.6990} = 10^1.10^{0.6990} = 10(5) = 50$

(b) We are given $10^{0.6990} = 5$ and by taking logs to the base 10 we obtain $0.6990 = \log 5$.

(c) From $10^{0.6990} = 5$ we can see that

$$100(5) = 100(10^{0.6990})$$

and so

$$500 = 10^2.10^{0.6990} = 10^{2.6990}$$

Taking logs to the base 10 gives

$$\log 500 = 2.6990$$

Self-assessment questions 16.1

1. The base of a logarithm is always a positive integer. True or false?

Exercise 16.1

1. Use a scientific calculator to evaluate

   (a) $\log 150$ (b) $\ln 150$ (c) $\log 0.316$
   (d) $\ln 0.1$

2. Write down the logarithmic form of the following:

   (a) $3^8 = 6561$ (b) $6^5 = 7776$
   (c) $2^{10} = 1024$ (d) $10^5 = 100000$
   (e) $4^7 = 16384$

   (f) $\left(\dfrac{1}{2}\right)^5 = 0.03125$

   (g) $12^3 = 1728$ (h) $9^4 = 6561$

3. Write the exponential form of the following:

   (a) $\log_6 1296 = 4$ (b) $\log_{15} 225 = 2$
   (c) $\log_8 512 = 3$ (d) $\log_7 2401 = 4$
   (e) $\log_3 243 = 5$ (f) $\log_6 216 = 3$
   (g) $\log_{20} 8000 = 3$ (h) $\log_{16} 4096 = 3$
   (i) $\log_2 4096 = 12$

4. Given $10^{0.4771} = 3$, evaluate the following:

   (a) $\log 3$ (b) $\log 300$ (c) $\log 0.03$

5. Given $\log 7 = 0.8451$ evaluate the following:

   (a) $10^{0.8451}$ (b) $\log 700$ (c) $\log 0.07$

16.2　Calculating logarithms to any base

In §16.1 we showed how equations of the form $y = a^x$ could be written in logarithmic form as $\log_a y = x$. The number a, called the **base**, is always positive, that is, $a > 0$. We also introduced the common bases 10 and e. Scientific calculators are programmed with logarithms to base 10 and to base e. Suppose that we wish to calculate logarithms to bases other than 10 or e. For example, in communication theory, logarithms to base 2 are commonly used. To calculate logarithms to base a we use one of the following two formulae.

KEY POINT

$$\log_a X = \frac{\log_{10} X}{\log_{10} a} \qquad \log_a X = \frac{\ln X}{\ln a}$$

Worked example

16.5　Evaluate (a) $\log_6 19$, (b) $\log_7 29$.

Solution　(a) We use the formula

$$\log_a X = \frac{\log_{10} X}{\log_{10} a}$$

Comparing $\log_6 19$ with $\log_a X$ we see $X = 19$ and $a = 6$. So

$$\log_6 19 = \frac{\log_{10} 19}{\log_{10} 6} = \frac{1.2788}{0.7782} = 1.6433$$

We could have equally well used the formula

$$\log_a X = \frac{\ln X}{\ln a}$$

With this formula we obtain

$$\log_6 19 = \frac{\ln 19}{\ln 6} = \frac{2.9444}{1.7918} = 1.6433$$

(b) Comparing $\log_a X$ with $\log_7 29$ we see $X = 29$ and $a = 7$. Hence

$$\log_7 29 = \frac{\log_{10} 29}{\log_{10} 7} = \frac{1.4624}{0.8451} = 1.7304$$

Alternatively we use

$$\log_7 29 = \frac{\ln 29}{\ln 7} = \frac{3.3673}{1.9459} = 1.7304$$

By considering the formula

$$\log_a X = \frac{\log_{10} X}{\log_{10} a}$$

with $X = a$ we obtain

$$\log_a a = \frac{\log_{10} a}{\log_{10} a} = 1$$

KEY POINT

$$\log_a a = 1$$

This same result could be derived by writing the logarithm form of $a = a^1$.

Exercise 16.2

1. Evaluate the following:

   (a) $\log_4 6$ (b) $\log_3 10$ (c) $\log_{20} 270$
   (d) $\log_5 0.65$ (e) $\log_2 100$ (f) $\log_2 0.03$
   (g) $\log_{100} 10$ (h) $\log_7 7$

2. Show that

   $2.3026 \, \log_{10} X = \ln X$

3. Evaluate the following

   (a) $\log_3 7 + \log_4 7 + \log_5 7$
   (b) $\log_8 4 + \log_8 0.25$ (c) $\log_{0.7} 2$
   (d) $\log_2 0.7$

16.3   Laws of logarithms

Logarithms obey several laws which we now examine. They are introduced via examples.

Worked example

16.6 Evaluate (a) $\log 7$, (b) $\log 12$, (c) $\log 84$ and $\log 7 + \log 12$. Comment on your findings.

Solution (a) $\log 7 = 0.8451$ (b) $\log 12 = 1.0792$

(c) $\log 84 = 1.9243$, and $\log 7 + \log 12 = 0.8451 + 1.0792 = 1.9243$

We note that $\log 7 + \log 12 = \log 84$.

Example 16.6 illustrates the **first law** of logarithms which states:

$$\log A + \log B = \log AB$$

This law holds true for any base. However, in any one calculation all bases must be the same.

Worked examples

16.7 Simplify to a single log term

(a) $\log 9 + \log x$

(b) $\log t + \log 4t$

(c) $\log 3x^2 + \log 2x$

Solution (a) $\log 9 + \log x = \log 9x$

(b) $\log t + \log 4t = \log(t.4t) = \log 4t^2$

(c) $\log 3x^2 + \log 2x = \log(3x^2.2x) = \log 6x^3$

16.8 Simplify

(a) $\log 7 + \log 3 + \log 2$

(b) $\log 3x + \log x + \log 4x$

Solution (a) We know $\log 7 + \log 3 = \log(7 \times 3) = \log 21$, and so

$$\log 7 + \log 3 + \log 2 = \log 21 + \log 2$$
$$= \log(21 \times 2) = \log 42$$

(b) We have

$$\log 3x + \log x = \log(3x.x) = \log 3x^2$$

and so

$$\log 3x + \log x + \log 4x = \log 3x^2 + \log 4x$$
$$= \log(3x^2.4x) = \log 12x^3$$

We now consider an example which introduces the second law of logarithms.

Worked example

16.9   (a) Evaluate $\log 12$, $\log 4$ and $\log 3$.

     (b) Compare the values of $\log 12 - \log 4$ and $\log 3$.

Solution   (a) $\log 12 = 1.0792$, $\log 4 = 0.6021$, $\log 3 = 0.4771$.

     (b) From part (a),

$$\log 12 - \log 4 = 1.0792 - 0.6021 = 0.4771$$

and also

$$\log 3 = 0.4771$$

We note that $\log 12 - \log 4 = \log 3$.

This example illustrates the **second law** of logarithms, which states:

KEY POINT

$$\log A - \log B = \log\left(\frac{A}{B}\right)$$

Worked examples

16.10   Use the second law of logarithms to simplify the following to a single log term:

(a) $\log 20 - \log 10$ (b) $\log 500 - \log 75$

(c) $\log 4x^3 - \log 2x$ (d) $\log 5y^3 - \log y$

Solution   (a) Using the second law of logarithms we have

$$\log 20 - \log 10 = \log\left(\frac{20}{10}\right) = \log 2$$

(b) $\log 500 - \log 75 = \log\left(\frac{500}{75}\right) = \log\left(\frac{20}{3}\right)$

(c) $\log 4x^3 - \log 2x = \log\left(\frac{4x^3}{2x}\right) = \log 2x^2$

(d) $\log 5y^3 - \log y = \log\left(\frac{5y^3}{y}\right) = \log 5y^2$

16.11   Simplify

(a) $\log 20 + \log 3 - \log 6$

(b) $\log 18 - \log 24 + \log 2$

Solution   (a) Using the first law of logarithms we see that

$$\log 20 + \log 3 = \log 60$$

and so

$$\log 20 + \log 3 - \log 6 = \log 60 - \log 6$$

Using the second law of logarithms we see that

$$\log 60 - \log 6 = \log\left(\frac{60}{6}\right) = \log 10$$

Hence

$$\log 20 + \log 3 - \log 6 = \log 10$$

(b) $\log 18 - \log 24 + \log 2 = \log\left(\dfrac{18}{24}\right) + \log 2$

$$= \log\left(\frac{3}{4}\right) + \log 2$$

$$= \log\left(\frac{3}{4} \times 2\right)$$

$$= \log 1.5$$

16.12   Simplify

(a) $\log 2 + \log 3x - \log 2x$

(b) $\log 5y^2 + \log 4y - \log 10y^2$

Solution   (a) $\log 2 + \log 3x - \log 2x = \log(2 \times 3x) - \log 2x$

$$= \log 6x - \log 2x$$

$$= \log\left(\frac{6x}{2x}\right) = \log 3$$

(b) $\log 5y^2 + \log 4y - \log 10y^2 = \log(5y^2 . 4y) - \log 10y^2$

$$= \log 20y^3 - \log 10y^2$$

$$= \log\left(\frac{20y^3}{10y^2}\right) = \log 2y$$

We consider a special case of the second law. Consider $\log A - \log A$. This is clearly 0. However, using the second law we may write

$$\log A - \log A = \log\left(\frac{A}{A}\right) = \log 1$$

Thus

KEY POINT

$$\boxed{\log 1 = 0}$$

In any base, the logarithm of 1 equals 0.

Finally we introduce the third law of logarithms.

Worked example

16.13 (a) Evaluate $\log 16$ and $\log 2$.

(b) Compare $\log 16$ and $4\log 2$.

Solution (a) $\log 16 = 1.204$, $\log 2 = 0.301$.

(b) $\log 16 = 1.204$, $4\log 2 = 1.204$. Hence we see that $4\log 2 = \log 16$.

Noting that $16 = 2^4$, example 16.13 suggests the **third law** of logarithms:

KEY POINT

$$\boxed{n\log A = \log A^n}$$

This law applies if n is integer, fractional, positive or negative.

Worked examples

16.14 Write the following as a single logarithmic expression:

(a) $3\log 2$ (b) $2\log 3$ (c) $4\log 3$

Solution (a) $3\log 2 = \log 2^3 = \log 8$

(b) $2\log 3 = \log 3^2 = \log 9$

(c) $4\log 3 = \log 3^4 = \log 81$

16.15 Write as a single log term

(a) $\frac{1}{2}\log 16$ (b) $-\log 4$ (c) $-2\log 2$ (d) $-\frac{1}{2}\log 0.5$

Solution (a) $\frac{1}{2}\log 16 = \log 16^{1/2} = \log \sqrt{16} = \log 4$

(b) $-\log 4 = -1.\log 4 = \log 4^{-1} = \log\left(\dfrac{1}{4}\right) = \log 0.25$

(c) $-2\log 2 = \log 2^{-2} = \log\left(\dfrac{1}{2^2}\right) = \log\left(\dfrac{1}{4}\right) = \log 0.25$

(d) $-\dfrac{1}{2}\log 0.5 = -\dfrac{1}{2}\log\left(\dfrac{1}{2}\right) = \log\left(\dfrac{1}{2}\right)^{-1/2} = \log 2^{1/2} = \log\sqrt{2}$

16.16 Simplify

(a) $3\log x - \log x^2$

(b) $3\log t^3 - 4\log t^2$

(c) $\log Y - 3\log 2Y + 2\log 4Y$

Solution (a) $3\log x - \log x^2 = \log x^3 - \log x^2$

$$= \log\left(\dfrac{x^3}{x^2}\right)$$

$$= \log x$$

(b) $3\log t^3 - 4\log t^2 = \log(t^3)^3 - \log(t^2)^4$

$$= \log t^9 - \log t^8$$

$$= \log\left(\dfrac{t^9}{t^8}\right)$$

$$= \log t$$

(c) $\log Y - 3\log 2Y + 2\log 4Y = \log Y - \log(2Y)^3 + \log(4Y)^2$

$$= \log Y - \log 8Y^3 + \log 16Y^2$$

$$= \log\left(\dfrac{Y.16Y^2}{8Y^3}\right)$$

$$= \log 2$$

16.17 Simplify

(a) $2\log 3x - \dfrac{1}{2}\log 16x^2$

(b) $\dfrac{3}{2}\log 4x^2 - \log\left(\dfrac{1}{x}\right)$

(c) $2\log\left(\dfrac{2}{x^2}\right) - 3\log\left(\dfrac{2}{x}\right)$

Solution   (a) $2\log 3x - \dfrac{1}{2}\log 16x^2 = \log(3x)^2 - \log(16x^2)^{1/2}$

$$= \log 9x^2 - \log 4x$$

$$= \log\left(\frac{9x^2}{4x}\right)$$

$$= \log\left(\frac{9x}{4}\right)$$

(b) $\dfrac{3}{2}\log 4x^2 - \log\left(\dfrac{1}{x}\right) = \log(4x^2)^{3/2} - \log(x^{-1})$

$$= \log 8x^3 + \log x$$

$$= \log 8x^4$$

(c) $2\log\left(\dfrac{2}{x^2}\right) - 3\log\left(\dfrac{2}{x}\right) = \log\left(\dfrac{2}{x^2}\right)^2 - \log\left(\dfrac{2}{x}\right)^3$

$$= \log\left(\frac{4}{x^4}\right) - \log\left(\frac{8}{x^3}\right)$$

$$= \log\left(\frac{4/x^4}{8/x^3}\right)$$

$$= \log\left(\frac{1}{2x}\right)$$

Self-assessment questions 16.3

1. State the three laws of logarithms.

Exercise 16.3

1. Write the following as a single log term using the laws of logarithms:

   (a) $\log 5 + \log 9$ (b) $\log 9 - \log 5$
   (c) $\log 5 - \log 9$ (d) $2 \log 5 + \log 1$
   (e) $2 \log 4 - 3 \log 2$ (f) $\log 64 - 2 \log 2$
   (g) $3 \log 4 + 2 \log 1 + \log 27 - 3 \log 12$

2. Simplify as much as possible:

   (a) $\log 3 + \log x$ (b) $\log 4 + \log 2x$
   (c) $\log 3X - \log 2X$ (d) $\log T^3 - \log T$
   (e) $\log 5X + \log 2X$

3. Simplify:

   (a) $3 \log X - \log X^2$ (b) $\log y - 2 \log \sqrt{y}$

   (c) $5 \log x^2 + 3 \log \dfrac{1}{x}$

   (d) $4 \log X - 3 \log X^2 + \log X^3$

   (e) $3 \log y^{1.4} + 2 \log y^{0.4} - \log y^{1.2}$

4. Simplify the following as much as possible by using the laws of logarithms:

   (a) $\log 4x - \log x$ (b) $\log t^3 + \log t^4$

   (c) $\log 2t - \log\left(\dfrac{t}{4}\right)$

   (d) $\log 2 + \log\left(\dfrac{3}{x}\right) - \log\left(\dfrac{x}{2}\right)$

   (e) $\log\left(\dfrac{t^2}{3}\right) + \log\left(\dfrac{6}{t}\right) - \log\left(\dfrac{1}{t}\right)$

   (f) $2 \log y - \log y^2$

   (g) $3 \log\left(\dfrac{1}{t}\right) + \log t^2$

   (h) $4 \log \sqrt{x} + 2 \log\left(\dfrac{1}{x}\right)$

   (i) $2 \log x + 3 \log t$ (j) $\log A - \dfrac{1}{2} \log 4A$

   (k) $\dfrac{\log 9x + \log 3x^2}{3}$

   (l) $\log xy + 2 \log\left(\dfrac{x}{y}\right) + 3 \log\left(\dfrac{y}{x}\right)$

   (m) $\log\left(\dfrac{A}{B}\right) - \log\left(\dfrac{B}{A}\right)$

   (n) $\log\left(\dfrac{2t}{3}\right) + \dfrac{1}{2} \log 9t - \log\left(\dfrac{1}{t}\right)$

5. Express as a single log term:

   $\log_{10} X + \ln X$

6. Simplify

   (a) $\log(9x - 3) - \log(3x - 1)$
   (b) $\log(x^2 - 1) - \log(x + 1)$
   (c) $\log(x^2 + 3x) - \log(x + 3)$

16.4   Solving equations with logarithms

This section illustrates the use of logarithms in solving certain types of equations. For reference we note from §16.1 the equivalence of

$$y = a^x \text{ and } \log_a y = x$$

and from §16.3 the laws of logarithms:

$$\log A + \log B = \log AB$$
$$\log A - \log B = \log\left(\frac{A}{B}\right)$$
$$n \log A = \log A^n$$

Worked examples

16.18   Solve the following equations

(a) $10^x = 59$ (b) $10^x = 0.37$ (c) $e^x = 100$ (d) $e^x = 0.5$

Solution   (a) $10^x = 59$

Taking logs to base 10 gives

$x = \log 59 = 1.7709$

(b) $10^x = 0.37$

Taking logs to base 10 gives

$x = \log 0.37 = -0.4318$

(c) $e^x = 100$

Taking logs to base e gives

$x = \ln 100 = 4.6052$

(d) $e^x = 0.5$

Taking logs to base e we have

$x = \ln 0.5 = -0.6931$

16.19   Solve the following equations

(a) $\log x = 1.76$ (b) $\ln x = -0.5$ (c) $\log(3x) = 0.76$

(d) $\ln\left(\frac{x}{2}\right) = 2.6$ (e) $\log(2x - 4) = 1.1$ (f) $\ln(7 - 3x) = 1.75$

Solution   We note that if $\log_a X = n$, then $X = a^n$.

(a) $\log x = 1.76$ and so $x = 10^{1.76}$. Using the 'x^y' button of a scientific calculator we find

$$x = 10^{1.76} = 57.5440$$

(b) $\ln x = -0.5$

$$x = e^{-0.5} = 0.6065$$

(c) $\log 3x = 0.76$

$$3x = 10^{0.76}$$

$$x = \frac{10^{0.76}}{3} = 1.9181$$

(d) $\ln\left(\frac{x}{2}\right) = 2.6$

$$\frac{x}{2} = e^{2.6}$$

$$x = 2e^{2.6} = 26.9275$$

(e) $\log(2x - 4) = 1.1$

$$2x - 4 = 10^{1.1}$$

$$2x = 10^{1.1} + 4$$

$$x = \frac{10^{1.1} + 4}{2} = 8.2946$$

(f) $\ln(7 - 3x) = 1.75$

$$7 - 3x = e^{1.75}$$

$$3x = 7 - e^{1.75}$$

$$x = \frac{7 - e^{1.75}}{3} = 0.4151$$

16.20   Solve the following:

(a) $e^{3+x}.e^x = 1000$ (b) $\ln\left(\frac{x}{3} + 1\right) + \ln\left(\frac{1}{x}\right) = -1$
(c) $3e^{x-1} = 75$ (d) $\log(x + 2) + \log(x - 2) = 1.3$

Solution   (a) Using the laws of logarithms we have

$$e^{3+x}.e^x = e^{3+2x}$$

Hence

$$e^{3+2x} = 1000$$

$$3 + 2x = \ln 1000$$

$$2x = \ln(1000) - 3$$

$$x = \frac{\ln(1000) - 3}{2} = 1.9539$$

(b) Using the laws of logarithms we may write

$$\ln\left(\frac{x}{3}+1\right) + \ln\left(\frac{1}{x}\right) = \ln\left(\frac{x}{3}+1\right)\frac{1}{x} = \ln\left(\frac{1}{3}+\frac{1}{x}\right)$$

So $\ln\left(\frac{1}{3}+\frac{1}{x}\right) = -1$

$$\frac{1}{3}+\frac{1}{x} = e^{-1}$$

$$\frac{1}{x} = e^{-1} - \frac{1}{3}$$

$$\frac{1}{x} = 0.0345$$

$$x = 28.947$$

(c) $3e^{x-1} = 75$

$e^{x-1} = 25$

$x - 1 = \ln 25$

$x = \ln 25 + 1 = 4.2189$

(d) Using the first law we may write

$$\log(x+2) + \log(x-2) = \log(x+2)(x-2) = \log(x^2 - 4)$$

Hence

$$\log(x^2 - 4) = 1.3$$

$$x^2 - 4 = 10^{1.3}$$

$$x^2 = 10^{1.3} + 4$$

$$x = \sqrt{10^{1.3} + 4} = 4.8941$$

Exercise 16.4

1. Solve the following equations, giving your answer to 4 d.p.

   (a) $\log x = 1.6000$ (b) $10^x = 75$

   (c) $\ln x = 1.2350$ (d) $e^x = 36$

2. Solve each of the following equations giving your answer to 4 d.p.

   (a) $\log(3t) = 1.8$ (b) $10^{2t} = 150$

   (c) $\ln(4t) = 2.8$ (d) $e^{3t} = 90$

3. Solve the following equations, giving your answer to 4 d.p.:

   (a) $\log x = 0.3940$ (b) $\ln x = 0.3940$
   (c) $10^y = 5.5$ (d) $e^z = 500$

   (e) $\log(3v) = 1.6512$ (f) $\ln\left(\frac{t}{6}\right) = 1$

   (g) $10^{2r+1} = 25$ (h) $e^{(2t-1)/3} = 7.6700$
   (i) $\log(4b^2) = 2.6987$

   (j) $\log\left(\frac{6}{2+t}\right) = 1.5$

   (k) $\ln(2r^3 + 1) = 3.0572$
   (l) $\ln(\log t) = -0.3$ (m) $10^{t^2-1} = 180$

   (n) $10^{3r^4} = 170\,000$

   (o) $\log(10^t) = 1.6$ (p) $\ln(e^x) = 20\,000$

4. Solve the following equations:

   (a) $e^{3x}.e^{2x} = 59$
   (b) $10^{3t}.10^{4-t} = 27$
   (c) $\log(5 - t) + \log(5 + t) = 1.2$
   (d) $\log x + \ln x = 4$

16.5	**Properties and graph of the logarithm function**

Values of $\log x$ and $\ln x$ are given in Table 16.1 and graphs of the functions $y = \log x$ and $y = \ln x$ are illustrated in Figure 16.1. The following properties are noted from the graphs.

(a) As x increases, the values of $\log x$ and $\ln x$ increase.
(b) $\log 1 = \ln 1 = 0$.
(c) As x approaches 0 the values of $\log x$ and $\ln x$ increase negatively.
(d) When $x < 1$, the values of $\log x$ and $\ln x$ are negative.

Table 16.1

x	0.01	0.1	0.5	1	2	5	10	100
$\log x$	-2	-1	-0.30	0	0.30	0.70	1	2
$\ln x$	-4.61	-2.30	-0.69	0	0.69	1.61	2.30	4.61

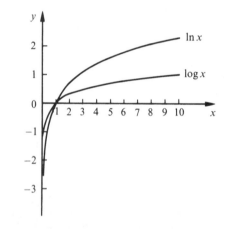

Figure 16.1
Graphs of $y = \log x$ and
$y = \ln x$

Self-assessment questions 16.5

1. State two properties which are common to both $y = \log x$ and $y = \ln x$.
2. It is possible to find a value of x such that the value of $\log x$ exceeds $100\,000\,000$. True or false?

Exercise 16.5

1. (a) Plot a graph of $y = \log x^2$ for
   $0 < x \le 10$.
   (b) Plot a graph of $y = \log(1/x)$ for
   $0 < x \le 10$.

2. (a) Plot a graph of $y = \log x$ for
   $0 < x \le 3$.
   (b) Plot on the same axes $y = 1 - x/3$
   for $0 \le x \le 3$.
   (c) Use your graphs to find an
   approximate solution to the equation

$$\log x = 1 - \frac{x}{3}$$

3. (a) Plot $y = \log x$ for $0 < x \le 7$
   (b) On the same axes, plot $y = 0.5 - 0.1x$
   (c) Use your graphs to obtain an
   approximate solution to

$$0.5 - 0.1x = \log x$$

Test and assignment exercises 16

1. Use a scientific calculator to evaluate

   (a) $\log 107\,000$ (b) $\ln 0.0371$ (c) $\log 0.1$
   (d) $\ln 150$

2. Evaluate

   (a) $\log_7 20$ (b) $\log_{16} 100$ (c) $\log_2 60$
   (d) $\log_8 150$ (e) $\log_6 4$

3. Simplify the following expressions to a
   single log term:

   (a) $\log 7 + \log t$ (b) $\log t - \log x$
   (c) $\log x + 2\log y$ (d) $3\log t + 4\log r$
   (e) $\frac{1}{2}\log 9v^4 - 3\log 1$

   (f) $\log(a+b) + \log(a-b)$
   (g) $\frac{2}{3}\log x + \frac{1}{3}\log xy^3$

4. Solve the following logarithmic
   equations, giving your answer to 4 d.p.:

   (a) $\log 7x = 2.9$ (b) $\ln 2x = 1.5$
   (c) $\ln x + \ln 2x = 3.6$

   (d) $\ln\left(\frac{x^2}{2}\right) - \ln x = 0.7$

   (e) $\log(4t - 3) = 0.9$

   (f) $\log 3t + 3\log t = 2$

   (g) $3\ln\left(\frac{2}{x}\right) - \frac{1}{2}\ln x = -1$

5. (a) Plot $y = \log x$ for $0 < x \le 3$
   (b) On the same axes plot $y = \sin x$
   where x is measured in radians.
   (c) Hence find an approximate solution
   to $\log x = \sin x$.

6. Solve the following equations, giving
   your answer correct to 4 d.p.

   (a) $10^{x+1} = 70$ (b) $e^{2x+3} = 500$

   (c) $3(10^{2x}) = 750$ (d) $4(e^{-x+2}) = 1000$

   (e) $\frac{4}{6 + e^{3x}} = 0.1500$

7. By substituting $z = e^x$ solve the
   equation

   $$e^{2x} - 9e^x + 14 = 0$$

8. By substituting $z = 10^x$ solve the
   following equation, giving your answer
   correct to 4 d.p.

   $$10^{2x} - 9(10^x) + 20 = 0$$

9. Solve the following equations:

   (a) $\log(x - 1) + \log(x + 1) = 2$

   (b) $\log(10 + 10^x) = 2$

   (c) $\frac{\log 2x}{\log x} = 2$

10. Solve

   (a) $8(10^{-x}) + 10^x = 6$

   (b) $10^{3x} - 4(10^x) = 0$

17 Angles

17.1 Measuring angles

θ is the Greek letter 'theta' commonly used to denote angles.

We use angles to measure the amount by which a line has been turned. An angle is commonly denoted by θ. In Figure 17.1 the angle between BC and BA is θ. We write $\angle ABC = \theta$. We could also write $\angle CBA = \theta$. The two units for measuring angle are the **degree** and the **radian**. We examine each in turn.

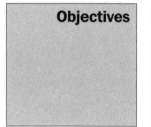

Figure 17.1
The angle ABC is θ

Degrees

A full revolution is 360 degrees, denoted 360°. To put it another way, 1 degree (1°) is $\frac{1}{360}$th of a full revolution. In Figure 17.2, AB is rotated through half a revolution, that is, 180°, to the position AC. CAB is a straight line and so 180° is sometimes called a **straight line angle**.

Figure 17.2
180° is a straight line
angle

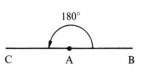

Figure 17.3 shows an angle of 90°. This is the angle in a corner. An angle of 90° is called a **right angle**. A triangle containing a right angle is a **right-angled triangle**. The side opposite the right angle is called the **hypotenuse**. An angle between 0° and 90° is an **acute angle**. An angle between 90° and 180° is an **obtuse angle**. An angle greater than 180° is a **reflex angle**.

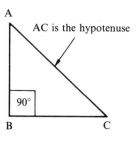

Figure 17.3
An angle of 90° is a right
angle

Radians

The other unit of angle is the radian and this too is based on examining a circle. Radian is short for 'radius angle'. Consider Figure 17.4 which shows a circle of radius *r*, centre O and an arc AB. We say that the arc **AB** subtends an angle AOB at the centre of the circle. Clearly if AB is short then ∠AOB will be small; if AB is long then ∠AOB will be a large angle. If arc AB has a length equal to 1 radius then we say that ∠AOB is 1 **radian**.

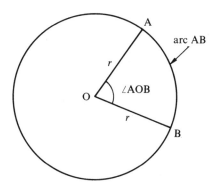

Figure 17.4
Arc AB subtends ∠AOB
at the centre of the circle

1 radian is the angle subtended at the centre of a circle by an arc length of 1 radius.

If the angle is in degrees we use the degree symbol, °; otherwise assume the angle is in radians. Hence we abbreviate 1 radian to 1. From the definition we see that an arc length $2r$, that is, 2 radii, subtends an angle of 2, an arc length of $3r$ subtends an angle of 3 and so on. In particular, an arc length of $2\pi r$ subtends an angle of 2π. We note that the circumference of the circle is $2\pi r$ and a full circumference subtends a full revolution, that is, 360°, at the centre. Hence

$$2\pi = 360°$$

and so

$$\pi = 180°$$

Your scientific calculator is pre-programmed to give a value for π. Check that you can use it correctly.

Note that π is a constant whose value is approximately 3.142.

This equation allows us to convert degrees to radians and radians to degrees.

Worked examples

17.1 Convert to degrees

(a) $\dfrac{\pi}{2}$ (b) $\dfrac{\pi}{4}$ (c) $\dfrac{\pi}{3}$ (d) 0.7π (e) 1 (f) 3 (g) 1.3

Solution (a) We have

$$\pi = 180°$$

and so

$$\frac{\pi}{2} = \frac{180°}{2} = 90°$$

Hence $\dfrac{\pi}{2}$ is a right angle.

(b) $\pi = 180°$

$$\frac{\pi}{4} = \frac{180°}{4}$$
$$= 45°$$

(c) $\pi = 180°$

$$\frac{\pi}{3} = \frac{180°}{3}$$
$$= 60°$$

(d) $\pi = 180°$

$$0.7\pi = 0.7 \times 180°$$
$$= 126°$$

(e) $\pi = 180°$

$1 = \dfrac{180°}{\pi}$

$= 57.3°$

(g) $1 = \dfrac{180°}{\pi}$

$1.3 = 1.3 \times \dfrac{180°}{\pi}$

$= 74.5°$

(f) $\pi = 180°$

$1 = \dfrac{180°}{\pi}$

$3 = 3 \times \dfrac{180°}{\pi}$

$= 171.9°$

17.2 Convert to radians

(a) 72°   (b) 120°   (c) 12°   (d) 200°

Solution (a) $180° = \pi$

$1° = \dfrac{\pi}{180}$

$72° = 72 \times \dfrac{\pi}{180}$

$= \dfrac{2\pi}{5} = 1.26$

(b) $180° = \pi$

$1° = \dfrac{\pi}{180}$

$120° = 120 \times \dfrac{\pi}{180}$

$= \dfrac{2\pi}{3} = 2.09$

(c) $180° = \pi$

$1° = \dfrac{\pi}{180}$

$12° = 12 \times \dfrac{\pi}{180}$

$= \dfrac{\pi}{15} = 0.21$

(d) $180° = \pi$

$1° = \dfrac{\pi}{180}$

$200° = 200 \times \dfrac{\pi}{180}$

$= \dfrac{10\pi}{9} = 3.49$

Note the following commonly met angles:

KEY POINT

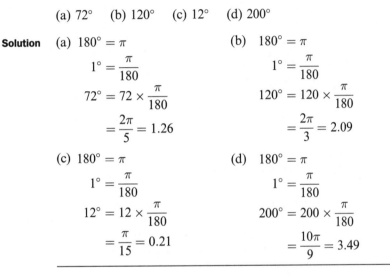

$30° = \dfrac{\pi}{6}$ radians $45° = \dfrac{\pi}{4}$ radians

$60° = \dfrac{\pi}{3}$ radians $90° = \dfrac{\pi}{2}$ radians

$135° = \dfrac{3\pi}{4}$ radians $180° = \pi$ radians

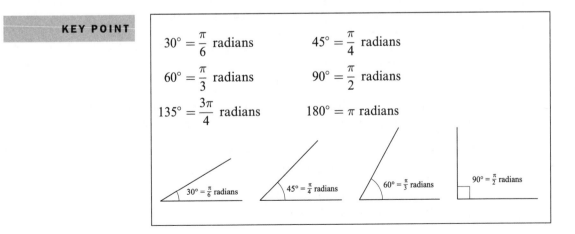

Your calculator should be able to work with angles measured in both radians and degrees. Usually the MODE button allows you to select the appropriate measure.

Self-assessment questions 17.1

1. Define the units of angle: degree and radian.
2. How are the degree and the radian related?

Exercise 17.1

1. Convert to radians

   (a) 240° (b) 300° (c) 400° (d) 37°
   (e) 1000°

2. Convert to degrees

   (a) $\dfrac{\pi}{10}$ (b) $\dfrac{2\pi}{9}$ (c) 2 (d) 3.46 (e) 1.75

3. Convert the following angles in radians to degrees:

   (a) 4π (b) 3.5π (c) 5 (d) 1.56

4. Convert the following angles in degrees to radians, expressing your answer as a multiple of π:

   (a) 504° (b) 216° (c) 420° (d) 126° (e) 324°

17.2	**Length of an arc**

In §17.1 we saw the connection between the length of an arc and the angle it subtends at the centre. Knowing the angle subtended by an arc we are able to calculate the length of the arc.

An arc of length 1 × radius subtends an angle of 1. An arc of length 2 × radius subtends an angle of 2. Clearly an arc of length θ × radius subtends an angle of θ.

KEY POINT

An angle θ is subtended by an arc of length $r\theta$. Note that θ must be in radians. Hence arc length $= r\theta$.

Worked examples

17.3 A circle has a radius of 8 cm. An angle of 1.4 is subtended at the centre by an arc. Calculate the length of the arc.

Solution An angle of 1 is subtended by an arc of length 1 radius. So an angle of 1.4 is subtended by an arc of length 1.4 × radius. Now

$$1.4 \times \text{radius} = 1.4 \times 8 = 11.2$$

The length of the arc is 11.2 cm.

17.4   A circle has radius 6 cm. An arc, AB, of the circle measures 9 cm. Calculate the angle subtended by AB at the centre of the circle.

Solution   Using arc length $= r\theta$ we have

$$9 = 6\theta$$

$$\theta = \frac{9}{6} = 1.5$$

Hence AB subtends an angle of 1.5 at the centre.

17.5   A circle has a radius of 18 cm. An arc AB subtends an angle of 142° at the centre of the circle. Calculate the length of the arc AB.

Solution   We convert 142° to radians,

$$180° = \pi$$

$$142° = \frac{142 \times \pi}{180}$$

An angle of $\dfrac{142\pi}{180}$ is subtended by an arc of length $\dfrac{142\pi}{180} \times$ radius.

$$\frac{142\pi}{180} \times \text{ radius} = \frac{142\pi}{180} \times 18 = 44.61$$

The length of the arc is 44.61 cm.

Self-assessment questions 17.2

1. Describe the connection between the angle subtended at the centre of a circle and arc length.

Exercise 17.2

1. A circle has a radius of 15 cm. Calculate the length of arc which subtends an angle of

   (a) 2 (b) 3 (c) 1.2 (d) 100° (e) 217°

2. A circle has radius 12 cm. Calculate the angle subtended at the centre by an arc of length

   (a) 12 cm (b) 6 cm (c) 24 cm (d) 15 cm (e) 2 cm

3. A circle has a radius of 18 cm. Calculate the length of arc which subtends an angle of:

   (a) 1.5 (b) 1.1 (c) $\frac{\pi}{3}$ (d) 1.76

4. A circle has a radius of 24 cm. Calculate the angle subtended at the centre by an arc of length:

   (a) 18 cm (b) 30 cm (c) 20 cm (d) 60 cm

17.3   Area of a sector

In Figure 17.5 the shaded area is a **sector**. Suppose the radius of the circle is r and the angle subtended at the centre is θ. We can calculate the area of the sector in terms of r and θ. When the angle subtended is 2π, the sector occupies the full circle and so has an area of πr^2. When the angle subtended is π, the sector is a semi-circle and so the sector has an area of $\pi r^2/2$. Thus, if the angle subtended is 1 the area of the sector is $\pi r^2/2\pi = r^2/2$. Finally, if the angle subtended is θ, the area of the sector is $r^2\theta/2$.

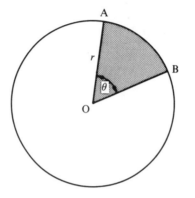

Figure 17.5
The sector AOB is shaded

KEY POINT

$$\text{Area of sector } = \frac{r^2\theta}{2}$$

Note that this formula is only valid when the angle θ is measured in radians.

Worked examples

17.6   A circle has a radius of 22 cm. A sector subtends an angle of 1.2. Calculate the area of the sector.

Solution   Here $r = 22$ and $\theta = 1.2$. Hence

$$\text{area of sector } = \frac{(22)^2(1.2)}{2} = 290.4$$

The area of the sector is 290.4 cm^2.

17.7   A circle has a radius of 10 cm. The area of a sector is 170 cm^2. Calculate the angle at the centre.

Solution   The area of sector $= \dfrac{r^2\theta}{2}$

Substituting the values given we obtain

$$170 = \frac{(10)^2\theta}{2}$$

$$\theta = \frac{2 \times 170}{10^2}$$

$$= 3.4$$

The angle is 3.4.

Self-assessment questions 17.3

1. Explain what is meant by a sector of a circle. Describe how to calculate the area of a sector.

Exercise 17.3

1. A circle has a radius of 9 cm. Calculate the area of the sectors which subtend angles of

   (a) 1.5 (b) 2 (c) 100° (d) 215°

2. A circle has a radius of 16 cm. Calculate the angle at the centre when a sector has an area of

   (a) 100 cm² (b) 5 cm² (c) 520 cm²

3. A circle has a radius of 18 cm. Calculate the area of a sector whose angle is

   (a) 1.5 (b) 2.2 (c) 120° (d) 217°

4. A circle, centre O, has a radius of 25 cm. An arc AB has length 17 cm. Calculate the area of the sector AOB.

5. A circle, centre O, radius 12 cm, has a sector AOB of area 370 cm². Calculate the length of the arc AB.

6. A circle, centre O, has a sector AOB. The arc length of AB is 16 cm and the angle subtended at O is 1.2. Calculate the area of the sector AOB.

Test and assignment exercises 17

1. Convert to radians

   (a) 73° (b) 196° (c) 1000°

2. Convert to degrees

   (a) 1.75 (b) 0.004 (c) 7.9

3. A circle, centre O, has radius 22 cm. An arc AB has length 35 cm.

   (a) Calculate the angle at O subtended by AB.

   (b) Calculate the area of the sector AOB.

4. A circle, centre O, has radius 42 cm. Arc AB subtends an angle of 235° at O.

   (a) Calculate the length of the arc AB.

   (b) Calculate the area of the sector AOB.

5. A circle, centre O, has radius 20 cm. The area of sector AOB is 320 cm².

   (a) Calculate the angle at O subtended by the arc AB.

   (b) Calculate the length of the arc AB.

6. A sector AOB has an area of 250 cm² and the angle at the centre, O, is 75°.

   (a) Calculate the area of the circle.

   (b) Calculate the radius of the circle.

7. A circle, centre O, has a radius of 18 cm. An arc AB has length 22 cm. Calculate the area of sector AOB.

8. A circle, centre O, has a sector AOB. The angle subtended at O is 108°, and the arc AB has length 11 cm. Calculate the area of the sector AOB.

9. A circle, centre O, radius 14 cm, has a sector AOB of area 290 cm². Calculate the length of the arc AB.

10. Figure 17.6 shows two concentric circles of radii 5 cm and 10 cm. Arc AB has length 15 cm. Calculate the shaded area.

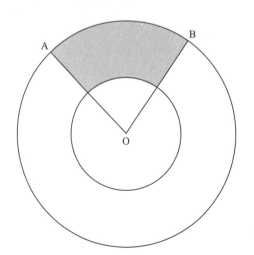

Figure 17.6
Figure for Test and assignment exercises 17, Q10

18 Introduction to trigonometry

Objectives	This chapter
	• introduces terminology associated with a right-angled triangle
	• defines the trigonometrical ratios sine, cosine, and tangent with reference to the lengths of the sides of a right-angled triangle
	• explains how to calculate an angle given one of its trigonometrical ratios

18.1 The trigonometrical ratios

Consider the right-angled triangle ABC as shown in Figure 18.1. There is a right-angle at B.

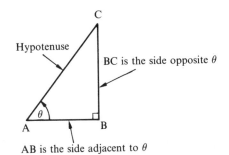

Figure 18.1
A right-angled triangle ABC

The side joining points A and B is referred to as AB. The side joining points A and C is referred to as AC. The angle at A, made by the sides AB and AC is written ∠BAC, ∠A, or simply as A. In Figure 18.1 we have labelled this angle θ.

The side opposite the right-angle is always called the **hypotenuse**. So, in Figure 18.1 the hypotenuse is AC. The side **opposite** θ is BC. The remaining side, AB is said to be **adjacent** to θ.

If we know the lengths BC and AC we can calculate $\frac{BC}{AC}$. This is known as the **sine** of θ, or simply $\sin\theta$. Similarly, we call $\frac{AB}{AC}$ the **cosine** of θ, written $\cos\theta$. Finally, $\frac{BC}{AB}$ is known as the **tangent** of θ, written $\tan\theta$. Sine, cosine and tangent are known as the **trigonometrical ratios**.

KEY POINT

$$\sin\theta = \frac{\text{side opposite to } \theta}{\text{hypotenuse}} = \frac{BC}{AC}$$

$$\cos\theta = \frac{\text{side adjacent to } \theta}{\text{hypotenuse}} = \frac{AB}{AC}$$

$$\tan\theta = \frac{\text{side opposite to } \theta}{\text{side adjacent to } \theta} = \frac{BC}{AB}$$

Note that all the trigonometrical ratios are defined as the ratio of two lengths and so they themselves have no units. Also, since the hypotenuse is always the longest side of a triangle, then $\sin\theta$ and $\cos\theta$ can never be greater than 1.

Worked examples

18.1   Calculate $\sin\theta, \cos\theta$ and $\tan\theta$ for $\triangle ABC$ as shown in Figure 18.2.

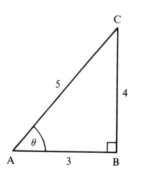

Figure 18.2
$\triangle ABC$ for Example 18.1

Solution

$$\sin\theta = \frac{BC}{AC} = \frac{4}{5} = 0.8$$

$$\cos\theta = \frac{AB}{AC} = \frac{3}{5} = 0.6$$

$$\tan\theta = \frac{BC}{AB} = \frac{4}{3} = 1.3333$$

18.2   Using the triangles shown in Figure 18.3 write down expressions for

(a) $\sin 45°$, $\cos 45°$, $\tan 45°$

(b) $\sin 60°$, $\cos 60°$, $\tan 60°$

(c) $\sin 30°$, $\cos 30°$, $\tan 30°$

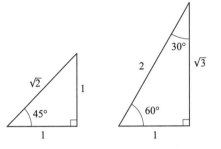

Figure 18.3

Solution   (a) $\sin 45° = \frac{1}{\sqrt{2}}$, $\cos 45° = \frac{1}{\sqrt{2}}$, $\tan 45° = \frac{1}{1} = 1$

(b) $\sin 60° = \frac{\sqrt{3}}{2}$, $\cos 60° = \frac{1}{2}$, $\tan 60° = \frac{\sqrt{3}}{1} = \sqrt{3}$

(c) $\sin 30° = \frac{1}{2}$, $\cos 30° = \frac{\sqrt{3}}{2}$, $\tan 30° = \frac{1}{\sqrt{3}}$

Notice that we have left our solutions in their exact forms using square roots if necessary rather than given their decimal approximations. These, so called, **surd forms** are frequently used.

Recall that an angle may be measured in degrees or in radians. An angle in degrees has the symbol °. Otherwise assume that the angle is measured in radians. If we know an angle, it is not necessary to draw a triangle to calculate its trigonometrical ratios. These ratios can be found directly using a scientific calculator. Your calculator must be instructed to work either in degrees or radians. Usually the MODE button is used to select the required units.

Worked examples

18.3   Use a scientific calculator to evaluate

(a) $\sin 65°$ (b) $\cos 17°$ (c) $\tan 50°$ (d) $\sin 1$ (e) $\cos 1.5$ (f) $\tan 0.5$

Solution   For (a), (b) and (c) ensure your calculator is set to DEGREES, and not to RADIANS or GRADS.

(a) $\sin 65° = 0.9063$ (b) $\cos 17° = 0.9563$ (c) $\tan 50° = 1.1918$

Now, change the MODE of your calculator in order to work in radians. Check that

(d) $\sin 1 = 0.8415$ (e) $\cos 1.5 = 0.0707$ (f) $\tan 0.5 = 0.5463$

18.4 Prove

$$\frac{\sin \theta}{\cos \theta} = \tan \theta$$

Solution From Figure 18.1, we have

$$\sin \theta = \frac{BC}{AC} \qquad \cos \theta = \frac{AB}{AC}$$

and so

$$\frac{\sin \theta}{\cos \theta} = \frac{BC/AC}{AB/AC} = \frac{BC}{AC} \times \frac{AC}{AB} = \frac{BC}{AB}$$

But

$$\tan \theta = \frac{BC}{AB}$$

and so

$$\frac{\sin \theta}{\cos \theta} = \tan \theta$$

Self-assessment questions 18.1

1. Define the trigonometrical ratios sine, cosine and tangent with reference to a right-angled triangle.
2. Explain why the trigonometrical ratios have no units.

Exercise 18.1

1. Evaluate

   (a) tan 30° (b) sin 20° (c) cos 75°
   (d) sin 1.2 (e) cos 0.89 (f) tan $\pi/4$

2. (a) Evaluate sin 70° and cos 20°.
   (b) Evaluate sin 25° and cos 65°.
   (c) Prove $\sin \theta = \cos(90° - \theta)$.
   (d) Prove that $\cos \theta = \sin(90° - \theta)$.

3. A right-angled triangle ABC has $\angle CBA$ = 90°, AB = 3 cm, BC = 3 cm and

   AC = $\sqrt{18}$ cm. $\angle CAB = \theta$. Without using a calculator find (a) sin θ, (b) cos θ, (c) tan θ.

4. A right-angled triangle XYZ has a right-angle at Y, XY = 10 cm, YZ = 4 cm and XZ = $\sqrt{116}$ cm. If $\angle ZXY = \theta$ and $\angle YZX = \alpha$, find

   (a) sin θ (b) cos θ (c) tan θ (d) sin α
   (e) cos α (f) tan α

18.2   Finding an angle given one of its trigonometrical ratios

Given an angle θ, we can use a scientific calculator to find $\sin\theta$, $\cos\theta$ and $\tan\theta$. Often we require to reverse the process, that is, given a value of $\sin\theta$, $\cos\theta$ or $\tan\theta$ we need to find the corresponding values of θ.

If

$$\sin\theta = x$$

The superscript -1 does not denote a power. It is a notation for the inverse of the trigonometrical ratios.

we write

$$\theta = \sin^{-1} x$$

and this is read as 'θ is the inverse sine of x'. This is the same as saying that θ is the angle whose sine is x. Similarly, $\cos^{-1} y$ is the angle whose cosine is y and $\tan^{-1} z$ is the angle whose tangent is z.

Most scientific calculators have inverse sine, inverse cosine and inverse tangent values programmed in. The buttons are usually denoted as $\sin^{-1}$, $\cos^{-1}$, and $\tan^{-1}$.

Worked examples

18.5   Use a scientific calculator to evaluate

(a) $\sin^{-1} 0.5$ (b) $\cos^{-1} 0.3$ (c) $\tan^{-1} 2$

Solution   (a) Using the $\sin^{-1}$ button we have

$$\sin^{-1} 0.5 = 30°$$

This is another way of saying that $\sin 30° = 0.5$.

(b) Using a scientific calculator we see

$$\cos^{-1} 0.3 = 72.5°$$

that is,

$$\cos 72.5° = 0.3$$

(c) Using a calculator we see $\tan^{-1} 2 = 63.4°$ and hence $\tan 63.4° = 2$.

18.6   Calculate the angles θ in Figure 18.4(a), (b) and (c).

(a)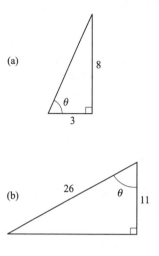

(b)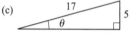

(c)

Figure 18.4

Solution   (a) In Figure 18.4(a) we are given the length of the side opposite θ and the length of the adjacent side. Hence we can make use of the tangent ratio. We write

$$\tan \theta = \frac{8}{3} = 2.6667$$

Therefore θ is the angle whose tangent is 2.6667, written

$$\theta = \tan^{-1} 2.6667$$

Using a calculator we can find the angle θ in either degrees or radians. If your calculator is set to degree mode, verify that $\theta = 69.44°$. Check also that you can obtain the equivalent answer in radians, that is 1.212.

(b) In Figure 18.4(b) we are given the length of the side adjacent to θ and the length of the hypotenuse. Hence we use the cosine ratio:

$$\cos \theta = \frac{11}{26} = 0.4231$$

Therefore

$$\theta = \cos^{-1} 0.4231$$

Using a calculator check that $\theta = 64.97°$

(c)   In Figure 18.4(c) $\sin\theta = \frac{5}{17}$. Hence

$$\theta = \sin^{-1}\frac{5}{17} = 17.10°$$

Exercise 18.2

1. Using a calculator find the following angles in degrees:

   (a) $\sin^{-1} 0.8$ (b) $\cos^{-1} 0.2$ (c) $\tan^{-1} 1.3$

2. Using a calculator find the following angles in radians:

   (a) $\sin^{-1} 0.63$ (b) $\cos^{-1} 0.25$ (c) $\tan^{-1} 2.3$

3. Find the angle θ in each of the right-angled triangles shown in Figure 18.5 giving your answer in degrees.

4. Find the angle θ in each of the right-angled triangles shown in Figure 18.6 giving your answer in radians.

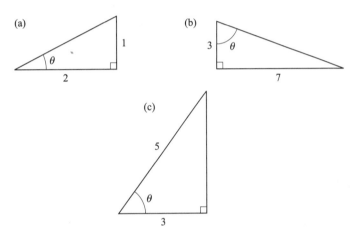

Figure 18.5

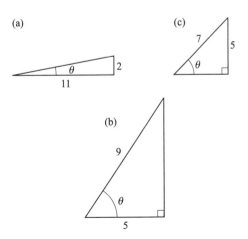

Figure 18.6

Test and assignment exercises 18

1. Use a calculator to evaluate

   (a) $\sin 23°$ (b) $\cos 52°$ (c) $\tan 77°$
   (d) $\sin 0.1$ (e) $\cos 1.1$ (f) $\tan 0.9$
   (g) $\tan \dfrac{\pi}{6}$ (h) $\tan \dfrac{\pi}{4}$ (i) $\cos \dfrac{\pi}{3}$

2. A right-angled triangle PQR has a right-angle at Q. $PR = 5$, $PQ = 3$, $QR = 4$, and $\angle PRQ = \theta$. Find $\sin \theta$, $\cos \theta$ and $\tan \theta$.

3. Using a calculator find the following angles in degrees:

   (a) $\sin^{-1} 0.2$ (b) $\cos^{-1} 0.5$ (c) $\tan^{-1} 2.4$

4. Using a calculator find the following angles in radians:

   (a) $\sin^{-1} 0.33$ (b) $\cos^{-1} 0.67$ (c) $\tan^{-1} 0.3$

5. Find the angles α and β in Figure 18.7

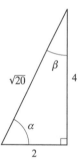

Figure 18.7

19 The trigonometrical functions and their graphs

Objectives

This chapter

- extends the definition of trigonometrical ratios to angles greater than 90°
- introduces the trigonometrical functions and their graphs

19.1 Extended definition of the trigonometrical ratios

In §18.1 we used a right-angled triangle in order to define the three trigonometrical ratios. The angle θ is thus limited to a maximum value of 90°. To give meaning to the trigonometrical ratios of angles greater than 90° we introduce an extended definition.

Quadrants

We begin by introducing the idea of quadrants.

The x and y axes divide the plane into four quadrants as shown in Figure 19.1. The origin is O. We consider an arm OC which can rotate into any of the quadrants. In Figure 19.1 the arm is in quadrant 1. We measure the anticlockwise angle from the positive x axis to the arm and call this angle θ. When the arm is in quadrant 1, then $0° \leq \theta \leq 90°$, when in the second quadrant $90° \leq \theta \leq 180°$, when in the third quadrant $180° \leq \theta \leq 270°$ and when in the fourth quadrant $270° \leq \theta \leq 360°$.

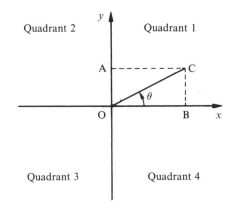

Figure 19.1
The plane is divided into four quadrants

On occasions, angles are measured in a clockwise direction from the positive x axis. In such cases these angles are conventionally taken to be negative. Figure 19.2 shows angles of $-60°$ and $-120°$. Note that for $-60°$ the arm is in the same position as for 300°. Similarly, for $-120°$ the arm is in the same position as for 240°.

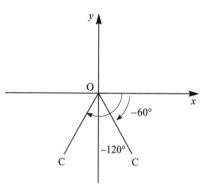

Figure 19.2
The negative angles
of $-60°$ and $-120°$

Projections

We consider the x and y projections of the arm OC. These are illustrated in Figure 19.3. We label the x projection OB and the y projection OA. One or both of these projections can be negative, depending upon the position of the arm OC. However, the length of the arm itself is always considered to be positive.

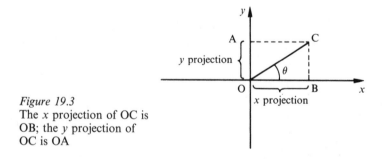

Figure 19.3
The x projection of OC is
OB; the y projection of
OC is OA

The trigonometrical ratios are now defined as

KEY POINT

$$\sin \theta = \frac{y \text{ projection of arm OC}}{\text{OC}} = \frac{\text{OA}}{\text{OC}}$$

$$\cos \theta = \frac{x \text{ projection of arm OC}}{\text{OC}} = \frac{\text{OB}}{\text{OC}}$$

$$\tan \theta = \frac{y \text{ projection of arm OC}}{x \text{ projection of arm OC}} = \frac{\text{OA}}{\text{OB}}$$

Note that the extended definition is in terms of projections and so θ is not limited to a maximum value of 90°.

Projections in the first quadrant

Consider the arm OC in the first quadrant as shown in Figure 19.3. From the right-angled triangle OCB we have

$$\sin \theta = \frac{\text{BC}}{\text{OC}} \qquad \cos \theta = \frac{\text{OB}}{\text{OC}} \qquad \tan \theta = \frac{\text{BC}}{\text{OB}}$$

Alternatively, using the extended definition we could also write

$$\sin \theta = \frac{\text{OA}}{\text{OC}} \qquad \cos \theta = \frac{\text{OB}}{\text{OC}} \qquad \tan \theta = \frac{\text{OA}}{\text{OB}}$$

Noting that OA = BC, we see that the two definitions are in agreement when $0° \leq \theta \leq 90°$.

Projections in the second quadrant

We now consider the arm in the second quadrant as shown in Figure 19.4. The x projection of OC is onto the negative part of the x axis; the y projection of OC is onto the positive part of the y axis. Hence $\sin \theta$ is positive, whereas $\cos \theta$ and $\tan \theta$ are negative.

Figure 19.4
When OC is in the second quadrant the x projection is negative

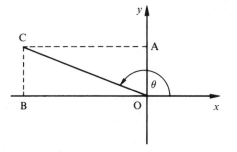

Projections in the third quadrant

When the arm is in the third quadrant as shown in Figure 19.5, the x and y projections are both negative. Hence for $180° < \theta < 270°$ $\sin \theta$ and $\cos \theta$ are negative and $\tan \theta$ is positive.

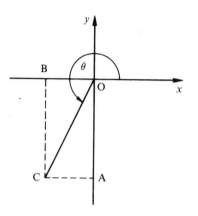

Figure 19.5
Both x and y projections are negative

Projections in the fourth quadrant

Finally the arm OC is rotated into the fourth quadrant as shown in Figure 19.6. The x projection is positive, the y projection is negative and so $\sin \theta$ and $\tan \theta$ are negative and $\cos \theta$ is positive.

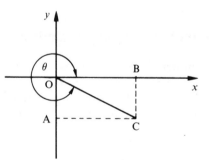

Figure 19.6
The x projection is positive and the y projection is negative

By looking at $\sin\theta, \cos\theta$ and $\tan\theta$ in the four quadrants we see that:

	Quadrant 1	Quadrant 2	Quadrant 3	Quadrant 4
$\sin\theta$	positive	positive	negative	negative
$\cos\theta$	positive	negative	negative	positive
$\tan\theta$	positive	negative	positive	negative

We have examined the trigonometrical ratios as θ varies from the first quadrant to the fourth quadrant, that is, from $0°$ to $360°$. It is possible for θ to have values outside the range $0°$ to $360°$. Adding or subtracting $360°$ to an angle is equivalent to rotating the arm through a complete revolution. This will leave its position unchanged. Hence adding or subtracting $360°$ to an angle will leave the trigonometrical ratios unaltered. We state this mathematically as

$$\sin\theta = \sin(\theta + 360°) = \sin(\theta - 360°)$$
$$\cos\theta = \cos(\theta + 360°) = \cos(\theta - 360°)$$
$$\tan\theta = \tan(\theta + 360°) = \tan(\theta - 360°)$$

Worked examples

19.1   Find $\tan\theta$ in Figure 19.7.

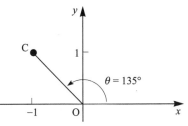

Figure 19.7

Solution   Angle θ is in the second quadrant. Hence its tangent will be negative. We can find its value from

$$\tan\theta = \frac{y\text{ projection}}{x\text{ projection}}$$
$$= \frac{1}{-1}$$
$$= -1$$

19.2   Find $\sin \theta$ in Figure 19.8.

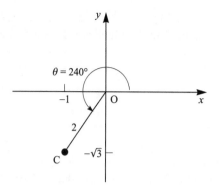

Figure 19.8

Solution   Angle θ is in the third quadrant. Hence its sine will be negative. We can find its value from

$$\sin \theta = \frac{y \text{ projection}}{\text{OC}}$$

$$= \frac{-\sqrt{3}}{2}$$

$$= -\frac{\sqrt{3}}{2}$$

19.3   Find $\cos \theta$ in Figure 19.9.

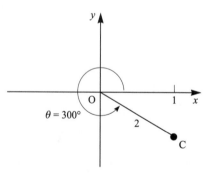

Figure 19.9

Solution   Angle θ is in the fourth quadrant. Hence its cosine will be positive. We can find its value from

$$\cos \theta = \frac{x \text{ projection}}{\text{OC}}$$

$$= \frac{1}{2}$$

Self-assessment questions 19.1

1. State the sign of $\sin\theta$, $\cos\theta$ and $\tan\theta$ in each of the four quadrants.

Exercise 19.1

1. An angle α is such that $\sin\alpha > 0$ and $\cos\alpha < 0$. In which quadrant does α lie?

2. An angle β is such that $\sin\beta < 0$. In which quadrants is it possible for β to lie?

3. The x projection of a rotating arm is negative. Which quadrants could the arm be in?

4. The y projection of a rotating arm is negative. Which quadrants could the arm be in?

5. Referring to Figure 19.10, state the sine, cosine and tangent of θ.

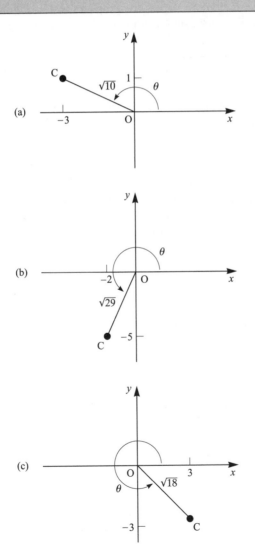

Figure 19.10

Trigonometrical functions and their graphs

Having introduced the ratios $\sin\theta$, $\cos\theta$ and $\tan\theta$, we are ready to consider the functions $y = \sin\theta$, $y = \cos\theta$ and $y = \tan\theta$. The independent variable is θ, and for every value of θ the output $\sin\theta$, $\cos\theta$ or $\tan\theta$ can be found. The graphs of these functions are illustrated in this section.

The sine function, $y = \sin\theta$

We can plot the function $y = \sin\theta$ by drawing up a table of values, as for example, in Table 19.1. Plotting these values and joining them with a smooth curve produces the graph shown in Figure 19.11. It is possible to plot $y = \sin\theta$ using a graphics calculator or a graph-plotting package.

Table 19.1

θ	0°	30°	60°	90°	120°	150°	180°
$\sin\theta$	0	0.5	0.8660	1	0.8660	0.5	0

θ	210°	240°	270°	300°	330°	360°
$\sin\theta$	−0.5	−0.8660	−1	−0.8660	−0.5	0

Recall from §19.1 that adding or subtracting 360° to an angle does not alter the trigonometrical ratio of that angle. Hence if we extend the values of θ below 0° and above 360°, the values of $\sin\theta$ are simply repeated. A graph of $y = \sin\theta$ for a larger domain of θ is shown in Figure 19.12. The pattern is repeated every 360°.

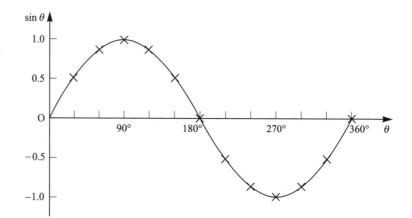

Figure 19.11
The function $y = \sin\theta$

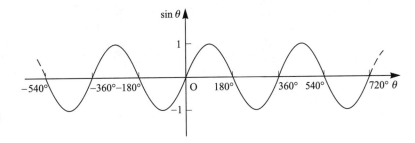

Figure 19.12
The values of sin θ repeat
every 360°

The cosine function, $y = \cos \theta$

Table 19.2 gives values of θ and $\cos \theta$. Plotting these values and joining them with a smooth curve produces the graph of $y = \cos \theta$ shown in Figure 19.13.

Table 19.2

θ	0°	30°	60°	90°	120°	150°	180°
$\cos \theta$	1	0.8660	0.5	0	−0.5	−0.8660	−1

θ	210°	240°	270°	300°	330°	360°
$\cos \theta$	−0.8660	−0.5	0	0.5	0.8660	1

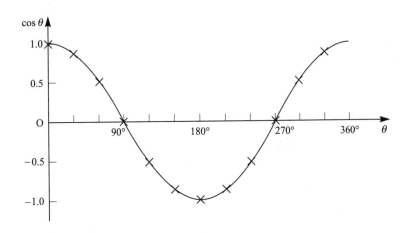

Figure 19.13
The function $y = \cos \theta$

Extending the values of θ beyond 0° and 360° produces the graph shown in Figure 19.14 in which we see that the values are repeated every 360°.

Note the similarity between $y = \sin \theta$ and $y = \cos \theta$. The two graphs are identical apart from the starting point.

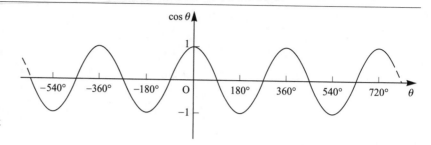

Figure 19.14
The values of $\cos\theta$ repeat
every 360°

The tangent function, $y = \tan\theta$

By constructing a table of values and plotting points a graph of $y = \tan\theta$
may be drawn. Figure 19.15 shows a graph of $y = \tan\theta$ as θ varies from
0° to 360°. Extending the values of θ produces Figure 19.16. Note that the
pattern is repeated every 180°. The values of $\tan\theta$ extend from minus
infinity to plus infinity.

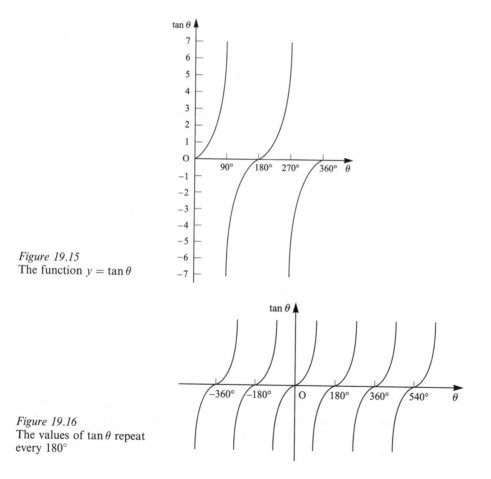

Figure 19.15
The function $y = \tan\theta$

Figure 19.16
The values of $\tan\theta$ repeat
every 180°

Worked examples

19.4   Draw $y = \sin 2\theta$ for $0° \leq \theta \leq 180°$.

Solution   Values of θ and $\sin 2\theta$ are given in Table 19.3. A graph of $y = \sin 2\theta$ is shown in Figure 19.17.

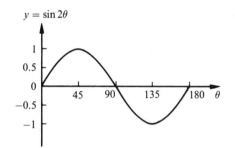

Figure 19.17
A graph of $y = \sin 2\theta$

Table 19.3
Values of θ and corresponding values of $\sin 2\theta$

θ	0	15	30	45	60	75	90	105
2θ	0	30	60	90	120	150	180	210
$\sin 2\theta$	0	0.5000	0.8660	1	0.8660	0.5000	0	−0.5000

θ	120	135	150	165	180
2θ	240	270	300	330	360
$\sin 2\theta$	−0.8660	−1	−0.8660	−0.5000	0

19.5   Draw a graph of $y = \cos(\theta + 30°)$ for $0° \leq \theta \leq 360°$.

Solution   A table of values is drawn up in Table 19.4. Figure 19.18 shows a graph of $y = \cos(\theta + 30°)$.

Table 19.4
Values of θ and corresponding values of $\cos(\theta + 30°)$

θ	0	30	60	90	120	150	180	210
$\theta + 30$	30	60	90	120	150	180	210	240
$\cos(\theta + 30)$	0.8660	0.5000	0	−0.5000	−0.8660	−1	−0.8660	−0.5000

θ	240	270	300	330	360
$\theta + 30$	270	300	330	360	390
$\cos(\theta + 30)$	0	0.5000	0.8660	1	0.8660

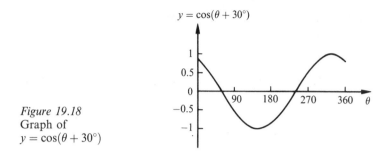

Figure 19.18
Graph of
$y = \cos(\theta + 30°)$

Self-assessment questions 19.2

1. State two properties which $y = \sin \theta$ and $y = \cos \theta$ have in common.
2. State three properties which distinguish $y = \tan \theta$ from either $y = \sin \theta$ or $y = \cos \theta$.
3. State the maximum and minimum values of the functions $\sin \theta$ and $\cos \theta$. At what values of θ do these maximum and minimum values occur?
4. With reference to the appropriate graphs explain why $\sin(-\theta) = -\sin \theta$ and $\cos(-\theta) = \cos \theta$.

Exercise 19.2

1. By observing the graphs of $\sin \theta$ and $\cos \theta$, write down the sine and cosine of the following angles:

   (a) 0° (b) 90° (c) 180° (d) 270° (e) 360°

   These values occur frequently and should be memorized.

2. Draw the following graphs:

   (a) $y = \sin(\theta + 45°)$ for $0° \le \theta \le 360°$

   (b) $y = 3\cos\left(\dfrac{\theta}{2}\right)$ for $0° \le \theta \le 720°$

   (c) $y = 2\tan(\theta + 60°)$ for $0° \le \theta \le 360°$

Test and assignment exercises 19

1. State which quadrant α lies in if $\alpha =$

   (a) 30° (b) −60° (c) −280°
   (d) 430° (e) 760°

2. State which quadrant θ lies in if $\theta =$

   (a) $\dfrac{\pi}{3}$ (b) $-\pi/6$ (c) $\dfrac{3\pi}{4}$ (d) 2.07 (e) $-\dfrac{3\pi}{4}$

3. Referring to Figure 19.19, state the sine, cosine and tangent of θ.

4. Use a graphics calculator or computer graph plotting package to draw the following graphs:

   (a) $y = \dfrac{1}{2}\sin 2\theta$ for $0° \le \theta \le 360°$

   (b) $y = 3\tan\left(\dfrac{\theta}{2}\right)$ for $0° \le \theta \le 720°$

   (c) $y = \cos(\theta - 90°)$ for $0° \le \theta \le 360°$

5. In which quadrant does the angle α lie, given

   (a) $\tan\alpha > 0$ and $\cos\alpha < 0$
   (b) $\sin\alpha < 0$ and $\tan\alpha < 0$
   (c) $\sin\alpha > 0$ and $\cos\alpha < 0$
   (d) $\tan\alpha < 0$ and $\cos\alpha < 0$

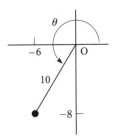

Figure 19.19

20 Trigonometrical identities and equations

Objectives	This chapter
	• explains what is meant by a trigonometrical identity and gives a table of important identities
	• shows how identities can be used to relate the trigonometrical ratios of angles in the second, third and fourth quadrants to those in the first
	• shows how complicated trigonometrical expressions can be simplified using the identities
	• explains how to solve equations involving trigonometrical functions

20.1   Trigonometrical identities

At first sight, an **identity** looks like an equation. The crucial and important difference is that the left-hand side and right-hand side of an identity are equal for *all* values of the variable involved. On the other hand, an equation contains one or more unknown quantities which must be found before the left-hand and right-hand sides are equal. We have already seen in Example 18.4 that

$$\frac{\sin \theta}{\cos \theta} = \tan \theta$$

This is an identity since $\frac{\sin \theta}{\cos \theta}$ and $\tan \theta$ have the same value whatever the value of θ. Table 20.1 lists some more trigonometrical identities. In the table the symbols A and B stand for any angle we choose.

Table 20.1
Common trigonometrical
identities

$$\frac{\sin A}{\cos A} = \tan A$$

$$\sin(A + B) = \sin A \cos B + \sin B \cos A$$

$$\sin(A - B) = \sin A \cos B - \sin B \cos A$$

$$\sin 2A = 2 \sin A \cos A$$

$$\cos(A + B) = \cos A \cos B - \sin A \sin B$$

$$\cos(A - B) = \cos A \cos B + \sin A \sin B$$

$$\cos 2A = (\cos A)^2 - (\sin A)^2 = \cos^2 A - \sin^2 A$$

$$\tan(A + B) = \frac{\tan A + \tan B}{1 - \tan A \tan B}$$

$$\tan(A - B) = \frac{\tan A - \tan B}{1 + \tan A \tan B}$$

$$\sin A + \sin B = 2 \sin\left(\frac{A + B}{2}\right) \cos\left(\frac{A - B}{2}\right)$$

$$\sin A - \sin B = 2 \sin\left(\frac{A - B}{2}\right) \cos\left(\frac{A + B}{2}\right)$$

$$\cos A + \cos B = 2 \cos\left(\frac{A + B}{2}\right) \cos\left(\frac{A - B}{2}\right)$$

$$\cos A - \cos B = -2 \sin\left(\frac{A - B}{2}\right) \sin\left(\frac{A + B}{2}\right)$$

$$\sin A = -\sin(-A)$$

$$\cos A = \cos(-A)$$

$$\tan A = -\tan(-A)$$

Note that $(\cos A)^2$ is usually written as $\cos^2 A$ and $(\sin A)^2$ is written as $\sin^2 A$.

There is another identity which deserves special mention:

KEY POINT $\boxed{\sin^2 A + \cos^2 A = 1}$

This identity shows that $\sin A$ and $\cos A$ are closely related. Knowing $\sin A$ the identity can be used to calculate $\cos A$ and vice versa.

Worked example

20.1   (a) Show that $\sin\theta = \sin(180° - \theta)$

(b) Show that $\cos\theta = -\cos(180° - \theta)$

(c) From parts (a) and (b) deduce that $\tan\theta = -\tan(180° - \theta)$

Solution   (a) Consider $\sin(180° - \theta)$. Because the right-hand side is the sine of the difference of two angles we use the identity for $\sin(A - B)$:

$$\sin(A - B) = \sin A \cos B - \sin B \cos A$$

Letting $A = 180°$ and $B = \theta$ we obtain

$$\sin(180° - \theta) = \sin 180° \cos\theta - \sin\theta \cos 180°$$
$$= 0.\cos\theta - \sin\theta(-1)$$
$$= \sin\theta$$

Hence $\sin\theta = \sin(180° - \theta)$.

(b) Consider $\cos(180° - \theta)$. Here we are dealing with the cosine of the difference of two angles. We use the identity

$$\cos(A - B) = \cos A \cos B + \sin A \sin B$$

Letting $A = 180°$ and $B = \theta$ we obtain

$$\cos(180° - \theta) = \cos 180° \cos\theta + \sin 180° \sin\theta$$
$$= -1(\cos\theta) + 0(\sin\theta)$$
$$= -\cos\theta$$

Hence $\cos\theta = -\cos(180° - \theta)$.

(c) We note from Table 20.1 that $\tan\theta = \frac{\sin\theta}{\cos\theta}$ and so using the results of parts (a) and (b),

$$\tan\theta = \frac{\sin\theta}{\cos\theta}$$
$$= \frac{\sin(180° - \theta)}{-\cos(180° - \theta)}$$
$$= -\tan(180° - \theta)$$

Because the results obtained in Example 20.1 are true for any angle θ, these too are trigonometrical identities. We will use them when solving trigonometrical equations in §20.2.

KEY POINT

$\sin\theta = \sin(180° - \theta)$

$\cos\theta = -\cos(180° - \theta)$

$\tan\theta = -\tan(180° - \theta)$

Worked example

20.2 Use a calculator to evaluate the following trigonometrical ratios and hence verify the identities obtained in Worked example 20.1.

(a) $\sin 30°$ and $\sin 150°$

(b) $\cos 70°$ and $\cos 110°$

(c) $\tan 30°$ and $\tan 150°$

Solution (a) Using a calculator verify that $\sin 30° = 0.5$. Similarly, $\sin 150° = 0.5$. We see that

$$\sin 30° = \sin 150° = \sin(180° - 30°)$$

This verifies the first of the identities for the case when $\theta = 30°$.

(b) $\cos 70° = 0.3420$ and $\cos 110° = -0.3420$. Hence

$$\cos 70° = -\cos 110° = -\cos(180° - 70°)$$

This verifies the second of the identities for the case when $\theta = 70°$.

(c) $\tan 30° = 0.5774$, and $\tan 150° = -0.5774$. So

$$\tan 30° = -\tan 150° = -\tan(180° - 30°)$$

This verifies the third of the identities for the case when $\theta = 30°$.

Using a similar analysis to Worked example 20.1 it is possible to obtain the following identities:

KEY POINT

$$\sin \theta = -\sin(\theta - 180°)$$
$$\cos \theta = -\cos(\theta - 180°)$$
$$\tan \theta = \tan(\theta - 180°)$$
$$\sin \theta = -\sin(360° - \theta)$$
$$\cos \theta = \cos(360° - \theta)$$
$$\tan \theta = -\tan(360° - \theta)$$

Worked examples

20.3   Evaluate the following trigonometrical ratios using a calculator and hence verify the results in the previous Key Point.

(a) $\sin 30°$ and $\sin 210°$

(b) $\cos 40°$ and $\cos 220°$

(c) $\tan 50°$ and $\tan 230°$

Solution   (a) $\sin 30° = 0.5$ and $\sin 210° = -0.5$. Note that $\sin 210° = -\sin 30°$.

(b) $\cos 40° = 0.7660$ and $\cos 220° = -0.7660$. Note that $\cos 220° = -\cos 40°$.

(c) $\tan 50° = 1.1918$ and $\tan 230° = 1.1918$. Note that $\tan 50° = \tan 230°$.

20.4   Evaluate the following trigonometrical ratios using a calculator and hence verify the results in the previous Key Point.

(a) $\sin 70°$ and $\sin 290°$

(b) $\cos 40°$ and $\cos 320°$

(c) $\tan 20°$ and $\tan 340°$

Solution   (a) $\sin 70° = 0.9397$, $\sin 290° = -0.9397$ and so $\sin 290° = -\sin 70°$.

(b) $\cos 40° = 0.7660$, $\cos 320° = 0.7660$ and so $\cos 40° = \cos 320°$.

(c) $\tan 20° = 0.3640$, $\tan 340° = -0.3640$ and so $\tan 340° = -\tan 20°$.

The remaining Worked examples in this section illustrate the use of a variety of identities in simplifying trigonometrical expressions. It is not always obvious which identity to use, especially for the inexperienced. The best advice is to try to experience a range of examples and work through a large number of exercises.

Worked examples

20.5   Simplify

$$1 - \sin A \cos A \tan A$$

Solution

$$1 - \sin A \cos A \tan A = 1 - \sin A \cos A \left(\frac{\sin A}{\cos A} \right)$$

$$= 1 - \sin^2 A$$

$$= \cos^2 A \qquad \text{since } \cos^2 A + \sin^2 A = 1$$

20.6   Show

$$\cos^4 A - \sin^4 A = \cos 2A$$

Solution   We note that

$$\cos^4 A - \sin^4 A = (\cos^2 A + \sin^2 A)(\cos^2 A - \sin^2 A)$$
$$= 1(\cos^2 A - \sin^2 A) \quad \text{using } \cos^2 A + \sin^2 A = 1$$
$$= \cos 2A \quad \text{using Table 20.1}$$

20.7   Show

$$\frac{\sin 6\theta - \sin 4\theta}{\sin \theta} = 2 \cos 5\theta$$

Solution   We note the identity

$$\sin A - \sin B = 2 \sin \left(\frac{A - B}{2}\right) \cos \left(\frac{A + B}{2}\right)$$

Letting $A = 6\theta$, $B = 4\theta$ we obtain

$$\sin 6\theta - \sin 4\theta = 2 \sin \theta \cos 5\theta$$

and so

$$\frac{\sin 6\theta - \sin 4\theta}{\sin \theta} = 2 \cos 5\theta$$

Exercise 20.1

1. In each case state value(s) of θ in the range $0°$ to $360°$ so that the following are true.

   (a) $\sin \theta = \sin 50°$ (b) $\cos \theta = -\cos 40°$
   (c) $\tan \theta = \tan 20°$ (d) $\sin \theta = -\sin 70°$
   (e) $\cos \theta = \cos 10°$ (f) $\tan \theta = -\tan 80°$
   (g) $\sin \theta = \sin 0°$ (h) $\cos \theta = \cos 90°$

2. State values of θ in the range $0°$ to $360°$ so that

   (a) $\cos \theta = \cos 20°$
   (b) $\sin \theta = -\sin 10°$
   (c) $\tan \theta = -\tan 40°$

3. Show:

   $$\cos 2A = 2 \cos^2 A - 1$$

4. Show:

   $$\tan^2 A + 1 = \frac{1}{\cos^2 A}$$

5. Simplify

   $$\frac{\cos 2\theta + \cos 8\theta}{2 \cos 3\theta}$$

6. Use the trigonometrical identities to expand and simplify if possible

   (a) $\cos(270° - \theta)$ (b) $\cos(270° + \theta)$
   (c) $\sin(270° + \theta)$ (d) $\tan(135° + \theta)$
   (e) $\sin(270° - \theta)$

continued

7. Noting that $\tan 45° = 1$, simplify

$$\frac{1 - \tan A}{1 + \tan A}$$

8. Show that

$$\frac{1 - \cos 2\theta + \sin 2\theta}{1 + \cos 2\theta + \sin 2\theta}$$

can be simplified to $\tan \theta$.

9. Simplify

$$\frac{\sin 4\theta + \sin 2\theta}{\cos 4\theta + \cos 2\theta}$$

10. Show that

$$\frac{\tan A}{\tan^2 A + 1} = \frac{1}{2} \sin 2A$$

11. Show

(a) $\dfrac{1}{\cos A} - \cos A = \sin A \tan A$

(b) $\dfrac{1}{\sin A} - \sin A = \dfrac{\cos A}{\tan A}$

12. By using the trigonometrical identity for $\sin(A + B)$ with $A = 2\theta$ and $B = \theta$ show

$$\sin 3\theta = 3 \sin \theta - 4 \sin^3 \theta$$

20.2   Solutions of trigonometrical equations

Trigonometrical equations are equations involving the trigonometrical ratios. This section shows how to solve trigonometrical equations.

We will need to make use of the inverse trigonometrical functions $\sin^{-1} x$, $\cos^{-1} x$ and $\tan^{-1} x$. The inverses of the trigonometrical ratios were first introduced in §18.2. Recall that since $\sin 30° = 0.5$, we write $30° = \sin^{-1} 0.5$, and say $30°$ is the angle whose sine is 0.5. Referring back to Worked example 20.2 you will see that $\sin 150°$ is also equal to 0.5, and so it is also true that $150° = \sin^{-1} 0.5$.

Hence the inverse sine of a number can yield more than one answer. This is also true for inverse cosine and inverse tangent. Scientific calculators will simply give one value. The other values must be deduced from knowledge of the functions $y = \sin \theta$, $y = \cos \theta$ and $y = \tan \theta$.

Worked examples

20.8   Find all values of $\sin^{-1} 0.4$ in the range $0°$ to $360°$.

Solution   Let $\theta = \sin^{-1} 0.4$, that is, $\sin \theta = 0.4$. Since $\sin \theta$ is positive then one value is in the first quadrant and another value is in the second quadrant. Using a scientific calculator we see

$$\theta = \sin^{-1} 0.4 = 23.6°$$

The formula
$$\sin\theta = \sin(180° - \theta)$$
was derived on p. 212.

This is the value in the first quadrant. The value in the second quadrant is found using $\sin\theta = \sin(180° - \theta)$. Hence the value in the second quadrant is $180° - 23.6° = 156.4°$.

20.9 Find all values of θ in the range $0°$ to $360°$ such that
(a) $\theta = \cos^{-1}(-0.5)$, (b) $\theta = \tan^{-1} 1$.

Solution (a) We have $\theta = \cos^{-1}(-0.5)$, that is, $\cos\theta = -0.5$. As $\cos\theta$ is negative then θ must be in the second and third quadrants. Using a calculator we find $\cos^{-1}(-0.5) = 120°$. This is the value of θ in the second quadrant. We now seek the value of θ in the third quadrant.

   We have, from the Key Point on page 213,

$$\cos(\theta - 180°) = -\cos\theta$$

We are given $\cos\theta = -0.5$. So

$$\cos(\theta - 180°) = 0.5$$

Now, since θ is in the third quadrant, $\theta - 180°$ must be an acute angle whose cosine equals 0.5. That is

$$\theta - 180° = 60°$$
$$\theta = 240°$$

This is the value of θ in the third quadrant. The required solutions are thus $\theta = 120°, 240°$.

(b) We are given $\theta = \tan^{-1} 1$, that is, $\tan\theta = 1$. Since $\tan\theta$ is positive, there is a value of θ in the first and third quadrants. Using a calculator we find $\tan^{-1} 1 = 45°$. This is the value of θ in the first quadrant.

   We have, from the Key Point on page 213,

$$\tan(\theta - 180°) = \tan\theta$$

We are given $\tan\theta = 1$.

So $\tan(\theta - 180°) = 1$.

Now $\theta - 180°$ must be an acute angle with tangent equal to 1.

That is $\theta - 180° = 45°$

$$\theta = 225°$$

The required values of θ are $45°$ and $225°$.

20.10   Find all values of θ in the range $0°$ to $360°$ such that $\theta = \sin^{-1}(-0.5)$.

Solution   We know $\sin\theta = -0.5$ and so θ must be in the third and fourth quadrants. Using a calculator we find

$$\sin^{-1}(-0.5) = -30°$$

We require solutions in the range $0°$ to $360°$. Recall that adding $360°$ to an angle leaves the values of the trigonometrical ratios unaltered. So

$$\sin^{-1}(-0.5) = -30° + 360° = 330°$$

We have found the value of θ in the fourth quadrant.

We now seek the value in the third quadrant. We have, from the Key Point on page 213,

$$\sin(\theta - 180°) = -\sin\theta$$

We are given $\sin\theta = -0.5$.

Therefore $\sin(\theta - 180°) = 0.5$.

Now $\theta - 180°$ must be an acute angle with sine equal to 0.5.

That is, $\theta - 180° = 30°$

$$\theta = 210°$$

The required values of θ are $210°$ and $330°$.

20.11   Find all values of θ in the range $0°$ to $360°$ such that $\sin 2\theta = 0.5$.

Solution   We make the substitution $z = 2\theta$. As θ varies from $0°$ to $360°$ then z varies from $0°$ to $720°$. Hence the problem as given is equivalent to finding all values of z in the range $0°$ to $720°$ such that $\sin z = 0.5$.

$$\sin z = 0.5$$
$$z = \sin^{-1} 0.5$$
$$= 30°$$

Also

$$\sin(180° - 30°) = \sin 30°$$
$$\sin 150° = \sin 30° = 0.5$$
$$\sin^{-1} 0.5 = 150°$$

The values of z in the range $0°$ to $360°$ are $30°$ and $150°$. Recalling that adding $360°$ to an angle leaves its trigonometrical ratios unaltered we see that $30° + 360° = 390°$ and $150° + 360° = 510°$ are values of z in the range $360°$ to $720°$. The solutions for z are thus

$$z = 30°, 150°, 390°, 510°$$

and hence the required values of θ are given by

$$\theta = \frac{z}{2} = 15°, 75°, 195°, 255°$$

20.12   Solve $\cos 3\theta = -0.5$ for $0° \le \theta \le 360°$.

Solution   We substitute $z = 3\theta$. As θ varies from $0°$ to $360°$ then z varies from $0°$ to $1080°$. Hence the problem is to find values of z in the range $0°$ to $1080°$ such that $\cos z = -0.5$.

We begin by finding values of z in the range $0°$ to $360°$. Using Worked example 20.9(a) we see that

$$z = 120°, 240°$$

To find values of z in the range $360°$ to $720°$ we add $360°$ to each of these solutions. This gives

$$z = 120° + 360° = 480° \quad \text{and} \quad z = 240° + 360° = 600°$$

To find values of z in the range $720°$ to $1080°$ we add $360°$ to these solutions. This gives

$$z = 840°, 960°$$

Hence

$$z = 120°, 240°, 480°, 600°, 840°, 960°$$

Finally, using $\theta = z/3$ we obtain values of θ:

$$\theta = \frac{z}{3} = 40°, 80°, 160°, 200°, 280°, 320°$$

20.13   Solve

$$\tan(2\theta + 20°) = 0.3 \qquad 0° \le \theta \le 360°$$

Solution   We substitute $z = 2\theta + 20°$. As θ varies from $0°$ to $360°$ then z varies from $20°$ to $740°$.

Firstly we solve

$$\tan z = 0.3 \qquad 0° \le z \le 360°$$

This leads to $z = 16.7°, 196.7°$. Values of z in the range $360°$ to $720°$ are

$$z = 16.7° + 360°, 196.7° + 360°$$
$$= 376.7°, 556.7°$$

By adding a further $360°$ values of z in the range $720°$ to $1080°$ are found. These are

$$z = 736.7°, 916.7°$$

Hence values of z in the range $0°$ to $1080°$ are

$$z = 16.7°, 196.7°, 376.7°, 556.7°, 736.7°, 916.7°$$

Values of z in the range $20°$ to $740°$ are thus

$$z = 196.7°, 376.7°, 556.7°, 736.7°$$

The values of θ in the range $0°$ to $360°$ are found using $\theta = (z - 20°)/2$.

$$\theta = \frac{z - 20°}{2} = 88.35°, 178.35°, 268.35°, 358.35°$$

Self-assessment questions 20.2

1. Can you explain why the inverses of the trigonometrical functions have several values for a single value of θ?

Exercise 20.2

1. Find all values in the range $0°$ to $360°$ of

   (a) $\sin^{-1} 0.9$ (b) $\cos^{-1} 0.45$ (c) $\tan^{-1} 1.3$
   (d) $\sin^{-1}(-0.6)$ (e) $\cos^{-1}(-0.75)$
   (f) $\tan^{-1}(-0.3)$ (g) $\sin^{-1} 1$ (h) $\cos^{-1}(-1)$

   (d) $\sin\left(\dfrac{\theta}{2}\right) = -0.5$

   (e) $\cos(2\theta - 30°) = -0.5$
   (f) $\tan(3\theta - 20°) = 0.25$
   (g) $\sin 2\theta = 2\cos 2\theta$

2. Solve the following trigonometrical equations for $0° \leq \theta \leq 360°$:

   (a) $\sin(2\theta + 50°) = 0.5$
   (b) $\cos(\theta + 110°) = 0.3$

   (c) $\tan\left(\dfrac{\theta}{2}\right) = 1$

3. Find values of θ in the range $0°$ to $360°$ such that

   (a) $\cos(\theta - 100°) = 0.3126$
   (b) $\tan(2\theta + 20°) = -1$
   (c) $\sin(\frac{2\theta}{3} + 30°) = -0.4325$

Test and assignment exercises 20

1. Solve the following trigonometrical equations for $0° \leq \theta \leq 360°$:

   (a) $13 \sin 2\theta = 5$ (b) $\cos\left(\dfrac{\theta}{3}\right) = -0.7$

   (c) $\tan(\theta - 110°) = 1.5$
   (d) $\cos 3\theta = 2 \sin 3\theta$

   (e) $\sin\left(\dfrac{\theta}{2} - 30°\right) = -0.65$

2. State all values in the range $0°$ to $360°$ of

   (a) $\sin^{-1} 0.85$ (b) $\cos^{-1}(-0.25)$
   (c) $\tan^{-1}(-1.25)$ (d) $\sin^{-1}(-0.4)$

   (e) $\cos^{-1}\left(\dfrac{1}{3}\right)$ (f) $\tan^{-1} 1.7$

3. Show
   $$\cos 2A = 1 - 2 \sin^2 A$$

4. Simplify
   $$\sin^4 A + 2 \sin^2 A \cos^2 A + \cos^4 A$$

5. Simplify $\dfrac{\sin 5\theta + \sin \theta}{\cos 5\theta + \cos \theta}$

6. (a) Show
   $$\cos 3A = \cos 2A \cos A - \sin 2A \sin A$$

   (b) Hence show that
   $$\cos 3A = 4 \cos^3 A - 3 \cos A$$

7. Solve the following equations for $0° \leq \theta \leq 360°$.

   (a) $\sin 2\theta = -0.6$
   (b) $\cos(\theta + 40°) = -0.25$
   (c) $\tan\left(\dfrac{2\theta}{3}\right) = 1.3$

8. State all values in the range $0°$ to $360°$ for the following:

   (a) $\sin^{-1}(-0.6500)$
   (b) $\cos^{-1}(-0.2500)$
   (c) $\tan^{-1}(1.2500)$

9. State values of θ in the range $0°$ to $360°$ for which:

   (a) $\tan \theta = \tan 25°$
   (b) $\cos \theta = -\cos 30°$
   (c) $\sin \theta + \sin 80° = 0$
   (d) $\tan \theta + \tan 15° = 0$

21 Solution of triangles

Objectives

This chapter

- explains the terms 'scalene', 'isosceles', 'equilateral' and 'right-angled' as applied to triangles
- states Pythagoras' theorem and shows how it can be used in the solution of right-angled triangles
- states the sine rule and the cosine rule and shows how they are used to solve triangles

A triangle is solved when all its angles and the lengths of all its sides are known. Before looking at the various rules used to solve triangles we define the various kinds of triangle.

21.1   Types of triangle

A **scalene triangle** is one in which all the sides are of different length. In a scalene triangle, all the angles are different too. An **isosceles triangle** is one in which two of the sides are of equal length. Figure 21.1 shows an isosceles triangle with $AB = AC$. In an isosceles triangle, there are also two equal angles. In Figure 21.1, $\angle ABC = \angle ACB$. An **equilateral triangle** has three equal sides and three equal angles each 60°. A **right-angled triangle** is one containing a right-angle, that is, an angle of 90°. The side opposite the right-angle is called the **hypotenuse**. Figure 21.2 shows a right-angled triangle with a right-angle at C. The hypotenuse is AB.

In **any** triangle we note

KEY POINT

The sum of the three angles is always 180°

Recall from §17.1 that an acute angle is less than 90° and an obtuse angle is greater than 90° and less than 180°.

Other properties include

(a) The longest side is opposite the largest angle.
(b) The shortest side is opposite the smallest angle.
(c) A triangle contains either three acute angles or two acute and one obtuse angle.

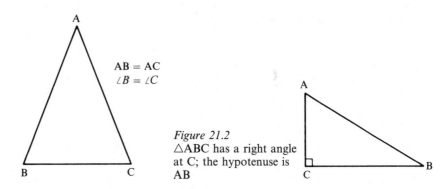

Figure 21.1
An isosceles triangle has two equal sides and two equal angles

$AB = AC$
$\angle B = \angle C$

Figure 21.2
$\triangle ABC$ has a right angle at C; the hypotenuse is AB

In $\triangle ABC$, as a shorthand, we often refer to $\angle ABC$ as $\angle B$, or more simply as B, $\angle ACB$ as C and so on. The sides also have a shorthand notation. In any triangle ABC, AB is always opposite C, AC is always opposite B and BC is always opposite A. Hence we refer to AB as c, AC as b and BC as a. Figure 21.3 illustrates this. We say that A is **included** by AC and AB, B is included by AB and BC and C is included by AC and BC.

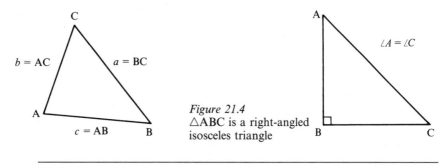

Figure 21.3
$\triangle ABC$ with $a = BC$, $b = AC$, $c = AB$

$b = AC$ $a = BC$ $c = AB$

Figure 21.4
$\triangle ABC$ is a right-angled isosceles triangle

$\angle A = \angle C$

Worked example

21.1 State all the angles of a right-angled isosceles triangle.

Solution Let $\triangle ABC$ be a right-angled isosceles triangle with a right-angle at B. Figure 21.4 illustrates the situation. The angles sum to 180° and $A = C$ because the triangle is isosceles.

$$A + B + C = 180$$
$$A + 90 + C = 180$$
$$A + C = 90$$
$$2A = 90$$
$$A = 45$$

Then $C = 45$ also. The angles are $A = 45°, B = 90°, C = 45°$.

Self-assessment questions 21.1

1. Explain the terms scalene, isosceles and equilateral when applied to triangles.
2. What is the sum of the angles of a triangle?
3. Explain the meaning of the term hypotenuse.
4. Can a triangle containing an obtuse angle possess a hypotenuse?

Exercise 21.1

1. In △ABC, $A = 70°$ and $B = 42°$. Calculate C.

2. In △ABC, $A = 42°$ and B is twice C. Calculate B and C.

3. An isosceles triangle ABC has $C = 114°$. Calculate A and B.

4. In △ABC, $A = 50°$ and $C = 58°$. State (a) the longest side, (b) the shortest side.

5. The smallest angle of an isosceles triangle is 40°. Calculate the angles of the triangle.

21.2   Pythagoras' theorem

One of the oldest theorems in the study of triangles is due to the ancient Greek mathematician Pythagoras. It applies to right-angled triangles. Consider any right-angled triangle ABC as shown in Figure 21.5.

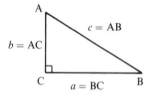

Figure 21.5
A right-angled triangle

KEY POINT

Pythagoras' theorem states $c^2 = a^2 + b^2$

A **theorem** is an important mathematical result which although not obvious, is possible to prove.

In words we have: the square of the hypotenuse (c^2) equals the sum of the squares of the other two sides ($a^2 + b^2$). Note that as Pythagoras' theorem refers to a hypotenuse, it can be applied only to right-angled triangles.

Worked examples

21.2   In Figure 21.5, AC = 3 cm and BC = 4 cm. Calculate the length of the hypotenuse AB.

Solution   We have $a = BC = 4$ and $b = AC = 3$:

$$a^2 = 4^2 = 16 \quad b^2 = 3^2 = 9$$

We apply Pythagoras' theorem:

$$c^2 = a^2 + b^2 = 16 + 9 = 25$$
$$c = 5$$

The hypotenuse, AB, is 5 cm.

21.3   In Figure 21.5, $AC = 14$ cm and the hypotenuse is 22 cm. Calculate the length of BC.

Solution   We have $b = AC = 14$ and $c = AB = 22$. We need to find the length of BC, that is, a. We apply Pythagoras' theorem:

$$c^2 = a^2 + b^2$$
$$22^2 = a^2 + 14^2$$
$$484 = a^2 + 196$$
$$a^2 = 288$$
$$a = \sqrt{288} = 16.97$$

The length of BC is 16.97 cm.

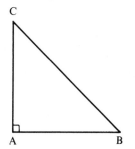

Figure 21.6
Right-angled triangle for
Example 21.4

21.4   ABC is a right-angled isosceles triangle with $A = 90°$. If $BC = 12$ cm, calculate the lengths of the other sides.

Solution   Figure 21.6 illustrates $\triangle ABC$. The triangle is isosceles and so $AC = AB$, that is, $b = c$. We are told that the hypotenuse, BC, is 12 cm, that is, $a = 12$. Applying Pythagoras' theorem gives

$$a^2 = b^2 + c^2$$

Substituting $a = 12$ and $b = c$ we get

$$12^2 = b^2 + b^2$$
$$144 = 2b^2$$
$$b^2 = 72$$
$$b = 8.49$$

The sides AC and AB are both 8.49 cm.

Self-assessment questions 21.2

1. State Pythagoras' theorem.
2. Explain what is meant by the hypotenuse of a triangle.

Exercise 21.2

1. △ABC has $C = 90°$. If AB = 30 cm and AC = 17 cm calculate the length of BC.

2. △CDE has $D = 90°$. Given CD = 1.2 m and DE = 1.7 m calculate the length of CE.

3. △ABC has a right-angle at B. Given AC:AB is 2:1 calculate AC:BC.

4. △XYZ has $X = 90°$. If XY = 10 cm and YZ = 2XZ calculate the lengths of XZ and YZ.

5. △LMN has LN as hypotenuse. Given LN = 30 cm and LM = 26 cm calculate the length of MN.

21.3   Solution of right-angled triangles

Recall that a triangle is solved when all angles and all lengths have been found. We can use Pythagoras' theorem and the trigonometrical ratios to solve right-angled triangles.

Worked examples

21.5   In △ABC, $B = 90°$, AB = 7 cm and BC = 4 cm. Solve △ABC.

Solution   Figure 21.7 shows the information given. We use Pythagoras' theorem to find AC:

$$b^2 = a^2 + c^2$$
$$= 16 + 49 = 65$$
$$b = \sqrt{65} = 8.06$$

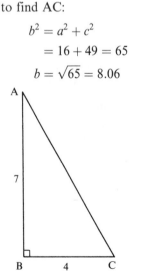

Figure 21.7
△ABC for Example 21.5

We use the trigonometrical ratios to find C:

$$\tan C = \frac{AB}{BC} = \frac{7}{4} = 1.75$$
$$C = \tan^{-1}(1.75) = 60.26°$$

The sum of the angles is $180°$:

$$A + B + C = 180$$
$$A + 90 + 60.26 = 180$$
$$A = 29.74$$

The solution to $\triangle ABC$ is

$$
\begin{array}{ll}
A = 29.74° & BC = 4 \text{ cm} \\
B = 90° & AC = 8.06 \text{ cm} \\
C = 60.26° & AB = 7 \text{ cm}
\end{array}
$$

Note that the longest side is opposite the largest angle; the shortest side is opposite the smallest angle.

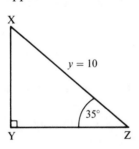

Figure 21.8
$\triangle XYZ$ for Example 21.6

21.6   In $\triangle XYZ$, $Y = 90°$, $Z = 35°$ and $XZ = 10$ cm. Solve $\triangle XYZ$.

Solution   Figure 21.8 illustrates the information given. We find X using the fact that the sum of the angles is $180°$:

$$X + Y + Z = 180$$
$$X + 90 + 35 = 180$$
$$X = 55$$

We find the length XY using the sine ratio:

$$\sin Z = \frac{XY}{XZ}$$

$$\sin 35° = \frac{XY}{10}$$

$$XY = 10 \sin 35° = 5.74$$

We find the length YZ using the cosine ratio:

$$\cos Z = \frac{YZ}{XZ}$$

$$\cos 35° = \frac{YZ}{10}$$

$$YZ = 10 \cos 35° = 8.19$$

The triangle is now completely solved.

$X = 55°$ $YZ = 8.19$ cm
$Y = 90°$ $XZ = 10$ cm
$Z = 35°$ $XY = 5.74$ cm

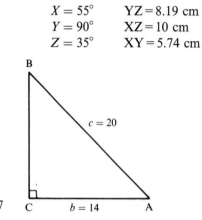

Figure 21.9
$\triangle ABC$ for Example 21.7

21.7 In $\triangle ABC$, $C = 90°$, $AB = 20$ cm and $AC = 14$ cm. Solve $\triangle ABC$.

Solution Figure 21.9 illustrates the given information. We use Pythagoras' theorem to calculate BC:

$$c^2 = a^2 + b^2$$

$$400 = a^2 + 196$$

$$a^2 = 204$$

$$a = \sqrt{204} = 14.28$$

We use the cosine ratio to calculate A:

$$\cos A = \frac{AC}{AB} = \frac{14}{20} = 0.7$$

$$A = \cos^{-1}(0.7) = 45.57°$$

Since the angles sum to $180°$ then we have

$$B = 180 - C - A$$

$$= 180 - 90 - 45.57$$

$$= 44.43$$

The solution is

$$A = 45.57°$$
$$B = 44.43°$$
$$C = 90°$$

BC = 14.28 cm
AC = 14 cm
AB = 20 cm

Self-assessment questions 21.3

1. Explain what is meant by 'solving a triangle'.

Exercise 21.3

1. $\triangle$ABC has $C = 90°$, $A = 37°$ and AC = 36 cm. Solve $\triangle$ABC.

2. $\triangle$CDE has $E = 90°$, CE = 14 cm and DE = 21 cm. Solve $\triangle$CDE.

3. $\triangle$XYZ has $X = 90°$, $Y = 26°$ and YZ = 45 mm. Solve $\triangle$XYZ.

4. $\triangle$IJK has $J = 90°$, IJ = 15 cm and JK = 27 cm. Solve $\triangle$IJK.

5. In $\triangle$PQR, $P = 62°$, $R = 28°$ and PR = 22 cm. Solve $\triangle$PQR.

6. In $\triangle$RST, $R = 70°$, $S = 20°$ and RT = 12 cm. Solve $\triangle$RST.

21.4   The sine rule

In §21.3 we saw how to solve right-angled triangles. Many triangles do not contain a right-angle and in such cases we need to use other techniques. One such technique is the **sine rule**.

Consider **any** triangle ABC as shown in Figure 21.3. Recall the notation $a =$ BC, $b =$ AC, $c =$ AB.

KEY POINT

The sine rule states

$$\frac{a}{\sin A} = \frac{b}{\sin B} = \frac{c}{\sin C}$$

It must be stressed that the sine rule can be applied to any triangle. The sine rule is used when we are given either (a) two angles and one side or (b) two sides and a non-included angle.

Worked examples

21.8   In $\triangle ABC$, $A = 30°$, $B = 84°$ and $AC = 19$ cm. Solve $\triangle ABC$.

Solution   The information is illustrated in Figure 21.10. We are given two angles and a side and so the sine rule can be used. We can immediately find C since the angles sum to 180°.

$$C = 180° - A - B = 180° - 30° - 84° = 66°$$

We know $A = 30°$, $B = 84°$, $C = 66°$ and $b = AC = 19$. The sine rule states

$$\frac{a}{\sin A} = \frac{b}{\sin B} = \frac{c}{\sin C}$$

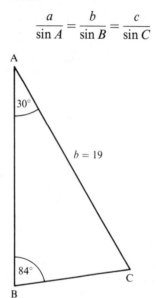

Figure 21.10
$\triangle ABC$ for Example 21.8

Substituting in values gives

$$\frac{a}{\sin 30°} = \frac{19}{\sin 84°} = \frac{c}{\sin 66°}$$

Hence

$$a = \frac{19\sin 30°}{\sin 84°} = 9.55 \qquad c = \frac{19\sin 66°}{\sin 84°} = 17.45$$

The solution is

$$
\begin{aligned}
A &= 30° & a &= BC = 9.55 \text{ cm}\\
B &= 84° & b &= AC = 19 \text{ cm}\\
C &= 66° & c &= AB = 17.45 \text{ cm}
\end{aligned}
$$

21.9   In $\triangle ABC$, $B = 42°$, $AB = 12$ cm and $AC = 17$ cm. Solve $\triangle ABC$.

Solution   Figure 21.11 illustrates the situation. We know two sides and a non-included angle and so it is appropriate to use the sine rule. We are given $c = AB = 12$, $b = AC = 17$ and $B = 42°$. The sine rule states

$$\frac{a}{\sin A} = \frac{b}{\sin B} = \frac{c}{\sin C}$$

Substituting in the known values gives

$$\frac{a}{\sin A} = \frac{17}{\sin 42°} = \frac{12}{\sin C}$$

Using the equation

$$\frac{17}{\sin 42°} = \frac{12}{\sin C}$$

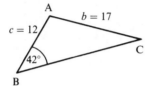

Figure 21.11
$\triangle ABC$ for Example 21.9

we see that

$$\sin C = \frac{12 \sin 42°}{17} = 0.4723$$

$$C = \sin^{-1}(0.4723) = 28.19° \text{ or } 151.81°$$

As $B = 42°$, then the value 151.81° for C must be rejected as the sum of the angles must be 180°. So $C = 28.19°$. Then

$$A = 180° - B - C = 180° - 42° - 28.19° = 109.81°$$

Using

$$\frac{a}{\sin A} = \frac{17}{\sin 42°}$$

we see that

$$a = \frac{17 \sin A}{\sin 42°}$$

$$= \frac{17 \sin 109.81°}{\sin 42°} = 23.90$$

The solution is

$$
\begin{aligned}
A &= 109.81° & a &= BC = 23.90 \text{ cm} \\
B &= 42° & b &= AC = 17 \text{ cm} \\
C &= 28.19° & c &= AB = 12 \text{ cm}
\end{aligned}
$$

21.10   In $\triangle ABC$, $AB = 23$ cm, $BC = 30$ cm and $C = 40°$. Solve $\triangle ABC$.

Solution   We are told the length of two sides and a non-included angle and so the sine rule can be applied. We have $c = AB = 23$, $a = BC = 30$ and $C = 40°$. The sine rule states

$$\frac{a}{\sin A} = \frac{b}{\sin B} = \frac{c}{\sin C}$$

Substituting in the known values gives

$$\frac{30}{\sin A} = \frac{b}{\sin B} = \frac{23}{\sin 40°}$$

Using the equation

$$\frac{30}{\sin A} = \frac{23}{\sin 40°}$$

we have

$$\sin A = \frac{30 \sin 40°}{23} = 0.8384$$

$$A = \sin^{-1}(0.8384)$$

$$= 56.97° \text{ or } 123.03°$$

There is no reason to reject either of these values and so there are two possible solutions for A. We consider each in turn.

$A = 56.97°$ Here

$$B = 180° - 40° - 56.97° = 83.03°$$

We can now use

$$\frac{b}{\sin B} = \frac{23}{\sin 40°}$$

$$b = \frac{23 \sin B}{\sin 40°} = \frac{23 \sin 83.03°}{\sin 40°}$$

$$= 35.52$$

$A = 123.03°$ In this case

$$B = 180° - 40° - 123.03° = 16.97°$$

We have

$$b = \frac{23 \sin B}{\sin 40°} = \frac{23 \sin 16.97°}{\sin 40°} = 10.44$$

The two solutions are

$$A = 56.97^\circ \qquad a = BC = 30 \text{ cm}$$
$$B = 83.03^\circ \qquad b = AC = 35.52 \text{ cm}$$
$$C = 40^\circ \qquad c = AB = 23 \text{ cm}$$

and

$$A = 123.03^\circ \qquad a = BC = 30 \text{ cm}$$
$$B = 16.97^\circ \qquad b = AC = 10.44 \text{ cm}$$
$$C = 40^\circ \qquad c = AB = 23 \text{ cm}$$

They are illustrated in Figures 21.12(a) and 21.12(b).

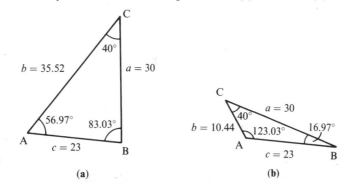

Figure 21.12
Solutions for Example
21.10

            (a)   **(b)**

Self-assessment questions 21.4

1. State the sine rule and the conditions under which it can be used to solve a triangle.

Exercise 21.4

1. Solve $\triangle ABC$ given

   (a) AB = 31 cm, AC = 24 cm, $C = 37^\circ$
   (b) AB = 19 cm, AC = 24 cm, $C = 37^\circ$
   (c) BC = 17 cm, $A = 17^\circ$, $C = 101^\circ$

   (d) $A = 53^\circ$, AB = 9.6 cm, BC = 8.9 cm
   (e) $A = 62^\circ$, AB = 12.2 cm, BC = 14.5 cm
   (f) $B = 36^\circ$, $C = 50^\circ$, AC = 11 cm

21.5   The cosine rule

The cosine rule also allows us to solve any triangle. With the usual notation the cosine rule states that:

POINT

$$a^2 = b^2 + c^2 - 2bc \cos A$$
$$b^2 = a^2 + c^2 - 2ac \cos B$$
$$c^2 = a^2 + b^2 - 2ab \cos C$$

This rule is used when we are given either (a) three sides or (b) two sides and the included angle.

Worked examples

21.11   Solve $\triangle ABC$ given $AB = 16$ cm, $AC = 23$ cm and $BC = 21$ cm.

Solution   We are told three sides and so the cosine rule can be used. We have $a = BC = 21$, $b = AC = 23$, $c = AB = 16$. Applying the cosine rule we have

$$a^2 = b^2 + c^2 - 2bc \cos A$$
$$21^2 = 23^2 + 16^2 - 2(23)(16) \cos A$$
$$736 \cos A = 344$$
$$\cos A = \frac{344}{736} = 0.4674$$
$$A = \cos^{-1}(0.4674) = 62.13°$$

We apply the cosine rule again:

$$b^2 = a^2 + c^2 - 2ac \cos B$$
$$23^2 = 21^2 + 16^2 - 2(21)(16) \cos B$$
$$672 \cos B = 168$$
$$\cos B = \frac{168}{672} = 0.25$$
$$B = \cos^{-1}(0.25) = 75.52°$$

Finally

$$C = 180° - A - B = 180° - 62.13° - 75.52° = 42.35°$$

The solution is

$$A = 62.13° \qquad a = BC = 21 \text{ cm}$$
$$B = 75.52° \qquad b = AC = 23 \text{ cm}$$
$$C = 42.35° \qquad c = AB = 16 \text{ cm}$$

21.12   Solve $\triangle ABC$ given $A = 51°$, AC = 14 cm and AB = 24 cm.

Solution   Figure 21.13 illustrates the situation. We are given two sides and the included angle and so the cosine rule can be used. We have $A = 51°$, $c = AB = 24$ and $b = AC = 14$. We use

$$a^2 = b^2 + c^2 - 2bc \cos A$$

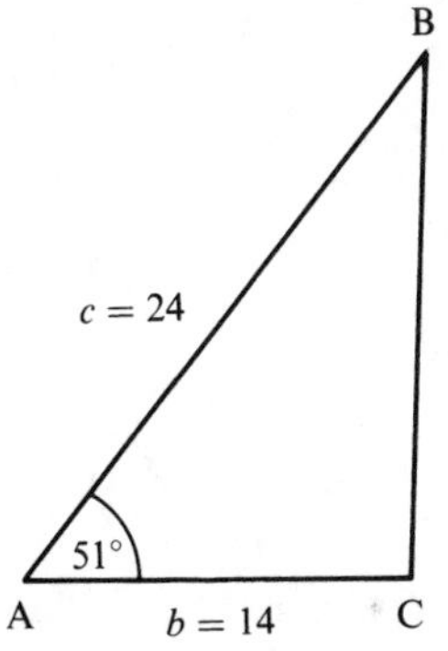

Figure 21.13
$\triangle ABC$ for Example 21.12

Substituting in the values given we have

$$a^2 = 14^2 + 24^2 - 2(14)(24) \cos 51°$$
$$= 196 + 576 - 672 \cos 51°$$
$$= 349.10$$
$$a = 18.68$$

We now use $b^2 = a^2 + c^2 - 2ac \cos B$ to find B:

$$14^2 = (18.68)^2 + 24^2 - 2(18.68)(24) \cos B$$
$$896.64 \cos B = 728.94$$
$$\cos B = \frac{728.94}{896.64} = 0.8130$$
$$B = \cos^{-1}(0.8130) = 35.61°$$

Finally
$$C = 180° - A - B = 180° - 51° - 35.61° = 93.39°$$

The solution is

$A = 51°$ $a = BC = 18.68$ cm
$B = 35.61°$ $b = AC = 14$ cm
$C = 93.39°$ $c = AB = 24$ cm

Self-assessment questions 21.5

1. State the cosine rule and the conditions under which it can be used to solve a triangle.

Exercise 21.5

1. Solve △ABC given

    (a) AC=119 cm, BC=86 cm and
        AB=53 cm
    (b) AB=42 cm, AC=30 cm and
        $A = 115°$

    (c) BC=74 cm, AC=93 cm and $C = 39°$
    (d) AB=1.9 cm, BC=3.6 cm and
        AC=2.7 cm
    (e) AB=29 cm, BC=39 cm and
        $B = 100°$

Test and assignment exercises 21

1. Solve △ABC given

    (a) $A = 45°, C = 57°$, BC=19 cm
    (b) AC=3.9 cm, AB=4.7 cm, $A = 64°$
    (c) AC=41 cm, AB=37 cm, $C = 50°$
    (d) AB=22 cm, BC=29 cm, $A = 37°$
    (e) AB=123 cm, BC=100 cm,
        AC=114 cm
    (f) AB=46 cm, $A = 33°, B = 76°$
    (g) BC=64 cm, AC=54 cm, $B = 42°$
    (h) BC=30 cm, AC=69 cm, $C = 97°$

2. △ABC has a right-angle at B. Solve
   △ABC given
    (a) AB=9 cm, BC=14 cm
    (b) BC=10 cm, AC=17 cm
    (c) AB=20 cm, $A = 40°$
    (d) BC=17 cm, $A = 45°$

    (e) AC=25 cm, $C = 27°$
    (f) $A = 52°$, AC=8.6 cm

3. Figure 21.14 illustrates △ABC. AC is
   112 cm, $\angle C = 47°$ and $\angle A = 31°$. From
   B, a line BD is drawn which is at right-
   angles to AC. Calculate the length of
   BD.

4. △ABC has $\angle A = 42°, \angle C = 59°$ and
   BC=17 cm. Calculate the length of the
   longest side of △ABC.

5. A circle, centre O, has radius 10 cm. An
   arc AB has length 17 cm. Calculate the
   length of the line AB.

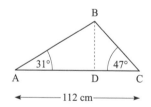

Figure 21.14

22 Matrices

Objectives	This chapter
	• explains what is meant by a matrix
	• shows how matrices can be added, subtracted and multiplied
	• explains what is meant by the determinant of a matrix
	• shows how to find the inverse of a matrix
	• shows how matrices can be used to solve simultaneous equations

22.1 What is a matrix?

A **matrix** is a set of numbers arranged in the form of a rectangle and enclosed in curved brackets. The plural of matrix is **matrices**. For example,

$$\begin{pmatrix} 1 & 2 & 3 \\ 3 & 4 & 6 \end{pmatrix} \qquad \begin{pmatrix} 1 \\ 2 \\ -4 \end{pmatrix} \qquad \begin{pmatrix} 1 & 1 & 2 \\ -3 & 4 & 5 \\ \frac{1}{2} & 2 & 1 \end{pmatrix}$$

are all matrices. Each number in a matrix is known as an **element**. To refer to a particular matrix we label it with a capital letter, so that we could write

$$A = \begin{pmatrix} 1 & 2 & 3 \\ 3 & 4 & 6 \end{pmatrix} \qquad B = \begin{pmatrix} 1 \\ 2 \\ -4 \end{pmatrix} \qquad C = \begin{pmatrix} 1 & 1 & 2 \\ -3 & 4 & 5 \\ \frac{1}{2} & 2 & 1 \end{pmatrix}$$

We refer to the **size** of a matrix by giving its number of rows and number of columns, in that order. So matrix A above has two rows and three columns – we say it is a 'two by three' matrix, and write this as '2×3'. Similarly B has size 3×1 and C has size 3×3. Some particular types of matrix occur so frequently that they have been given special names.

A **square** matrix has the same number of rows as columns.

$$\begin{pmatrix} 2 & -7 \\ -1 & 6 \end{pmatrix} \text{ and } \begin{pmatrix} 1 & 2 & -1 \\ 9 & 8 & 5 \\ 6 & -7 & 2 \end{pmatrix}$$

are both square matrices.

A **diagonal** matrix is a square matrix in which all the elements are 0 except those on the diagonal from the top left to the bottom right. This diagonal is called the **leading diagonal**. On the leading diagonal the elements can take any value including zero.

$$\begin{pmatrix} 7 & 0 \\ 0 & 9 \end{pmatrix}, \quad \begin{pmatrix} 9 & 0 \\ 0 & 0 \end{pmatrix} \text{ and } \begin{pmatrix} -1 & 0 & 0 \\ 0 & 8 & 0 \\ 0 & 0 & -17 \end{pmatrix}$$

are all diagonal matrices.

An **identity** matrix is a diagonal matrix, all the diagonal entries of which are equal to 1.

$$\begin{pmatrix} 1 & 0 \\ 0 & 1 \end{pmatrix}, \quad \begin{pmatrix} 1 & 0 & 0 \\ 0 & 1 & 0 \\ 0 & 0 & 1 \end{pmatrix} \text{ and } \begin{pmatrix} 1 & 0 & 0 & 0 \\ 0 & 1 & 0 & 0 \\ 0 & 0 & 1 & 0 \\ 0 & 0 & 0 & 1 \end{pmatrix}$$

are all identity matrices.

Self-assessment questions 22.1

1. Explain what is meant by a matrix.
2. Give an example of a 5×1 matrix and a 1×5 matrix.
3. Explain what is meant by a 'square' matrix.
4. Explain what is meant by the 'leading diagonal' of a square matrix.
5. Explain what is meant by a 'diagonal' matrix.
6. Explain what is meant by an 'identity' matrix.
7. Is it true that all identity matrices must be square?

Exercise 22.1

1. Give the size of each of the following matrices:

$$A = \begin{pmatrix} 1 & 0 & 0 & 2 \end{pmatrix} \quad B = \begin{pmatrix} 1 & 2 \\ 3 & 7 \\ 9 & 8 \\ 9 & -9 \end{pmatrix} \quad C = \begin{pmatrix} 1 \end{pmatrix} \quad D = \begin{pmatrix} 1 \\ -1 \\ 0 \\ 0 \\ 9 \end{pmatrix}$$

2. Classify each of the following matrices as square, diagonal or identity, as appropriate.

$$A = \begin{pmatrix} 3 & 4 \\ 2 & 9 \end{pmatrix} \qquad B = \begin{pmatrix} 1 & 0 & 0 \\ 0 & 1 & 0 \end{pmatrix} \qquad C = \begin{pmatrix} 1 & 0 & 0 \\ 0 & 1 & 0 \\ 0 & 0 & 1 \end{pmatrix}$$

$$D = \begin{pmatrix} 1 & 5 & 7 \\ 0 & 1 & 0 \end{pmatrix} \qquad E = (1 \quad 2 \quad 3 \quad 4) \qquad F = \begin{pmatrix} 9 & 0 \\ 0 & 8 \end{pmatrix} \qquad G = \begin{pmatrix} 0 & 8 \\ 9 & 0 \end{pmatrix} \qquad H = \begin{pmatrix} 0 & 1 \\ 1 & 0 \end{pmatrix}$$

3. How many elements are there in a matrix whose size is
   (a) 3×1 (b) 1×3 (c) $m \times n$ (d) $n \times n$?

22.2   Addition, subtraction and multiplication of matrices

Matrices can be added, subtracted and multiplied. In this section we shall see how to carry out these operations. However, they can never be divided. Two matrices having the same size can be added or subtracted by simply adding or subtracting the corresponding elements. For example,

$$\begin{pmatrix} 1 & 2 \\ 3 & 4 \end{pmatrix} + \begin{pmatrix} 5 & 2 \\ 1 & 0 \end{pmatrix} = \begin{pmatrix} 1+5 & 2+2 \\ 3+1 & 4+0 \end{pmatrix} = \begin{pmatrix} 6 & 4 \\ 4 & 4 \end{pmatrix}$$

Similarly,

$$\begin{pmatrix} 1 & 2 & 3 & 4 \\ 2 & 1 & 1 & 7 \end{pmatrix} - \begin{pmatrix} 0 & 1 & 3 & 9 \\ 7 & 0 & 0 & 1 \end{pmatrix} = \begin{pmatrix} 1-0 & 2-1 & 3-3 & 4-9 \\ 2-7 & 1-0 & 1-0 & 7-1 \end{pmatrix}$$

$$= \begin{pmatrix} 1 & 1 & 0 & -5 \\ -5 & 1 & 1 & 6 \end{pmatrix}$$

The matrices

$$\begin{pmatrix} 1 & 2 & 9 \\ -1 & \frac{1}{2} & 0 \end{pmatrix} \quad \text{and} \quad \begin{pmatrix} 1 \\ 0 \\ 7 \end{pmatrix}$$

can be neither added to nor subtracted from one another because they have different sizes.

A matrix is multiplied by a number by multiplying each element by that number. For example,

$$4\begin{pmatrix} 1 & 2 \\ 3 & -9 \end{pmatrix} = \begin{pmatrix} 4 \times 1 & 4 \times 2 \\ 4 \times 3 & 4 \times -9 \end{pmatrix} = \begin{pmatrix} 4 & 8 \\ 12 & -36 \end{pmatrix}$$

and

$$\frac{1}{4}\begin{pmatrix} 16 \\ 8 \end{pmatrix} = \begin{pmatrix} \frac{1}{4} \times 16 \\ \frac{1}{4} \times 8 \end{pmatrix} = \begin{pmatrix} 4 \\ 2 \end{pmatrix}$$

Worked example

22.1   If

$$A = \begin{pmatrix} 5 \\ 1 \\ 9 \end{pmatrix} \text{ and } B = \begin{pmatrix} 8 \\ 2 \\ -6 \end{pmatrix} \text{ find}$$

(a) $A + B$ (b) $A - B$ (c) $7A$ (d) $-\frac{1}{2}B$ (e) $3A + 2B$

Solution   (a)

$$A + B = \begin{pmatrix} 5 \\ 1 \\ 9 \end{pmatrix} + \begin{pmatrix} 8 \\ 2 \\ -6 \end{pmatrix} = \begin{pmatrix} 13 \\ 3 \\ 3 \end{pmatrix}$$

(b)

$$A - B = \begin{pmatrix} 5 \\ 1 \\ 9 \end{pmatrix} - \begin{pmatrix} 8 \\ 2 \\ -6 \end{pmatrix} = \begin{pmatrix} -3 \\ -1 \\ 15 \end{pmatrix}$$

(c)

$$7A = 7\begin{pmatrix} 5 \\ 1 \\ 9 \end{pmatrix} = \begin{pmatrix} 35 \\ 7 \\ 63 \end{pmatrix}$$

(d)

$$-\frac{1}{2}B = -\frac{1}{2}\begin{pmatrix} 8 \\ 2 \\ -6 \end{pmatrix} = \begin{pmatrix} -4 \\ -1 \\ 3 \end{pmatrix}$$

(e)

$$3A + 2B = 3\begin{pmatrix} 5 \\ 1 \\ 9 \end{pmatrix} + 2\begin{pmatrix} 8 \\ 2 \\ -6 \end{pmatrix} = \begin{pmatrix} 15 \\ 3 \\ 27 \end{pmatrix} + \begin{pmatrix} 16 \\ 4 \\ -12 \end{pmatrix} = \begin{pmatrix} 31 \\ 7 \\ 15 \end{pmatrix}$$

Two matrices can only be multiplied together if the number of columns in the first is the same as the number of rows in the second. The product of two such matrices is a matrix which has the same number of rows as the first matrix and the same number of columns as the second. In symbols, this states that if A has size $p \times q$ and B has size $q \times s$, then AB has size $p \times s$. The way that the multiplication is performed may seem strange at first. You will need to work through several examples to understand how it is done.

Worked examples

22.2 Find AB where

$$A = \begin{pmatrix} 1 & 4 \\ 6 & 3 \end{pmatrix} \quad \text{and} \quad B = \begin{pmatrix} 2 \\ 5 \end{pmatrix}$$

Solution The size of A is 2×2. The size of B is 2×1. Therefore the number of columns in the first matrix, A, equals the number of rows in the second matrix B. We can therefore find the product AB. The result will be a 2×1 matrix. Let us call it C. The working is as follows.

$$C = \begin{pmatrix} 1 & 4 \\ 6 & 3 \end{pmatrix} \begin{pmatrix} 2 \\ 5 \end{pmatrix} = \begin{pmatrix} 1 \times 2 + 4 \times 5 \\ 6 \times 2 + 3 \times 5 \end{pmatrix} = \begin{pmatrix} 22 \\ 27 \end{pmatrix}$$

The first row of A multiplies the first column of B to give

$$(1 \times 2) + (4 \times 5) = 2 + 20 = 22$$

This is the element in row 1, column 1 of C. The second row of A then multiplies the first column of B to give

$$(6 \times 2) + (3 \times 5) = 12 + 15 = 27$$

This is the element in row 2, column 1 of C.

22.3 Using the same A and B as the previous example, is it possible to find BA?

Solution Recall that B has size 2×1 and A has size 2×2. When written in the order BA, the number of columns in the first matrix is 1 whilst the number of rows in the second is 2. It is not therefore possible to find BA.

22.4 Find, if possible,

$$\begin{pmatrix} 1 & 4 & 9 \\ 2 & 0 & 1 \end{pmatrix} \begin{pmatrix} 1 & 9 \\ 8 & 7 \\ -7 & 3 \end{pmatrix}$$

Solution The number of columns in the first matrix is 3 and this is the same as the number of rows in the second. We can therefore perform the multiplication and the answer will have size 2×2.

$$\begin{pmatrix} 1 & 4 & 9 \\ 2 & 0 & 1 \end{pmatrix} \begin{pmatrix} 1 & 9 \\ 8 & 7 \\ -7 & 3 \end{pmatrix} = \begin{pmatrix} 1 \times 1 + 4 \times 8 + 9 \times -7 & 1 \times 9 + 4 \times 7 + 9 \times 3 \\ 2 \times 1 + 0 \times 8 + 1 \times -7 & 2 \times 9 + 0 \times 7 + 1 \times 3 \end{pmatrix}$$

$$= \begin{pmatrix} -30 & 64 \\ -5 & 21 \end{pmatrix}$$

22.5 Find $(1 \quad 5)\begin{pmatrix} 2 \\ 1 \end{pmatrix}$

Solution $(1 \quad 5)\begin{pmatrix} 2 \\ 1 \end{pmatrix} = (1 \times 2 + 5 \times 1) = (7)$

Note that in this example the result is a 1×1 matrix, that is, a single number.

22.6 Find $\begin{pmatrix} 3 & 2 \\ 5 & 3 \end{pmatrix}\begin{pmatrix} x \\ y \end{pmatrix}$

Solution $\begin{pmatrix} 3 & 2 \\ 5 & 3 \end{pmatrix}\begin{pmatrix} x \\ y \end{pmatrix} = \begin{pmatrix} 3x + 2y \\ 5x + 3y \end{pmatrix}$

22.7 Find IX where I is the 2×2 identity matrix

$$\begin{pmatrix} 1 & 0 \\ 0 & 1 \end{pmatrix} \quad \text{and} \quad X = \begin{pmatrix} a \\ b \end{pmatrix}$$

Solution $IX = \begin{pmatrix} 1 & 0 \\ 0 & 1 \end{pmatrix}\begin{pmatrix} a \\ b \end{pmatrix} = \begin{pmatrix} 1 \times a + 0 \times b \\ 0 \times a + 1 \times b \end{pmatrix} = \begin{pmatrix} a \\ b \end{pmatrix}$

Note that the effect of multiplying the matrix X by the identity matrix is to leave X unchanged. This is a very important and useful property of identity matrices which we will require later. It should be remembered. This property of identity matrices should remind you of the fact that multiplying a number by 1 leaves the number unchanged: for example, $7 \times 1 = 7$, $1 \times -9 = -9$. An identity matrix plays the same role for matrices as the number 1 does when dealing with ordinary arithmetic.

Self-assessment questions 22.2

1. Under what conditions can two matrices be added or subtracted?
2. Two different types of multiplication have been described. What are these two types and how do they differ?
3. If A has size $p \times q$ and B has size $r \times s$, under what conditions will the product AB exist? If AB does exist what will be its size? Under what conditions will the product BA exist and what will be its size?
4. Suppose A and B are two matrices. The products AB and BA both exist. What can you say about the sizes of A and B?

Exercise 22.2

1. If $M = \begin{pmatrix} 7 & 8 \\ 2 & -1 \end{pmatrix}$ and $N = \begin{pmatrix} 3 & 2 \\ -1 & 4 \end{pmatrix}$

   find MN and NM. Comment upon your two answers.

2. If

   $A = \begin{pmatrix} 1 & -7 \\ -1 & 2 \end{pmatrix}$ $B = \begin{pmatrix} 3 & 7 & 2 \\ -1 & 0 & -10 \end{pmatrix}$

   $C = \begin{pmatrix} 1 & 0 & -1 & 1 \end{pmatrix}$

   $D = \begin{pmatrix} 7 \\ 9 \\ 8 \\ 1 \end{pmatrix}$ $E = \begin{pmatrix} 1/2 \\ 0 \\ 0 \\ 1 \end{pmatrix}$

   find, if possible,

   (a) $3A$ (b) $A + B$ (c) $A + C$ (d) $5D$ (e) BC
   (f) CB (g) AC (h) CA (i) $5A - 2C$
   (j) $D - E$ (k) BA (l) AB (m) AA (that is, A times itself, usually written A^2).

3. I is the 2×2 identity matrix

   $\begin{pmatrix} 1 & 0 \\ 0 & 1 \end{pmatrix}$

   If

   $A = \begin{pmatrix} 3 & 7 \\ -1 & -2 \end{pmatrix}$ $B = \begin{pmatrix} x \\ y \end{pmatrix}$

   and $C = \begin{pmatrix} 3 & 9 & 5 \\ 4 & 2 & 8 \end{pmatrix}$

   find, where possible,

   (a) IA (b) AI (c) BI (d) IB (e) CI (f) IC. Comment upon your answers.

4. If

   $A = \begin{pmatrix} a & b \\ c & d \end{pmatrix}$ and $B = \begin{pmatrix} e & f \\ g & h \end{pmatrix}$

   find

   (a) $A + B$ (b) $A - B$ (c) AB (d) BA

5. Find

   $\begin{pmatrix} -7 & 5 \\ -2 & -1 \end{pmatrix} \begin{pmatrix} x \\ y \end{pmatrix}$

6. Find

   $\begin{pmatrix} 9 & -5 \\ 3 & 8 \end{pmatrix} \begin{pmatrix} a \\ b \end{pmatrix}$

7. Given

   $A = \begin{pmatrix} 2 & 3 & -1 \\ 4 & 0 & 1 \end{pmatrix}$ $B = \begin{pmatrix} 1 & 2 \\ 3 & -2 \end{pmatrix}$

   find
   (a) BA (b) B^2 (c) $B^2 A$

8. Given

   $A = \begin{pmatrix} 1 & 3 \\ -1 & 4 \end{pmatrix}$ $B = \begin{pmatrix} 0 & -1 \\ 3 & 2 \end{pmatrix}$

   find
   (a) AB (b) BA (c) A^2 (d) B^3

9. Given

   $A = \begin{pmatrix} 4 & 11 \\ 1 & 3 \end{pmatrix}$ and $B = \begin{pmatrix} 3 & -11 \\ -1 & 4 \end{pmatrix}$

   verify that $AB = I$ where I is the 2×2 identity matrix.

10. Calculate

    $\begin{pmatrix} 3 & 0 \\ 0 & 2 \end{pmatrix} \begin{pmatrix} a & b \\ c & d \end{pmatrix}$

    Comment upon the effect of multiplying a matrix by

    $\begin{pmatrix} 3 & 0 \\ 0 & 2 \end{pmatrix}$

22.3   The inverse of a 2 × 2 matrix

Consider the 2×2 matrix

$$A = \begin{pmatrix} a & b \\ c & d \end{pmatrix}$$

An important matrix which is related to A is known as the **inverse** of A and is given the symbol A^{-1}. Here, the superscript '-1' should not be read as a power, but is meant purely as a notation for the inverse matrix. A^{-1} can be found from the following formula:

KEY POINT

If $A = \begin{pmatrix} a & b \\ c & d \end{pmatrix}$ then $A^{-1} = \dfrac{1}{ad - bc} \begin{pmatrix} d & -b \\ -c & a \end{pmatrix}$

This formula states that

- the elements on the leading diagonal are interchanged
- the remaining elements change sign
- the resulting matrix is multiplied by $\dfrac{1}{ad - bc}$.

The inverse matrix has the property that:

KEY POINT

$$A A^{-1} = A^{-1} A = I$$

that is, when a 2×2 matrix and its inverse are multiplied together the result is the identity matrix. Consider the following example.

Worked example

22.8   Find the inverse of the matrix

$$A = \begin{pmatrix} 6 & 5 \\ 2 & 2 \end{pmatrix}$$

and verify that $A A^{-1} = A^{-1} A = I$.

Solution   Using the formula for the inverse we find

$$A^{-1} = \frac{1}{(6)(2) - (5)(2)} \begin{pmatrix} 2 & -5 \\ -2 & 6 \end{pmatrix} = \frac{1}{2} \begin{pmatrix} 2 & -5 \\ -2 & 6 \end{pmatrix} = \begin{pmatrix} 1 & -\frac{5}{2} \\ -1 & 3 \end{pmatrix}$$

The inverse of $\begin{pmatrix} 6 & 5 \\ 2 & 2 \end{pmatrix}$

is therefore $\begin{pmatrix} 1 & -\frac{5}{2} \\ -1 & 3 \end{pmatrix}$

Evaluating $A A^{-1}$

we find

$$\begin{pmatrix} 6 & 5 \\ 2 & 2 \end{pmatrix} \begin{pmatrix} 1 & -\frac{5}{2} \\ -1 & 3 \end{pmatrix} = \begin{pmatrix} 6 \times 1 + 5 \times -1 & 6 \times (-\frac{5}{2}) + 5 \times 3 \\ 2 \times 1 + 2 \times -1 & 2 \times (-\frac{5}{2}) + 2 \times 3 \end{pmatrix} = \begin{pmatrix} 1 & 0 \\ 0 & 1 \end{pmatrix}$$

which is the 2 × 2 identity matrix. Also evaluating $A^{-1}A$ we find

$$\begin{pmatrix} 1 & -\frac{5}{2} \\ -1 & 3 \end{pmatrix} \begin{pmatrix} 6 & 5 \\ 2 & 2 \end{pmatrix} = \begin{pmatrix} 1 \times 6 + (-\frac{5}{2}) \times 2 & 1 \times 5 + (-\frac{5}{2}) \times 2 \\ -1 \times 6 + 3 \times 2 & -1 \times 5 + 3 \times 2 \end{pmatrix} = \begin{pmatrix} 1 & 0 \\ 0 & 1 \end{pmatrix}$$

which is the 2 × 2 identity matrix. We have shown that
$A A^{-1} = A^{-1} A = I$.

The quantity $ad - bc$ in the formula for the inverse is known as the **determinant** of the matrix A. We often write it as $|A|$, the vertical bars indicating that we mean the determinant of A and not the matrix itself.

KEY POINT

If $A = \begin{pmatrix} a & b \\ c & d \end{pmatrix}$ then its determinant equals

$$|A| = \begin{vmatrix} a & b \\ c & d \end{vmatrix} = ad - bc$$

Worked examples

22.9   Find the value of the determinant

$$\begin{vmatrix} 8 & 7 \\ 9 & 2 \end{vmatrix}$$

Solution   The determinant is given by $(8)(2) - (7)(9) = 16 - 63 = -47$.

22.10　Find the determinant of each of the following matrices:

$$\text{(a) } C = \begin{pmatrix} 7 & 2 \\ 4 & 9 \end{pmatrix} \quad \text{(b) } D = \begin{pmatrix} 4 & -2 \\ 9 & -1 \end{pmatrix} \quad \text{(c) } I = \begin{pmatrix} 1 & 0 \\ 0 & 1 \end{pmatrix} \quad \text{(d) } E = \begin{pmatrix} 4 & 2 \\ 2 & 1 \end{pmatrix}$$

Solution　(a) $|C| = (7)(9) - (2)(4) = 63 - 8 = 55$. Note that the determinant is always a single number.

(b) $|D| = (4)(-1) - (-2)(9) = -4 + 18 = 14$.

(c) $|I| = (1)(1) - (0)(0) = 1 - 0 = 1$. It is always true that the determinant of an identity matrix is 1.

(d) $|E| = (4)(1) - (2)(2) = 4 - 4 = 0$. In this example the determinant of the matrix E is zero.

As we have seen in the previous example, on some occasions a matrix may be such that $ad - bc = 0$, that is, its determinant is zero. Such a matrix is called **singular.** When a matrix is singular it cannot have an inverse. This is because it is impossible to evaluate the quantity $1/(ad - bc)$ which appears in the formula for the inverse; the quantity $1/0$ has no meaning in mathematics.

Worked example

22.11　Show that the matrix

$$P = \begin{pmatrix} 6 & -2 \\ -24 & 8 \end{pmatrix}$$

has no inverse.

Solution　We first find the determinant of P:

$$|P| = (6)(8) - (-2)(-24) = 48 - 48 = 0$$

The determinant is zero and so the matrix P is singular. It does not have an inverse.

Self-assessment questions 22.3

1. Explain what is meant by the inverse of a 2×2 matrix. Under what conditions will this inverse exist?
2. Explain what is meant by a singular matrix.
3. A diagonal matrix has the elements on its leading diagonal equal to 7 and -9. What is the value of its determinant?

Exercise 22.3

1. Evaluate the following determinants:

(a) $\begin{vmatrix} 9 & 7 \\ -1 & -1 \end{vmatrix}$ (b) $\begin{vmatrix} 8 & 7 \\ 2 & -1 \end{vmatrix}$ (c) $\begin{vmatrix} 0 & 1 \\ -1 & 0 \end{vmatrix}$ (d) $\begin{vmatrix} 8 & 0 \\ 0 & 11 \end{vmatrix}$

2. Which of the following matrices are singular?

(a) $A = \begin{pmatrix} 9 & 5 \\ -2 & 0 \end{pmatrix}$ (b) $B = \begin{pmatrix} 6 & 45 \\ -2 & 15 \end{pmatrix}$ (c) $C = \begin{pmatrix} 6 & 45 \\ 2 & 15 \end{pmatrix}$ (d) $D = \begin{pmatrix} 9 & 5 \\ 0 & 1 \end{pmatrix}$

(e) $E = \begin{pmatrix} a & b \\ a & b \end{pmatrix}$

3. By multiplying the matrices

$\begin{pmatrix} 4 & -9 \\ -3 & 7 \end{pmatrix}$ and $\begin{pmatrix} 7 & 9 \\ 3 & 4 \end{pmatrix}$ together show that each one is the inverse of the other.

4. Find, if it exists, the inverse of each of the following matrices:

(a) $\begin{pmatrix} 2 & 4 \\ 6 & 10 \end{pmatrix}$ (b) $\begin{pmatrix} 1 & 0 \\ 0 & 1 \end{pmatrix}$ (c) $\begin{pmatrix} -1 & 3 \\ 2 & 2 \end{pmatrix}$ (d) $\begin{pmatrix} 8 & 4 \\ 2 & 1 \end{pmatrix}$

5. Find the inverse of each of the following matrices

(a) $\begin{pmatrix} 9 & 2 \\ 4 & 1 \end{pmatrix}$ (b) $\begin{pmatrix} 3 & 2 \\ 5 & 4 \end{pmatrix}$ (c) $\begin{pmatrix} -2 & 1 \\ 5 & -3 \end{pmatrix}$ (d) $\begin{pmatrix} 4 & -2 \\ 9 & 5 \end{pmatrix}$ (e) $\begin{pmatrix} -2 & -3 \\ -4 & -5 \end{pmatrix}$

22.4 Application of matrices to solving simultaneous equations

Matrices can be used to solve simultaneous equations. Suppose we wish to solve the simultaneous equations

$$3x + 2y = -3$$
$$5x + 3y = -4$$

First of all we rewrite them using matrices as

$$\begin{pmatrix} 3 & 2 \\ 5 & 3 \end{pmatrix} \begin{pmatrix} x \\ y \end{pmatrix} = \begin{pmatrix} -3 \\ -4 \end{pmatrix}$$

Writing

$$A = \begin{pmatrix} 3 & 2 \\ 5 & 3 \end{pmatrix}, \quad X = \begin{pmatrix} x \\ y \end{pmatrix} \quad \text{and} \quad B = \begin{pmatrix} -3 \\ -4 \end{pmatrix}$$

we have $AX = B$. Note that we are trying to find x and y; in other words we must find X. Now, if

$$AX = B$$

then, provided A^{-1} exists, we multiply both sides by A^{-1} to obtain

$$A^{-1}AX = A^{-1}B$$

But $A^{-1}A = I$, so this becomes

$$IX = A^{-1}B$$

However, multiplying X by the identity matrix leaves X unaltered, so we can write

$$X = A^{-1}B$$

In other words if we multiply B by the inverse of A we will have X as required. Now, given

$$A = \begin{pmatrix} 3 & 2 \\ 5 & 3 \end{pmatrix}$$

then

$$A^{-1} = \frac{1}{(3)(3) - (2)(5)} \begin{pmatrix} 3 & -2 \\ -5 & 3 \end{pmatrix} = \frac{1}{-1} \begin{pmatrix} 3 & -2 \\ -5 & 3 \end{pmatrix} = \begin{pmatrix} -3 & 2 \\ 5 & -3 \end{pmatrix}$$

Finally,

$$X = A^{-1}B = \begin{pmatrix} -3 & 2 \\ 5 & -3 \end{pmatrix} \begin{pmatrix} -3 \\ -4 \end{pmatrix} = \begin{pmatrix} 1 \\ -3 \end{pmatrix}$$

Therefore the solution of the simultaneous equations is $x = 1$ and $y = -3$.

We note that the crucial step is

KEY POINT

$$AX = B$$
$$X = A^{-1}B \quad \text{provided } A^{-1} \text{ exists}$$

Worked example

22.12   Solve the simultaneous equations

$$x + 2y = 13$$
$$2x - 5y = 8$$

Solution   First of all we rewrite the equations using matrices as

$$\begin{pmatrix} 1 & 2 \\ 2 & -5 \end{pmatrix} \begin{pmatrix} x \\ y \end{pmatrix} = \begin{pmatrix} 13 \\ 8 \end{pmatrix}$$

Writing

$$A = \begin{pmatrix} 1 & 2 \\ 2 & -5 \end{pmatrix}, \quad X = \begin{pmatrix} x \\ y \end{pmatrix} \quad \text{and} \quad B = \begin{pmatrix} 13 \\ 8 \end{pmatrix}$$

we have $AX = B$. We must now find the matrix X. If $AX = B$ then multiplying both sides by A^{-1} gives $A^{-1}AX = A^{-1}B$. But $A^{-1}A = I$, so this becomes $IX = A^{-1}B$. It follows that

$$X = A^{-1}B$$

We must multiply B by the inverse of A. The inverse of A is

$$A^{-1} = \frac{1}{(1)(-5) - (2)(2)} \begin{pmatrix} -5 & -2 \\ -2 & 1 \end{pmatrix} = \frac{1}{-9} \begin{pmatrix} -5 & -2 \\ -2 & 1 \end{pmatrix}$$

Finally,

$$X = A^{-1}B = \frac{1}{-9} \begin{pmatrix} -5 & -2 \\ -2 & 1 \end{pmatrix} \begin{pmatrix} 13 \\ 8 \end{pmatrix} = -\frac{1}{9} \begin{pmatrix} -81 \\ -18 \end{pmatrix} = \begin{pmatrix} 9 \\ 2 \end{pmatrix}$$

Therefore the solution of the simultaneous equations is $x = 9$ and $y = 2$.

Self-assessment questions 22.4

1. Explain how the simultaneous equations $ax + by = c$, $dx + ey = f$, where a, b, c, d, e and f are given numbers and x and y are the unknowns, can be written using matrices.

Exercise 22.4

1. Use matrices to solve the following simultaneous equations:
   (a) $x + y = 17$, $2x - y = 10$ (b) $2x - y = 10$, $x + 3y = -2$
   (c) $x + 2y = 15$, $3x - y = 10$

2. Try to solve the simultaneous equations $2x - 3y = 5$, $6x - 9y = 15$ using the matrix method. What do you find? Can you explain this?

3. Use matrices to solve each of the following simultaneous equations.
   (a) $3x - y = 3$ (b) $2x + 3y = 10$ (c) $x + \dfrac{y}{2} = 2$ (d) $\dfrac{x}{2} + 3y = 5$
   $\quad\ \ x + y = 5$ $\quad\ \ x - 2y = -9$ $\quad 4x - 3y = 18$ $\quad 2x + 5y = 6$

Test and assignment exercises 22

1. Express

$$\frac{1}{2}\begin{pmatrix} 6 & 6 \\ -18 & 4 \end{pmatrix} + 7\begin{pmatrix} 8 & 2 \\ 0 & 2 \end{pmatrix} \text{ as a single matrix.}$$

2. If

$$A = \begin{pmatrix} 7 & 7 & 1 \\ 2 & 0 & 9 \end{pmatrix} \text{ and } B = \begin{pmatrix} 9 \\ 3 \\ 6 \end{pmatrix}$$

find, if possible, AB and BA.

3. Find the value of the following determinants:

(a) $\begin{vmatrix} 9 & -7 \\ 8 & 8 \end{vmatrix}$ (b) $\begin{vmatrix} 9 & 0 \\ 0 & 7 \end{vmatrix}$ (c) $\begin{vmatrix} 0 & 11 \\ -7 & 0 \end{vmatrix}$ (d) $\begin{vmatrix} \alpha & \beta \\ \gamma & \delta \end{vmatrix}$

4. Use matrices to solve the following simultaneous equations:
   (a) $x + 2y = 15$, $x - 2y = -5$ (b) $7x + 8y = 56$, $2x - 2y = 16$
   (c) $2x - 3y = 1$, $x + 2y = -3$

5. Find the inverse of each of the following matrices, if it exists:

(a) $\begin{pmatrix} -1 & 0 \\ 0 & 1 \end{pmatrix}$ (b) $\begin{pmatrix} 10 & 7 \\ 6 & 4 \end{pmatrix}$ (c) $\begin{pmatrix} \frac{1}{2} & 1 \\ 3 & 7 \end{pmatrix}$

6. Show that each of the following matrices has no inverse:

(a) $\begin{pmatrix} 16 & 8 \\ 4 & 2 \end{pmatrix}$ (b) $\begin{pmatrix} a & b \\ a^2 & ab \end{pmatrix}$

7. If

$$A = \begin{pmatrix} 7 & 5 \\ 2 & -3 \end{pmatrix} \text{ and } B = \begin{pmatrix} -2 & 2 \\ 3 & 7 \end{pmatrix}$$

find AB and BA.

8. Solve the simultaneous equations $2x + 2y = 4$, $7x + 5y = 8$ using the matrix method.

9. Find AB and BA when

$$A = \begin{pmatrix} 1 & 2 & 1 \\ 3 & 4 & 0 \\ 0 & 1 & -1 \end{pmatrix} \quad \text{and} \quad B = \begin{pmatrix} 2 & 4 & 1 \\ 0 & 3 & 1 \\ 0 & 0 & 4 \end{pmatrix}$$

10. Given

$$A = \begin{pmatrix} 1 & 2 & -1 \\ 0 & 3 & 1 \\ 4 & 2 & -2 \end{pmatrix} \quad B = \begin{pmatrix} 3 & -1 \\ 4 & 1 \\ 0 & 2 \end{pmatrix}$$

find
(a) AB (b) A^2

23 Measurement

Objectives

This chapter

- revises common units of length and conversions between them
- revises the concepts of area and volume
- provides a summary of important formulas needed for calculating areas of common shapes and volumes of common solids
- revises common units of mass and conversions between them

23.1 Units of length, area and volume

The standard unit of **length** is the metre, abbreviated to m, although several smaller units are commonly used as well. The following relationships are given for future reference:

KEY POINT

> 1 metre (m) = 100 centimetres (cm)
> = 1000 millimetres (mm)

Note also that 1000 metres is equal to 1 kilometre (km).

Worked examples

23.1 Convert a length of 6.7 metres into centimetres.

Solution Because 1 m equals 100 cm then 6.7 metres equals $6.7 \times 100 = 670$ cm.

23.2   Convert a length of 24 mm into metres.

Solution   Because 1 m equals 1000 mm, then 1 mm equals $\frac{1}{1000}$ m. Therefore 24 mm equals

$$24 \times \frac{1}{1000} = 0.024 \, \text{m}$$

The **area** of a shape is measured by finding how many square units it contains. 1 square metre, written 1 m^2, is the area of a square having sides of length 1 metre. Similarly 1 square centimetre, written 1 cm^2, is the area of a square having sides of length 1 centimetre.

The **volume** of a solid is measured by finding how many cubic units it contains. 1 cubic metre, written 1 m^3, is the volume of a cube with sides of length 1 metre. Similarly, 1 cubic centimetre, written 1 cm^3, is the volume of a cube having sides of length 1 centimetre.

When measuring the volume of liquids, the litre is commonly used. One litre, that is, 1 l, is equal to 1000 cubic centimetres, that is,

$$1 \, l = 1000 \, \text{cm}^3$$

Note that 1000 cm^3 is the volume of a cube of side 10 cm, and so it is helpful to remember that such a cube holds 1 litre of liquid. Commonly millilitres are used. 1 millilitre (1 ml) equals 1 cubic centimetre, that is,

$$1 \, \text{ml} = 1 \, \text{cm}^3$$

KEY POINT

> 1 litre (l) = 1000 millilitres (ml)
>
> 1000 litres (l) = 1 m^3

Worked example

23.3   Convert 256 millilitres into litres.

Solution   Because 1 l = 1000 ml then 1 ml is equal to $\frac{1}{1000}$ l. Therefore

$$256 \, \text{ml} = 256 \times \frac{1}{1000} = 0.256 \, l$$

The areas of common shapes and volumes of common solids can be found using formulas. Table 23.1 summarizes a variety of such formulas. The calculation of an area or volume is then simply an exercise in substituting values into a formula.

Table 23.1
Formulae for areas and
volumes of common
shapes and solids

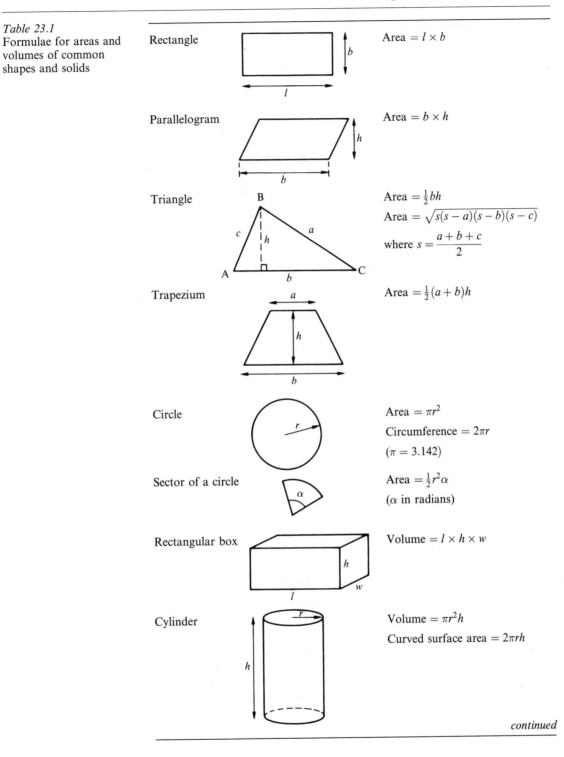

Rectangle

Area $= l \times b$

Parallelogram

Area $= b \times h$

Triangle

Area $= \frac{1}{2}bh$

Area $= \sqrt{s(s-a)(s-b)(s-c)}$

where $s = \dfrac{a+b+c}{2}$

Trapezium

Area $= \frac{1}{2}(a+b)h$

Circle

Area $= \pi r^2$

Circumference $= 2\pi r$

$(\pi = 3.142)$

Sector of a circle

Area $= \frac{1}{2}r^2\alpha$

$(\alpha$ in radians)

Rectangular box

Volume $= l \times h \times w$

Cylinder

Volume $= \pi r^2 h$

Curved surface area $= 2\pi rh$

continued

Table 23.1 (continued)   Cone

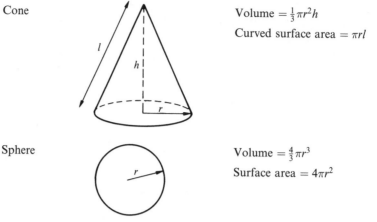

Volume $= \frac{1}{3}\pi r^2 h$

Curved surface area $= \pi r l$

Sphere

Volume $= \frac{4}{3}\pi r^3$

Surface area $= 4\pi r^2$

Worked examples

23.4 A square has sides of length 1 m. Calculate the area in

(a) m^2 (b) cm^2 (c) mm^2

Solution (a) When working in metres the area is $1 \times 1 = 1\,m^2$.

(b) When working in centimetres, each side is of length 100 cm. Therefore the area is $100 \times 100 = 10000 = 10^4\,cm^2$.

(c) When working in millimetres, each side is of length 1000 mm and so the area is $1000 \times 1000 = 10^3 \times 10^3 = 10^6\,mm^2$.

Note that in all three cases the areas are the same and so we can say
$$1\,m^2 = 10^4\,cm^2 = 10^6\,mm^2$$

23.5 A cubic box has sides of length 1 m. Calculate its volume in

(a) m^3 (b) cm^3 (c) mm^3

Solution (a) When working in metres the volume is $1 \times 1 \times 1 = 1\,m^3$.

(b) When working in centimetres, each side is of length 100 cm. Therefore the volume is $100 \times 100 \times 100 = 1000000 = 10^6\,cm^3$.

(c) When working in millimetres, each side is of length 1000 mm and so the volume is $1000 \times 1000 \times 1000 = 10^3 \times 10^3 \times 10^3 = 10^9\,mm^3$.

Note that in all three cases the volumes are the same and so we can say
$$1\,m^3 = 10^6\,cm^3 = 10^9\,mm^3$$

Exercise 23.1

1. Convert 16 cm into (a) mm, (b) m.

2. Change 356 mm into (a) cm, (b) m.

3. Change 156 m into km.

4. Find the area of a rectangle whose length is 14 cm and whose width is 7 cm.

5. A room has length 5 m and width 3.5 m. It is required to carpet this room at £4.99 per square metre. Calculate the cost of the carpet assuming there is no wastage due to cutting.

6. Find the area of a parallelogram which has a vertical height of 2 m and a base of length 8 m.

7. A triangle has sides of length 8 cm, 12 cm and 14 cm. Find its area.

8. Find the area of a trapezium whose parallel sides are 14 cm and 8 cm long and whose distance apart is 2 cm.

9. Find the area of a circle whose radius is 18 cm.

10. Find the area of a circle whose diameter is 10 cm.

11. Find the radius of the circle whose area is 26 cm^2.

12. A circular piece of cardboard of radius 70 cm has a circular piece of radius 25 cm removed from it. Calculate the area of the remaining cardboard.

13. Find the volume of a tin of beans of base radius 5 cm and height 14 cm. What is the tin's capacity in litres?

14. Find the volume of a sphere of radius 8 cm.

15. A sphere has volume 86 cm^3. Calculate its radius.

16. Calculate the area of a sector of a circle of angle 45° and radius 10 cm.

17. A garden swimming pool is designed in the form of a cylinder of base radius 3 m. Calculate how many litres of water are required to fill it to a depth of 1 m.

23.2   Units of mass

The standard unit of mass is the kilogram, abbreviated to kg. Sometimes smaller units, the gram (g) or milligram (mg), are used. A very sensitive balance may be able to measure in micrograms (μg). The relationship between these units is given as follows:

KEY POINT

> 1 kilogram (kg) = 1000 grams (g) = 1 000 000 milligrams (mg)
> 1 gram (g) = 1000 milligrams (mg) = 1 000 000 micrograms (μg)
> 1 milligram = 1000 micrograms (μg)

Note also that 1000 kilograms equals 1 tonne (t).

Worked examples

23.6   Convert 8.5 kg into grams.

Solution   1 kg is equal to 1000 g and so 8.5 kg must equal $8.5 \times 1000 = 8500$ g.

23.7   Convert 50 g to kilograms.

Solution   Because 1 kg equals 1000 g then 1 g equals $\frac{1}{1000}$ of a kilogram. Consequently,

$$50 \text{ g} = 50 \times \frac{1}{1000} = 0.05 \text{ kg}$$

Exercise 23.2

1. Convert 12 kg into (a) grams (b) milligrams.

2. Convert 168 mg into (a) grams (b) kilograms.

3. Convert 0.005 t into kilograms.

4. Convert 875 t into kilograms.

5. Convert 3500 kg into tonnes.

Test and assignment exercises 23

1. Change the following to grams:

   (a) 125 mg (b) 15 kg (c) 0.5 mg

2. Change the following lengths to centimetres:

   (a) 65 mm (b) 0.245 mm (c) 5 m

3. Convert 7500 m and 125.5 m into kilometres.

4. Convert 8 l into millilitres.

5. Convert 56 ml into litres.

6. A household water tank has dimensions 1 m × 0.6 m × 0.8 m. Calculate its capacity in litres assuming it can be filled to the brim.

7. Calculate the area of the sector of a circle of radius 25 cm and angle 60°.

8. A sector of a circle of radius 10 cm has area 186 cm². Calculate the angle of the sector.

9. Calculate the volume of a cylinder whose height is 1 m and whose base radius is 10 cm. (Be careful of the units used.)

10. A sphere has diameter 16 cm. Calculate its volume and surface area.

11. A trapezium has parallel sides of length y_1 and y_2. If these sides are a distance h apart find an expression for the area of the trapezium. Calculate this area when $y_1 = 7$ cm, $y_2 = 10$ cm and $h = 1$ cm.

12. A lorry weighs 7.5 t. Express this in kilograms.

24 Gradients of curves

Objectives

This chapter

- introduces a technique called differentiation for calculating the gradient of a curve at any point
- explains the terms 'maximum' and 'minimum' when applied to functions
- applies the technique of differentiation to locating maximum and minimum values of a function

24.1   The gradient function

Suppose we have a function, $y = f(x)$, and are interested in its slope, or gradient, at several points. For example, Figure 24.1 shows a graph of the function $y = 2x^2 + 3x$. Imagine tracing the graph from the left to the right. At point A the graph is falling rapidly. At point B the graph is falling but less rapidly than at A. Point C lies at the bottom of the dip. At point D the graph is rising. At E it is rising but more quickly than at D. The important point is that the gradient of the curve changes from point

Figure 24.1
Graph of $y = 2x^2 + 3x$

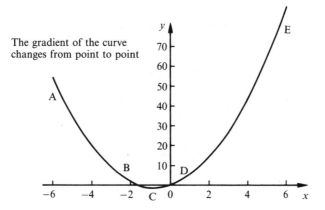

The gradient of the curve
changes from point to point

y' stands for the gradient function of *y* = *f*(*x*). We can also denote the gradient function by *f* '(*x*)

to point. The following section describes a mathematical technique for measuring the gradient at different points. We introduce another function called the **gradient function**, which we write as $\frac{dy}{dx}$. This is read as 'dee *y* by dee *x*'. We sometimes simplify this notation to *y'*, read as '*y* dash'. For almost all of the functions you will meet this gradient function can be found by using a formula, applying a rule or checking in a table. Knowing the gradient function we can find the gradient of the curve at any point. The following sections show how to find the gradient function of a number of common functions.

KEY POINT

Given a function $y = f(x)$ we denote its gradient function by $\frac{dy}{dx}$ or simply by *y'*.

24.2   Gradient function of *y* = *x^n*

For any function of the form $y = x^n$ the gradient function is found from the following formula:

KEY POINT

If $y = x^n$ then $y' = nx^{n-1}$

Worked examples

24.1   Find the gradient function of (a) $y = x^3$, (b) $y = x^4$.

Solution   (a) Comparing $y = x^3$ with $y = x^n$ we see that $n = 3$. Then
$y' = 3x^{3-1} = 3x^2$.

(b) Applying the formula with $n = 4$ we find that if $y = x^4$ then
$y' = 4x^{4-1} = 4x^3$.

24.2   Find the gradient function of (a) $y = x^2$, (b) $y = x$.

Solution   (a) Applying the formula with $n = 2$ we find that if $y = x^2$ then
$y' = 2x^{2-1} = 2x^1$. Because x^1 is simply *x* we find that the gradient function is $y' = 2x$.

(b) Applying the formula with $n = 1$ we find that if $y = x^1$ then
$y' = 1x^{1-1} = 1x^0$. Because x^0 is simply 1 we find that the gradient function is $y' = 1$.

Finding the gradient of a graph is now a simple matter. Once the gradient function has been found the gradient at any value of x is found by substituting that value into the gradient function. If, after carrying out this substitution, the result is negative, then the curve is falling. If the result is positive, the curve is rising. The size of the gradient function is a measure of how rapidly this fall or rise is taking place. We write $y'(x = 2)$ or simply $y'(2)$ to denote the value of the gradient function when $x = 2$.

Worked examples

24.3   Find the gradient of $y = x^2$ at the points where

(a) $x = -1$ (b) $x = 0$ (c) $x = 2$ (d) $x = 3$

Solution   From Example 24.2, or from the formula, we know that the gradient function of $y = x^2$ is given by $y' = 2x$.

(a) When $x = -1$ the gradient of the graph is then $y'(-1) = 2(-1) = -2$. The fact that the gradient is negative means that the curve is falling at the point.

(b) When $x = 0$ the gradient is $y'(0) = 2(0) = 0$. The gradient of the curve is zero at this point. This means that the curve is neither falling nor rising.

(c) When $x = 2$ the gradient is $y'(2) = 2(2) = 4$. The fact that the gradient is positive means that the curve is rising.

(d) When $x = 3$ the gradient is $y'(3) = 2(3) = 6$ and so the curve is rising here. Comparing this answer with that of part (c) we conclude that the curve is rising more rapidly at $x = 3$ than at $x = 2$ where the gradient was found to be 4.

   The graph of $y = x^2$ is shown in Figure 24.2. If we compare our results with the graph we see that the curve is indeed falling when $x = -1$ (point A) and is rising when $x = 2$ (point C), and $x = 3$ (point D). At the point where $x = 0$ (point B) the curve is neither rising nor falling.

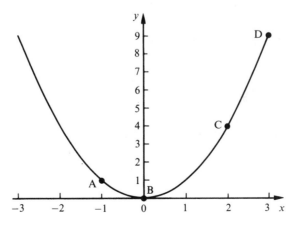

Figure 24.2
Graph of $y = x^2$

24.4   Find the gradient function of $y = x^{-3}$. Hence find the gradient of $y = x^{-3}$ when $x = 4$.

Solution   Using the formula with $n = -3$ we find that if $y = x^{-3}$ then $y' = -3x^{-3-1} = -3x^{-4}$. Because x^{-4} can also be written as $1/x^4$ we could write $y' = -3(1/x^4)$ or $-3/x^4$. When $x = 4$ we find $y'(4) = -3/4^4 = -\frac{3}{256}$. This number is very small and negative, which means that when $x = 4$ the curve is falling, but only slowly.

24.5   Find the gradient function of $y = 1$.

Solution   Before we use the formula to calculate the gradient function let us think about the graph of $y = 1$. Whatever the value of x, this function takes the value 1. Its graph must then be a horizontal line – it neither rises nor falls. We conclude that the gradient function must be zero, that is, $y' = 0$. To obtain the same result using the formula we must rewrite 1 as x^0. Then, using the formula with $n = 0$, we find that if $y = x^0$ then $y' = 0x^{0-1} = 0$.

The previous example illustrates an important result. Any constant function has a gradient function equal to zero because its graph is a horizontal line, and is therefore neither rising nor falling.

The technique introduced in this section has a very long history and as a consequence a wide variety of alternative names for the gradient function have emerged. For example, the gradient function is also called the **first derivative**, or simply the **derivative**. The process of obtaining this is also known as **differentiation.** Being asked to **differentiate** $y = x^5$, say, is equivalent to being asked to find its gradient function, y'. Furthermore, because the gradient function measures how rapidly a graph is changing, it is also referred to as the **rate of change** of y. When you meet these words you should realize that they are simply technical terms associated with finding the gradients of graphs.

It is possible to find the gradient function of a wide range of functions by simply referring to standard results. Some of these are given in Table 24.1. Note that for technical reasons, when finding the gradient functions of trigonometrical functions the angle x must always be measured in radians.

Worked examples

24.6   Use Table 24.1 to find the gradient function y' when y is

(a) $\sin x$ (b) $\sin 2x$ (c) $\cos 3x$ (d) e^x

Table 24.1 The gradient function of some common functions	$y = f(x)$	$y' = f'(x)$	Notes
	constant	0	
	x	1	
	x^2	$2x$	
	x^n	nx^{n-1}	
	e^x	e^x	
	e^{kx}	ke^{kx}	k is a constant
	$\sin x$	$\cos x$	
	$\cos x$	$-\sin x$	
	$\sin kx$	$k\cos kx$	k is a constant
	$\cos kx$	$-k\sin kx$	k is a constant
	$\ln kx$	$1/x$	k is a constant

Solution (a) Directly from the table we see that if $y = \sin x$ then its gradient function is given by $y' = \cos x$. This result occurs frequently and is worth remembering.

(b) Using the table and taking $k = 2$ we find that if $y = \sin 2x$ then $y' = 2\cos 2x$.

(c) Using the table and taking $k = 3$ we find that if $y = \cos 3x$ then $y' = -3\sin 3x$.

(d) Using the table we see directly that if $y = e^x$ then $y' = e^x$. Note that the exponential function e^x is very special because its gradient function y' is the same as y.

24.7 Find the gradient function of $y = e^{-x}$. Hence find the gradient of the graph of y at the point where $x = 1$.

Solution Noting that $e^{-x} = e^{-1x}$ and using Table 24.1 with $k = -1$ we find that if $y = e^{-x}$ then $y' = -1e^{-x} = -e^{-x}$. Using a calculator to evaluate this when $x = 1$ we find $y'(1) = -e^{-1} = -0.368$.

24.8 Find the gradient function of $y = \sin 4x$ where $x = 0.3$.

Solution Using Table 24.1 with $k = 4$ gives $y' = 4\cos 4x$. Remembering to measure x in radians we evaluate this when $x = 0.3$:

$$y'(0.3) = 4\cos 4(0.3) = 4\cos 1.2 = 1.4494$$

Self-assessment questions 24.2

1. What is the purpose of the gradient function?
2. State the formula for finding the gradient function when $y = x^n$.
3. State three alternative names for the gradient function.
4. State an alternative notation for y'.
5. What is the derivative of any constant? What is the graphical explanation of this answer?
6. When evaluating the derivative of the sine and cosine functions in which units must the angle x be measured – radians or degrees?
7. Which function is equal to its derivative?

Exercise 24.2

1. Find y' when y is given by

   (a) x^8 (b) x^7 (c) x^{-1} (d) x^{-5} (e) x^{13}
   (f) x^5 (g) x^{-2}

2. Find y' when y is given by

   (a) $x^{3/2}$ (b) $x^{5/2}$ (c) $x^{-1/2}$ (d) $x^{1/2}$ (e) $\sqrt{x}$
   (f) $x^{0.2}$

3. Sketch a graph of $y = 8$. State the gradient function, y'.

4. Find the first derivative of each of the following functions:

   (a) $y = x^{16}$ (b) $y = x^{0.5}$ (c) $y = x^{-3.5}$

   (d) $y = \dfrac{1}{x^{1/2}}$ (e) $y = \dfrac{1}{\sqrt{x}}$

5. Find the gradient of the graph of the function $y = x^4$ when

   (a) $x = -4$ (b) $x = -1$ (c) $x = 0$
   (d) $x = 1$ (e) $x = 4$

   Can you infer anything about the shape of the graph from this information?

6. Find the gradient of the graphs of each of the following functions at the points given:

   (a) $y = x^3$ at $x = -3$, at $x = 0$ and at $x = 3$
   (b) $y = x^4$ at $x = -2$ and at $x = 2$

   (c) $y = x^{1/2}$ at $x = 1$ and $x = 2$
   (d) $y = x^{-2}$ at $x = -2$ and $x = 2$

7. Sometimes a function is given in terms of variables other than x. This should pose no problems. Find the gradient functions of

   (a) $y = t^2$ (b) $y = t^{-3}$ (c) $y = t^{1/2}$
   (d) $y = t^{-1.5}$

8. Use Table 24.1 to find the gradient function y' when y is equal to

   (a) $\cos 7x$ (b) $\sin \frac{1}{2}x$ (c) $\cos \frac{x}{2}$
   (d) e^{4x} (e) e^{-3x}

9. Find the derivative of each of the following functions:

   (a) $y = \ln 3x$ (b) $y = \sin \frac{1}{3}x$
   (c) $y = e^{x/2}$

10. Find the gradient of the graph of the function $y = \sin x$ at the points where
    (a) $x = 0$ (b) $x = \pi/2$ (c) $x = \pi$

11. Table 24.1, although written in terms of the variable x, can still be applied when other variables are involved. Find y' when

    (a) $y = t^7$ (b) $y = \sin 3t$ (c) $y = e^{2t}$
    (d) $y = \ln 4t$ (e) $y = \cos \frac{t}{3}$

24.3   Some rules for finding gradient functions

Up to now the formulas we have used for finding gradient functions have applied to single terms. It is possible to apply these much more widely with the addition of three more rules. The first applies to the sum of two functions, the second to their difference and the third to a constant multiple of a function.

KEY POINT

Rule 1:   If $y = f(x) + g(x)$ then $y' = f'(x) + g'(x)$

In words this simply says that to find the gradient function of a sum of two functions we simply find the two gradient functions separately and add these together.

Worked example

24.9   Find the gradient function of $y = x^2 + x^4$.

Solution   The gradient function of x^2 is $2x$. The gradient function of x^4 is $4x^3$. Therefore,

$$\text{if } y = x^2 + x^4 \quad \text{then } y' = 2x + 4x^3$$

The second rule is really an extension of the first, but we state it here for clarity.

KEY POINT

Rule 2:   If $y = f(x) - g(x)$ then $y' = f'(x) - g'(x)$

Worked example

24.10   Find the gradient function of $y = x^5 - x^7$.

Solution   We find the gradient function of each term separately and subtract them. That is, $y' = 5x^4 - 7x^6$.

KEY POINT

Rule 3:   If $y = kf(x)$, where k is a number, then $y' = kf'(x)$

Worked examples

24.11   Find the gradient function of $y = 3x^2$.

Solution   This function is 3 times x^2. The gradient function of x^2 is $2x$. Therefore, using Rule 3, we find that

$$\text{if } y = 3x^2 \qquad \text{then } y' = 3(2x) = 6x$$

The rules can be applied at the same time. Consider the following example.

24.12   Find the derivative of $y = 4x^2 + 3x^{-3}$.

Solution   The derivative of $4x^2$ is $4(2x) = 8x$. The derivative of $3x^{-3}$ is $3(-3x^{-4}) = -9x^{-4}$. Therefore, if $y = 4x^2 + 3x^{-3}$, $y' = 8x - 9x^{-4}$.

24.13   Find the derivative of

(a) $y = 4 \sin t - 3 \cos 2t$ (b) $y = \dfrac{e^{2t}}{3} + 6 + \dfrac{\ln(2t)}{5}$

Solution   (a) We differentiate each quantity in turn using Table 24.1

$$y' = 4 \cos t - 3(-2 \sin 2t) = 4 \cos t + 6 \sin 2t$$

(b) Writing y as $\frac{1}{3}e^{2t} + 6 + \frac{1}{5}\ln(2t)$ we find

$$y' = \frac{2}{3}e^{2t} + 0 + \frac{1}{5}\left(\frac{1}{t}\right) = \frac{2e^{2t}}{3} + \frac{1}{5t}$$

Self-assessment questions 24.3

1. State the rules for finding the derivative of (a) $f(x) + g(x)$ (b) $f(x) - g(x)$ (c) $kf(x)$.

Exercise 24.3

1. Find the gradient function of each of the following:

(a) $y = x^6 - x^4$ (b) $y = 6x^2$

(c) $y = 9x^{-2}$ (d) $y = \frac{1}{2}x$

(e) $y = 2x^3 - 3x^2$

2. Find the gradient function of each of the following:

(a) $y = 2 \sin x$ (b) $y = 3 \sin 4x$
(c) $y = 7 \cos 9x + 3 \sin 4x$
(d) $y = e^{3x} - 5e^{-2x}$

3. Find the gradient function of $y = 3x^3 - 9x + 2$. Find the value of the gradient function at $x = 1$, $x = 0$ and $x = -1$. At which of these points is the curve steepest?

4. Write down a function which when differentiated gives $2x$. Are there any other functions you can differentiate to give $2x$?

5. Find the gradient of $y = 3 \sin 2t + 4 \cos 2t$ when $t = 2$. Remember to use radians.

continued

6. Find the gradient of $y = 2x - x^3 + e^{2x}$ when $x = 1$.

7. Find the gradient of $y = (2x - 1)^2$ when $x = 0$.

8. Show that the derivative of $f(x) = x^2 - 2x$ is 0 when $x = 1$.

9. Find the values of x such that the gradient of $y = \frac{x^3}{3} - x + 7$ is 0.

10. The function $y(x)$ is given by

$$y(x) = x + \cos x \qquad 0 \le x \le 2\pi$$

Find the value(s) of x where the gradient of y is 0.

24.4 Higher derivatives

In some applications, and in more advanced work, it is necessary to find the derivative of the gradient function itself. This is termed the **second derivative** and is written y'' and read 'y double dash'. Some books write y'' as $\dfrac{d^2y}{dx^2}$.

KEY POINT

y'' or $\dfrac{d^2y}{dx^2}$ is found by differentiating y'

It is a simple matter to find the second derivative by differentiating the first derivative. No new techniques or tables are required.

Worked examples

24.14 Find the first and second derivatives of $y = x^4$.

Solution From Table 24.1, if $y = x^4$ then $y' = 4x^3$. The second derivative is found by differentiating the first derivative. Therefore

$$\text{if } y' = 4x^3 \qquad \text{then } y'' = 4(3x^2) = 12x^2$$

24.15 Find the first and second derivatives of $y = 3x^2 - 7x + 2$.

Solution Using Table 24.1 we find $y' = 6x - 7$. The derivative of the constant 2 equals 0. Differentiating again we find $y'' = 6$, because the derivative of the constant -7 equals 0.

24.16 If $y = \sin x$, find

(a) $\dfrac{dy}{dx}$ (b) $\dfrac{d^2y}{dx^2}$

Solution   (a) Recall that dy/dx is the first derivative of y. From Table 24.1 this is given by $dy/dx = \cos x$.

(b) d^2y/dx^2 is the second derivative of y. This is found by differentiating the first derivative. From Table 24.1 we find that the derivative of $\cos x$ is $-\sin x$, and so $d^2y/dx^2 = -\sin x$.

Self-assessment questions 24.4

1. Explain how the second derivative of y is found.
2. Give two alternative notations for the second derivative of y.

Exercise 24.4

1. Find the first and second derivatives of the following functions:

   (a) $y = x^2$ (b) $y = 3x^8$ (c) $y = 9x^5$
   (d) $y = 3x^{1/2}$ (e) $y = 3x^4 + 5x^2$
   (f) $y = 9x^3 - 14x + 11$
   (g) $y = x^{1/2} + 4x^{-1/2}$

2. Find the first and second derivatives of

   (a) $y = \sin 3x + \cos 2x$ (b) $y = e^x + e^{-x}$

3. Find the second derivative of $y = \sin 4x$. Show that the sum of $16y$ and y'' is zero, that is $y'' + 16y = 0$.

4. If $y = x^4 - 2x^3 - 36x^2$, find the values of x for which $y'' = 0$.

5. Determine y'' given y is

   (a) $2 \sin 3x + 4 \cos 2x$
   (b) $\sin kt$ $\quad$ k constant
   (c) $\cos kt$ $\quad$ k constant
   (d) $A \sin kt + B \cos kt$
   $\quad$ A, B, k constants

6. Find y'' when y is given by

   (a) $6 + e^{3x}$ (b) $2 + 3x + 2e^{4x}$
   (c) $e^{2x} - e^{-2x}$ (d) $\frac{1}{e^x}$ (e) $e^x(e^x + 3)$

7. Given $y = 1000 + 200x + 6x^2 - x^3$, find the value(s) of x for which $y'' = 0$.

8. Given $y = 2e^x + \sin 2x$, evaluate y'' when $x = 1$.

9. Given $y = 2 \sin 4t - 5 \cos 2t$, evaluate y'' when $t = 1.2$.

24.5   Finding maximum and minimum points of a curve

Consider the graph sketched in Figure 24.3. There are a number of important points marked on this graph, all of which have one thing in common. At each point A, B, C and D, the gradient of the graph is zero. If you were travelling from the left to right then at these points the graph would appear to be flat. Points where the gradient is zero are known as **stationary points**. To the left of point A, the curve is rising; to its right the curve is falling. A point such as A is called a **maximum** turning point or simply a **maximum**. To the left of point C the curve is falling; to its right the curve is rising. A point such as C is called a **minimum turning point**, or simply a **minimum**. Points B and D are known as **points of inflexion**. At these points the slope of the curve is momentarily zero but then the curve continues rising or falling as before.

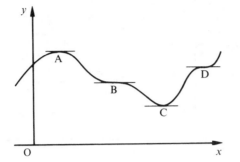

Figure 24.3
The curve has a maximum at A, a minimum at C and points of inflexion at B and D

Because at all these points the gradient is zero, they can be located by looking for values of x which make the gradient function zero.

KEY POINT

Stationary points are located by setting the gradient function equal to zero, that is, $y' = 0$.

Worked example

24.17   Find the stationary points of $y = 3x^2 - 6x + 8$.

Solution   We first determine the gradient function y' by differentiating y. This is found to be $y' = 6x - 6$. Stationary points occur when the gradient is zero, that is, when $y' = 6x - 6 = 0$. From this

$$6x - 6 = 0$$
$$6x = 6$$
$$x = 1$$

When $x = 1$ the gradient is zero, and therefore $x = 1$ is a stationary point. At this stage we cannot tell whether to expect a maximum, minimum or point of inflexion; all we know is that one of these occurs when $x = 1$. However, a sketch of the graph of $y = 3x^2 - 6x + 8$ is shown in Figure 24.4 which reveals that the point is a minimum.

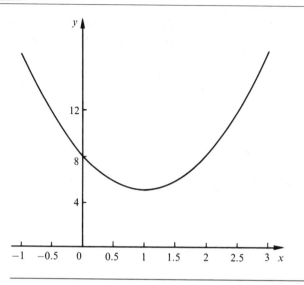

Figure 24.4
There is a minimum
turning point at $x = 1$

There are a number of ways to determine the nature of a stationary point once its location has been found. One way, as we have seen, is to sketch a graph of the function. Two alternative methods are now described.

Finding the nature of a stationary point by looking at the gradient on either side

At a stationary point we know that the gradient is zero. If this stationary point is a minimum, then we see from Figure 24.5 that, as we move from left to right, the gradient changes from negative to zero to positive. Therefore a little to the left of a minimum the gradient function will be negative; a little to the right it will be positive. Conversely, if the stationary point is a maximum, as we move from left to right, the gradient

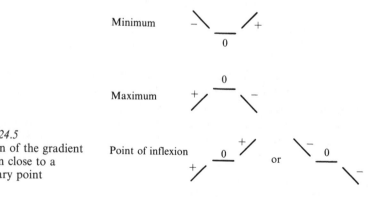

Figure 24.5
The sign of the gradient
function close to a
stationary point

changes from positive to zero to negative. A little to the left of a maximum the gradient function will be positive; a little to the right it will be negative.

The behaviour of the gradient function close to a point of inflexion is also shown. The sign of the gradient function is the same on both sides of a point of inflexion. With this information we can determine the nature of a stationary point without first plotting a graph.

Worked examples

24.18 Find the location and nature of the stationary points of $y = 2x^3 - 6x^2 - 18x$.

Solution In order to find the location of the stationary points we must calculate the gradient function. This is $y' = 6x^2 - 12x - 18$. At a stationary point $y' = 0$ and so the equation which must be solved is $6x^2 - 12x - 18 = 0$. This quadratic equation is solved as follows:

$$6x^2 - 12x - 18 = 0$$

factorizing:

$$6(x^2 - 2x - 3) = 0$$
$$6(x + 1)(x - 3) = 0$$
$$\text{so that } x = -1 \text{ and } 3$$

The stationary points are located at $x = -1$ and $x = 3$. We now find their nature.

When $x = -1$ A little to the left of $x = -1$, say at $x = -2$, we calculate the gradient of the graph using the gradient function. That is, $y'(-2) = 6(-2)^2 - 12(-2) - 18 = 24 + 24 - 18 = 30$. Therefore the graph is rising when $x = -2$. A little to the right of $x = -1$, say at $x = 0$, we calculate the gradient of the graph using the gradient function. That is, $y'(0) = 6(0)^2 - 12(0) - 18 = -18$. Therefore the graph is falling when $x = 0$. From this information we conclude that the turning point at $x = -1$ must be a maximum.

When $x = 3$ A little to the left of $x = 3$, say at $x = 2$, we calculate the gradient of the graph using the gradient function. That is, $y'(2) = 6(2)^2 - 12(2) - 18 = 24 - 24 - 18 = -18$. Therefore the graph is falling when $x = 2$. A little to the right of $x = 3$, say at $x = 4$, we calculate the gradient of the graph using the gradient function. That is, $y'(4) = 6(4)^2 - 12(4) - 18 = 96 - 48 - 18 = 30$. Therefore the graph is rising when $x = 4$. From this information we conclude that the turning point at $x = 3$ must be a minimum.

To show this behaviour a graph of $y = 2x^3 - 6x^2 - 18x$ is shown in Figure 24.6 where these turning points can be clearly seen.

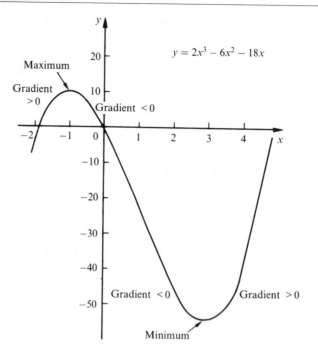

Figure 24.6
Graph of
$y = 2x^3 - 6x^2 - 18x$

24.19 Find the location and nature of the stationary points of
$y = x^3 - 3x^2 + 3x - 1$.

Solution First we find the gradient function: $y' = 3x^2 - 6x + 3$. At a stationary
point $y' = 0$ and so

$$3x^2 - 6x + 3 = 0$$
$$3(x^2 - 2x + 1) = 0$$
$$3(x - 1)(x - 1) = 0$$
$$\text{and so } x = 1$$

We conclude that there is only one stationary point, and this is at
$x = 1$. To determine its nature we look at the gradient function on
either side of this point. At $x = 0$, say, $y'(0) = 3$. At $x = 2$, say,
$y'(2) = 3(2^2) - 6(2) + 3 = 3$. The gradient function changes from
positive to zero to positive as we move through the stationary point.
We conclude that the stationary point is a point of inflexion. For
completeness a graph is sketched in Figure 24.7.

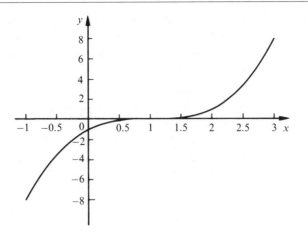

Figure 24.7
Graph of
$y = x^3 - 3x^2 + 3x - 1$
showing the point of
inflexion

The second derivative test for maximum and minimum points

A simple test exists which uses the second derivative for determining the nature of a stationary point once it has been located.

KEY POINT

> If y'' is positive at the stationary point, the point is a minimum.
> If y'' is negative at the stationary point, the point is a maximum.
> If y'' is equal to zero, this test does not tell us anything and the previous method should be used.

Worked example

24.20   Locate the stationary points of

$$y = \frac{x^3}{3} + \frac{x^2}{2} - 12x + 5$$

and determine their nature using the second derivative test.

Solution   We first find the gradient function: $y' = x^2 + x - 12$. Setting this equal to zero to locate the stationary points we find

$$x^2 + x - 12 = 0$$
$$(x + 4)(x - 3) = 0$$
$$x = -4 \text{ and } 3$$

There are two stationary points: at $x = -4$ and at $x = 3$. To apply the second derivative test we need to find y''. This is found by differentiating y' to give $y'' = 2x + 1$. We now evaluate this at each stationary point in turn.

$x = -4$ When $x = -4$, $y''(-4) = 2(-4) + 1 = -7$. This is negative and we conclude from the second derivative test that the point is a maximum point.

$x = 3$ When $x = 3$, $y''(3) = 2(3) + 1 = 7$. This is positive and we conclude from the second derivative test that the point is a minimum point. Remember, if in doubt, these results could be confirmed by sketching a graph of the function.

Self-assessment questions 24.5

1. State the three types of stationary point.
2. State the second derivative test for determining the nature of a stationary point.
3. Under what condition will the second derivative test fail?

Exercise 24.5

1. Determine the location and nature of any stationary point of the following functions:

   (a) $y = x^2 + 1$ (b) $y = -x^2$
   (c) $y = 2x^3 + 9x^2$ (d) $y = -2x^3 + 27x^2$
   (e) $y = x^3 - 3x^2 + 3x + 1$

2. Determine the location of the maximum and minimum points of $y = \sin x$ $0 \le x \le 2\pi$.

3. Find the maximum and minimum points of $y = e^x - x$.

4. Find the maximum and minimum points of $y = 2x^3 - 3x^2 - 12x + 1$.

5. Determine the location and nature of the stationary points of $y = -\frac{x^3}{3} + x$.

6. Determine the location and nature of the stationary points of $y = x^5 - 5x + 1$.

Test and assignment exercises 24

1. Find y' when y equals

   (a) $7x$ (b) x^{15} (c) $x^{3/2}$ (d) $3x^{-1}$ (e) $5x^4$
   (f) $16x^{-3}$

2. Find y' when y equals

   (a) $\dfrac{x+1}{x}$ (b) $3x^2 - 7x + 14$ (c) $\sin 3x$

   (d) $e^{5x} + e^{4x}$ (e) $3e^{3x}$

3. Find the first and second derivatives of

   (a) $f(x) = e^{-3x}$ (b) $f(x) = \ln 2x$

   (c) $f(x) = x + \dfrac{x^2}{2} + \dfrac{x^3}{3}$

   (d) $f(x) = -15$ (e) $f(x) = 3x$

4. Find y' when

   (a) $y = 3t - 7$ (b) $y = 16t^2 + 7t + 9$

5. Find the first derivative of each of the following functions:

   (a) $f(x) = x(x+1)$ (b) $f(x) = x(x-2)$
   (c) $f(x) = x^2(x+1)$
   (d) $f(x) = (x-2)(x+3)$
   (e) $f(t) = 3t^2 + 7t + 11$
   (f) $f(t) = 7(t-1)(t+1)$

6. Find the first and second derivatives of each of the following:

   (a) $2x^2$ (b) $4x^3$ (c) $8x^7 - 5x^2$ (d) $\dfrac{1}{x}$

7. Find the location and nature of the maxima and minima of
   $y = 2x^3 + 15x^2 - 84x$.

8. Find a function which when differentiated gives $\frac{1}{2}x^3$.

9. Find the gradient of y at the specified points:

   (a) $y = 3x + x^2$ at $x = -1$.

   (b) $y = 4\cos 3x + \dfrac{\sin 4x}{2}$ at $x = 0.5$.

   (c) $y = 3e^{2x} - 2e^{-3x}$ at $x = 0.1$.

10. Evaluate the second derivative of y at the specified point

    (a) $y = \frac{1}{2}\sin 4x$ $x = 1$
    (b) $y = e^{-2x} + x^2$ $x = 0$
    (c) $y = 3\ln 5x$ $x = 1$

11. Find the value(s) of t for which y' is zero

    (a) $y = t^3 - 12t^2 + 1$
    (b) $y = \ln(2t) - 2t^2$

12. Determine the location and nature of the stationary points of

    (a) $y = x^4 - 108x$
    (b) $y = \ln 2x + \frac{1}{x}$

25 Integration and areas under curves

Objectives

This chapter

- introduces the reverse process of differentiation which is called 'integration'
- explains what is meant by an 'indefinite integral' and a 'definite integral'
- shows how integration can be used to find the area under a curve

25.1 Introduction

In some applications it is necessary to find the area under a curve and above the x axis between two values of x, say a and b. Such an area is shown shaded in Figure 25.1. In this chapter we shall show how this area can be found using a technique known as **integration**. However, before we do this it is necessary to consider two types of integration: **indefinite** and **definite** integration.

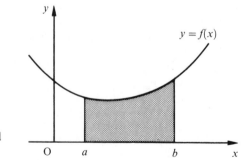

Figure 25.1
The shaded area is found by integration

25.2   Indefinite integration: the reverse of differentiation

Suppose we ask 'what function can we differentiate to give $2x$?' In fact, there are lots of correct answers to this question. We can use our knowledge of differentiation to find these. The function $y = x^2$ has derivative $dy/dx = 2x$, so clearly x^2 is one answer. Furthermore, $y = x^2 + 13$ also has derivative $2x$ as do $y = x^2 + 5$ and $y = x^2 - 7$, because differentiation of any constant equals zero. Indeed any function of the form $y = x^2 + c$ where c is a constant has derivative equal to $2x$. The answer to the original question 'what function can we differentiate to give $2x$?' is then any function of the form $y = x^2 + c$. Here we have started with the derivative of a function and then found the function itself. This is differentiation in reverse. We call this process **integration** or more precisely **indefinite integration**. We say that $2x$ has been **integrated** with respect to x to give $x^2 + c$. The quantity $x^2 + c$ is called the **indefinite integral** of $2x$. Both differentiation and indefinite integration are shown schematically in Figure 25.2.

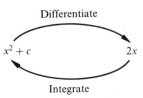

Figure 25.2
The indefinite integral of
$2x$ is $x^2 + c$

We now introduce a notation used to state this compactly. We write

$$\int 2x \, dx = x^2 + c$$

The $\int$ sign indicates that the reverse process to differentiation is to be performed. Following the function to be integrated is always a term of the form dx, which indicates the independent variable, in this case x. When finding indefinite integrals there will always be a constant known as the **constant of integration**. Every time we integrate, the constant of integration must be added.

As a further illustration, if $y = \sin x$ then we know from a table of derivatives that $y' = \cos x$. We can write this as

$$\int \cos x \, dx = \sin x + c$$

To find integrals it is often possible to use a table of derivatives such as that on page 261, and to read the table from right to left. However, it is more helpful to have a table of integrals such as that given in Table 25.1. As with the table of derivatives given in Chapter 24, when dealing with

Table 25.1
Table of indefinite
integrals

$f(x)$	$\int f(x)\, dx$, all '+c'	Notes
$2x$	x^2	
x	$\frac{1}{2}x^2$	
k, constant	kx	
x^n	$\frac{1}{n+1}x^{n+1}$	$n \neq -1$
$\frac{1}{x}$	$\ln x$	
e^x	e^x	
e^{kx}	$\frac{e^{kx}}{k}$	k is a constant
$\sin x$	$-\cos x$	
$\cos x$	$\sin x$	
$\sin kx$	$-\frac{\cos kx}{k}$	k is a constant
$\cos kx$	$\frac{\sin kx}{k}$	k is a constant

trigonometrical functions the angle x must always be measured in radians. The rule

$$\int x^n dx = \frac{1}{n+1}x^{n+1} + c$$

in the table applies for all values of n except $n = -1$.

Worked examples

25.1 Use the table to find the following indefinite integrals:

(a) $\int x^4\, dx$ (b) $\int x^{-2}\, dx$

Solution (a) From the table we note that

$$\int x^n\, dx = \frac{1}{n+1}x^{n+1} + c$$

Therefore, using this formula with $n = 4$, we find

$$\int x^4\, dx = \frac{1}{4+1}x^{4+1} + c = \frac{1}{5}x^5 + c$$

(b) Similarly, using the formula with $n = -2$ we find

$$\int x^{-2}dx = \frac{1}{-2+1}x^{-2+1} + c = \frac{1}{-1}x^{-1} + c = -x^{-1} + c$$

Note that because integration is the reverse of differentiation, in both cases the answer can be checked by differentiating.

25.2   Use Table 25.1 to find

(a) $\int e^x\, dx$ (b) $\int x^{-1}dx$.

Solution   (a) Directly from the table we see that $\int e^x dx = e^x + c$.

(b) Noting that $x^{-1} = 1/x$, Table 25.1 gives $\int x^{-1}dx = \ln x + c$. It is important to realize that the formula for integrating x^n cannot be used here; the formula specifically excludes use of the value $n = -1$.

25.3   Use Table 25.1 to find the following indefinite integrals:

(a) $\int e^{2x}dx$ (b) $\int \cos 3x\, dx$

Solution   (a) Using Table 25.1 we note that

$$\int e^{kx}dx = \frac{e^{kx}}{k} + c$$

Using this result with $k = 2$ we find

$$\int e^{2x}dx = \frac{e^{2x}}{2} + c$$

(b) Using Table 25.1 we note that

$$\int \cos kx dx = \frac{\sin kx}{k} + c$$

Using this result with $k = 3$ we find

$$\int \cos 3x dx = \frac{\sin 3x}{3} + c$$

25.4   Find $\int \sqrt{x}\, dx$.

Solution   Recall that $\sqrt{x}$ is the same as $x^{1/2}$. Therefore

$$\int \sqrt{x}dx = \int x^{1/2}dx = \frac{x^{3/2}}{3/2} + c = \frac{2}{3}x^{3/2} + c$$

Self-assessment questions 25.2

1. Explain what you understand by an 'indefinite integral'.
2. Why is it always necessary to add a constant of integration when finding an indefinite integral?

Exercise 25.2

1. Use Table 25.1 to find the following indefinite integrals:

   (a) $\int x \, dx$ (b) $\int 2x \, dx$ (c) $\int x^5 \, dx$

   (d) $\int x^7 \, dx$ (e) $\int x^{-5} \, dx$ (f) $\int x^{-1/2} \, dx$

   (g) $\int x^0 \, dx$ (h) $\int x^{-1/4} \, dx$

2. Find:

   (a) $\int 9 \, dx$ (b) $\int \frac{1}{2} \, dx$ (c) $\int -7 \, dx$

   (d) $\int 0.5 \, dx$

3. Use Table 25.1 to find the following integrals:

   (a) $\int \sin 5x \, dx$ (b) $\int \sin \frac{1}{2} x \, dx$

   (c) $\int \cos \frac{x}{2} \, dx$ (d) $\int e^{-3x} \, dx$

   (e) $\int \sin 0.25x \, dx$ (f) $\int e^{-0.5x} \, dx$

   (g) $\int e^{x/2} \, dx$

4. Find the following integrals:

   (a) $\int \cos 5x \, dx$ (b) $\int \sin 4x \, dx$

   (c) $\int e^{6t} \, dt$ (d) $\int \cos\left(\frac{t}{3}\right) \, dt$

5. Find the following integrals:

   (a) $\int \cos 3y \, dy$ (b) $\int \frac{1}{\sqrt{x}} \, dx$

   (c) $\int \sin\left(\frac{3t}{2}\right) \, dt$ (d) $\int \sin(-x) \, dx$

25.3   Some rules for finding other indefinite integrals

Up to now the formulae we have used for finding indefinite integrals have applied to single terms. More complicated functions can be integrated if we make use of three simple rules. The first applies to the sum of two functions, the second to their difference and the third to a constant multiple of a function.

KEY POINT

Rule 1:

$$\int f(x) + g(x) \, dx = \int f(x) \, dx + \int g(x) \, dx$$

In words, this simply says that to find the integral of a sum of two functions we simply find the integrals of the two separate parts and add these together.

Worked example

25.5   Find the integral $\int x^2 + x^3\, dx$.

Solution   From Table 25.1 the integral of x^2 is $x^3/3 + c_1$. The integral of x^3 is $x^4/4 + c_2$. Note that we have called the constants of integration c_1 and c_2. The integral of $x^2 + x^3$ is simply the sum of these separate integrals, that is,

$$\int x^2 + x^3\, dx = \frac{x^3}{3} + \frac{x^4}{4} + c_1 + c_2$$

The two constants can be combined into a single constant c, say, so that we would usually write

$$\int x^2 + x^3\, dx = \frac{x^3}{3} + \frac{x^4}{4} + c$$

The second rule is really an extension of the first, but we state it here for clarity.

KEY POINT

> *Rule 2:*
>
> $$\int f(x) - g(x)\, dx = \int f(x)\, dx - \int g(x)\, dx$$

In words, this says that to find the integral of the difference of two functions we simply find the integrals of the two separate parts and find their difference.

Worked examples

25.6   Find the integral $\int x^3 - x^6\, dx$.

Solution   From Table 25.1 the integral of x^3 is $x^4/4 + c_1$. The integral of x^6 is $x^7/7 + c_2$. Note once again that we have called the constants of integration c_1 and c_2. The integral of $x^3 - x^6$ is simply the difference of these separate integrals, that is,

$$\int x^3 - x^6\, dx = \frac{x^4}{4} - \frac{x^7}{7} + c_1 - c_2$$

As before the two constants can be combined into a single constant c, say, so that we would usually write

$$\int x^3 - x^6\, dx = \frac{x^4}{4} - \frac{x^7}{7} + c$$

25.7   Find $\int \sin 2x - \cos 2x \, dx$.

Solution   Using the second rule we can find the integral of each term separately. From Table 25.1 $\int \sin 2x \, dx = -(\cos 2x)/2 + c_1$ and $\int \cos 2x \, dx = (\sin 2x)/2 + c_2$. Therefore

$$\int \sin 2x - \cos 2x \, dx = \left(-\frac{\cos 2x}{2} + c_1 \right) - \left(\frac{\sin 2x}{2} + c_2 \right)$$

$$= -\frac{\cos 2x}{2} - \frac{\sin 2x}{2} + c$$

where we have combined the two constants into a single constant c.

KEY POINT

Rule 3: If k is a constant, then

$$\int k \times f(x) \, dx = k \times \int f(x) \, dx$$

In words, the integral of k times a function is simply k times the integral of the function. This means that a constant factor can be taken through the integral sign and written at the front, as in the following example.

Worked example

25.8   Find $\int 3x^4 \, dx$.

Solution   Using the third rule we can write

$$\int 3x^4 \, dx = 3 \times \int x^4 \, dx$$

taking the constant factor out. Now $\int x^4 \, dx = x^5/5 + c$ so that

$$\int 3x^4 \, dx = 3 \times \int x^4 \, dx = 3 \times \left(\frac{x^5}{5} + c \right)$$

$$= \frac{3x^5}{5} + 3c$$

It is common practice to write the constant $3c$ as simply another constant K, say, to give $\int 3x^4 \, dx = 3x^5/5 + K$.

Self-assessment questions 25.3

1. State the rules for finding the indefinite integrals $\int f(x) + g(x) \, dx$, $\int f(x) - g(x) \, dx$ and $\int k \times f(x) \, dx$ where k is a constant.

Exercise 25.3

1. Using Table 25.1 and the rules of integration given in this section, find

(a) $\int x^2 + x \, dx$ (b) $\int x^2 + x + 1 \, dx$

(c) $\int 4x^6 \, dx$ (d) $\int 5x + 7 \, dx$

(e) $\int \frac{1}{x} \, dx$ (f) $\int \frac{1}{x^2} \, dx$ (g) $\int \frac{3}{x} + \frac{7}{x^2} \, dx$

(h) $\int x^{1/2} \, dx$ (i) $\int 4x^{-1/2} \, dx$ (j) $\int \frac{1}{\sqrt{x}} \, dx$

(k) $\int 3x^3 + 7x^2 \, dx$ (l) $\int 2x^4 - 3x^2 \, dx$

(m) $\int x^5 - 7 \, dx$

2. Find the following integrals:

(a) $\int 3e^x \, dx$ (b) $\int \frac{1}{2} e^x \, dx$

(c) $\int 3e^x - 2e^{-x} \, dx$ (d) $\int 5e^{2x} \, dx$

(e) $\int \sin x - \cos 3x \, dx$

3. Integrate each of the following functions with respect to x:

(a) $3x^2 - 2e^x$ (b) $\frac{x}{2} + 1$

(c) $2(\sin 3x - 4\cos 3x)$

(d) $3\sin 2x + 5\cos 3x$ (e) $\frac{5}{x}$

4. Find $\int \frac{5}{t} - t^2 \, dt$

5. Find the following integrals:

(a) $\int 3\sin 2t + 4\cos 2t \, dt$

(b) $\int -\sin 3x - 2\cos 4x \, dx$

(c) $\int 1 + 2\sin\left(\frac{x}{2}\right) \, dx$

(d) $\int \frac{1}{2} - \frac{1}{3}\cos\left(\frac{x}{3}\right) \, dx$

6. Find the following integrals:

(a) $\int e^x(1 + e^x) \, dx$

(b) $\int (x+2)(x+3) \, dx$

(c) $\int 5\cos 2t - 2\sin 5t \, dt$

(d) $\int \frac{2}{x}(x+3) \, dx$

25.4   Definite integrals

There is a second form of integral known as a definite integral. This is written as

$$\int_a^b f(x) \, dx$$

and is called the **definite integral** of $f(x)$ between the **limits** $x = a$ and $x = b$. This definite integral looks just like the indefinite integral $\int f(x) \, dx$ except values a and b are placed below and above the integral sign. The value $x = a$ is called the **lower limit** and $x = b$ is called the **upper limit**.

Unlike the indefinite integral, which is a function depending upon x, the definite integral has a specific numerical value. To find this value we first find the indefinite integral of $f(x)$, and then calculate

its value at the upper limit $-$ its value at the lower limit

The resulting number is the value of the definite integral. To see how this value is obtained consider the following examples.

Worked examples

25.9   Evaluate the definite integral $\int_0^1 x^3 \, dx$.

Solution   We first find the indefinite integral $\int x^3 \, dx$. This is

$$\int x^3 \, dx = \frac{x^4}{4} + c$$

This is evaluated at the upper limit where $x = 1$, giving $\frac{1}{4} + c$, and at the lower limit where $x = 0$, giving $0 + c$, or simply c. The difference is found:

value at the upper limit $-$ value at the lower limit

$$= \left(\frac{1}{4} + c\right) - (c) = \frac{1}{4}$$

We note that the constant c has cancelled out; this will always happen and so for definite integrals it is usually omitted altogether. The calculation is written more compactly as

When evaluating definite integrals, the constant of integration may be omitted

$$\int_0^1 x^3 \, dx = \left[\frac{x^4}{4}\right]_0^1 \quad \text{the indefinite integral of } x^3 \text{ is written in square brackets}$$
$$= \frac{1}{4} - 0$$
$$= \frac{1}{4}$$

Note that the definite integral $\int_0^1 x^3 \, dx$ has a definite numerical value, $\frac{1}{4}$.

25.10   Find the definite integral $\int_{-1}^{1} 3x^2 + 7 \, dx$.

Solution

$$\int_{-1}^{1} 3x^2 + 7 \, dx = [x^3 + 7x]_{-1}^{1} \qquad \begin{array}{l} \text{(the indefinite integral of } 3x^2 + 7 \\ \text{is written in square brackets)} \end{array}$$

$$= (\text{value of } x^3 + 7x \text{ when } x = 1)$$
$$\quad - (\text{value of } x^3 + 7x \text{ when } x = -1)$$
$$= (1^3 + 7) - \{(-1)^3 - 7\}$$
$$= (8) - (-8)$$
$$= 16$$

25.11   Evaluate $\int_{0}^{1} e^x + 1 \, dx$.

Solution

$$\int_{0}^{1} e^x + 1 \, dx = [e^x + x]_{0}^{1}$$

$$= (e^1 + 1) - (e^0 + 0)$$
$$= e^1 + 1 - 1$$
$$= e^1$$
$$= 2.718$$

25.12   Find $\int_{\pi/4}^{\pi/2} \cos 2x \, dx$.

Solution   Note that the calculation must be carried out using radians and not degrees.

$$\int_{\pi/4}^{\pi/2} \cos 2x \, dx = \left[\frac{\sin 2x}{2} \right]_{\pi/4}^{\pi/2}$$

$$= \left(\frac{\sin \pi}{2} \right) - \left(\frac{\sin \pi/2}{2} \right)$$

$$= 0 - \frac{1}{2}$$

$$= -\frac{1}{2}$$

Self-assessment questions 25.4

1. Explain the important way in which a definite integral differs from an indefinite integral.

Exercise 25.4

1. Evaluate the following definite integrals:

(a) $\displaystyle\int_{-1}^{1} 7x\,dx$ (b) $\displaystyle\int_{0}^{3} 7x\,dx$ (c) $\displaystyle\int_{-1}^{1} x^2\,dx$

(d) $\displaystyle\int_{-1}^{1} (-x^2)\,dx$ (e) $\displaystyle\int_{0}^{2} x^2\,dx$

(f) $\displaystyle\int_{0}^{3} x^2 + 4x\,dx$ (g) $\displaystyle\int_{0}^{\pi} \cos x\,dx$

(h) $\displaystyle\int_{1}^{2} \frac{1}{x}\,dx$ (i) $\displaystyle\int_{0}^{\pi} -\sin x\,dx$ (j) $\displaystyle\int_{1}^{4} 13\,dx$

(k) $\displaystyle\int_{0}^{\pi} x^2 + \sin x\,dx$ (l) $\displaystyle\int_{1}^{3} 5x^2\,dx$

(m) $\displaystyle\int_{0}^{1} (-x)\,dx$ (n) $\displaystyle\int_{1}^{3} x^{-2}\,dx$

(o) $\displaystyle\int_{0}^{1} \sin 3t\,dt$

2. Evaluate the following definite integrals:

(a) $\displaystyle\int_{0}^{2} 3e^x + 1\,dx$ (b) $\displaystyle\int_{0}^{1} 2e^{2x} + \sin x\,dx$

(c) $\displaystyle\int_{1}^{2} 2\cos 2t - 3\sin 2t\,dt$

(d) $\displaystyle\int_{1}^{3} \frac{2}{x} + \frac{x}{2}\,dx$

3. Evaluate the following definite integrals:

(a) $\displaystyle\int_{0}^{2} x(2x+3)\,dx$

(b) $\displaystyle\int_{-1}^{1} 3e^{2x} - \frac{3}{e^{2x}}\,dx$

(c) $\displaystyle\int_{0}^{\pi} 2\sin\left(\frac{t}{2}\right) - 4\cos 2t\,dt$

(d) $\displaystyle\int_{1}^{2} \sin 3x - \frac{3}{x}\,dx$

4. Evaluate the following:

(a) $\displaystyle\int_{1}^{2} \frac{2}{x}\left(3 + \frac{1}{x}\right)dx$

(b) $\displaystyle\int_{0}^{1} e^{-2x}(e^x - 2e^{-x})\,dx$

(c) $\displaystyle\int_{0}^{0.5} \sin \pi t + \cos \pi t\,dt$

(d) $\displaystyle\int_{0}^{1} (2x-3)^2\,dx$

25.5   Areas under curves

As we stated at the beginning of this chapter integration can be used to find the area underneath a curve and above the x axis. Consider the shaded area in Figure 25.3 which lies under the curve $y = f(x)$, above the x axis, between the values $x = a$ and $x = b$. Provided all of the required area lies above the x axis, this area is given by the following formula.

KEY POINT

$$\text{area} = \int_{a}^{b} f(x)\,dx$$

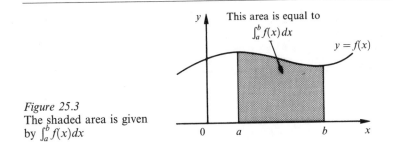

Figure 25.3
The shaded area is given
by $\int_a^b f(x)\,dx$

Worked examples

25.13 Find the area under $y = 7x + 3$ between $x = 0$ and $x = 2$.

Solution This area is shown in Figure 25.4. To find this area we must evaluate the definite integral $\int_0^2 7x + 3\,dx$:

$$\text{area} = \int_0^2 7x + 3\,dx = \left[\frac{7x^2}{2} + 3x\right]_0^2$$

$$= \left\{\frac{7(2)^2}{2} + 3(2)\right\} - (0)$$

$$= 20$$

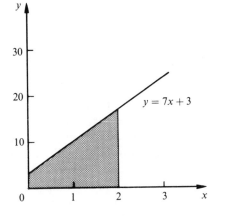

Figure 25.4
The shaded area is given
by $\int_0^2 7x + 3\,dx$

The required area is 20 square units. Convince yourself that this is correct by calculating the area of the shaded trapezium using the formula given in Chapter 23.

25.14   Find the area under the curve $y = x^2$ and above the x axis, between $x = 2$ and $x = 5$.

Solution   The required area is shown in Figure 25.5.

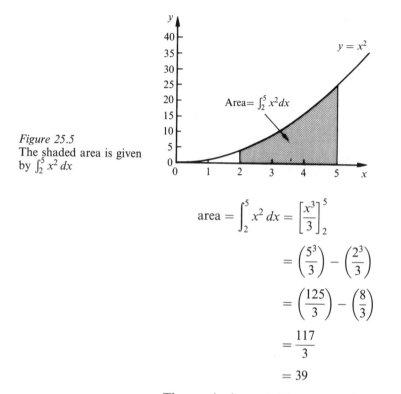

Figure 25.5
The shaded area is given by $\int_2^5 x^2 \, dx$

$$area = \int_2^5 x^2 \, dx = \left[\frac{x^3}{3}\right]_2^5$$

$$= \left(\frac{5^3}{3}\right) - \left(\frac{2^3}{3}\right)$$

$$= \left(\frac{125}{3}\right) - \left(\frac{8}{3}\right)$$

$$= \frac{117}{3}$$

$$= 39$$

The required area is 39 square units.

Note that the definite integral $\int_a^b f(x)dx$ only represents the area under $y = f(x)$ between $x = a$ and $x = b$ when all this area lies above the x axis. If between a and b the curve dips below the x axis, the result obtained will not be the area bounded by the curve. We shall see how to overcome this now.

Finding areas when some or all lie below the x axis

Consider the graph of $y = x^3$ shown in Figure 25.6. Suppose we want to know the area enclosed by the graph and the x axis between $x = -1$ and $x = 1$. Evaluating $\int_{-1}^1 x^3 \, dx$ we find:

$$\int_{-1}^{1} x^3 \, dx = \left[\frac{x^4}{4}\right]_{-1}^{1}$$

$$= \left\{\frac{(1)^4}{4}\right\} - \left\{\frac{(-1)^4}{4}\right\}$$

$$= \frac{1}{4} - \frac{1}{4}$$

$$= 0$$

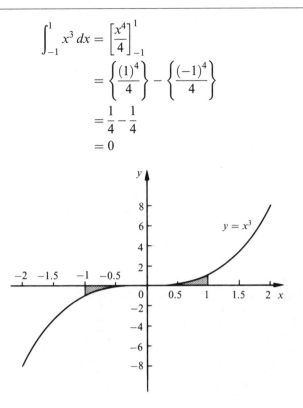

Figure 25.6
Graph of $y = x^3$

The result of finding the definite integral is 0, even though there is clearly a region enclosed by the curve and the x axis. The reason for this is that part of the curve lies below the x axis. Let us evaluate two separate integrals, one representing the area below the x axis and the other representing the area above the x axis. The area below the x axis is found from:

$$\int_{-1}^{0} x^3 \, dx = \left[\frac{x^4}{4}\right]_{-1}^{0}$$

$$= \left\{\frac{(0)^4}{4}\right\} - \left\{\frac{(-1)^4}{4}\right\}$$

$$= 0 - \left(\frac{1}{4}\right)$$

$$= -\frac{1}{4}$$

The result is negative, that is, $-\frac{1}{4}$. This means that the area bounded by the curve and the x axis between $x = -1$ and $x = 0$ is equal to $\frac{1}{4}$, the negative sign indicating that this area is below the x axis. The area above the x axis is found from:

$$\int_0^1 x^3\, dx = \left[\frac{x^4}{4}\right]_0^1$$

$$= \left\{\frac{(1)^4}{4}\right\} - \left\{\frac{(0)^4}{4}\right\}$$

$$= \frac{1}{4}$$

So the area between $x = 0$ and $x = 1$ is $\frac{1}{4}$. Note that we could have deduced this from the symmetry in the graph. The total area between $x = -1$ and $x = 1$ is therefore equal to $\frac{1}{4} + \frac{1}{4} = \frac{1}{2}$. The important point to note is that when some or all of the area lies below the x axis it is necessary to find the total area by finding the area of each part. It is therefore essential to draw a graph of the function to be integrated first.

Self-assessment questions 25.5

1. Explain how the area bounded by a curve and the x axis can be found using a definite integral. What precautions must be taken when doing this?

Exercise 25.5

1. Find the area bounded by the curve $y = x^3 + 3$, the x axis and the lines $x = 1$ and $x = 3$.

2. Find the area bounded by the curve $y = 4 - x^2$ and the x axis between $x = -1$ and $x = 4$.

3. Find the area under $y = e^{2t}$ from $t = 1$ to $t = 4$.

4. Find the area under $y = 1/t$ for $1 \le t \le 5$.

5. Find the area under $y = \sqrt{x}$ between $x = 1$ and $x = 2$.

6. Find the area under $y = \frac{1}{x}$ from $x = 1$ to $x = 4$.

7. Find the area under $y = \cos 2t$ from $t = 0$ to $t = 0.5$.

Test and assignment exercises 25

1. Find the following indefinite integrals:

   (a) $\int 3x\,dx$ (b) $\int 2x + 4\,dx$ (c) $\int x^2 + 7\,dx$

   (d) $\int 3x^2 + 7x - 7\,dx$ (e) $\int e^{1.5x}\,dx$

   (f) $\int \sin 5x - \cos 4x\,dx$ (g) $\int 5e^{-x}\,dx$

2. Evaluate the following definite integrals:

   (a) $\int_0^4 x^2\,dx$ (b) $\int_0^4 x^4\,dx$ (c) $\int_{-1}^2 3x^3\,dx$

   (d) $\int_0^1 x^2 + 3x + 4\,dx$ (e) $\int_{-2}^2 x^3 + 2\,dx$

   (f) $\int_{-3}^2 2x - 13\,dx$ (g) $\int_3^5 x + 2x + 3x^2\,dx$

   (h) $\int_2^7 5\,dx$ (i) $\int_1^2 11x^4\,dx$

3. Calculate the area bounded by the curve $y = x^2 - 4x + 3$ and the x axis from $x = 1$ to $x = 3$.

4. Sketch a graph of $y = \sin x$ from $x = 0$ to $x = \pi$. Calculate the area bounded by your curve and the x axis.

5. Find $\int 4e^{2t}\,dt$.

6. Find $\int 5 + x + x^2 + \dfrac{5}{x}\,dx$.

7. Find $\int_1^2 \dfrac{1}{x^2}\,dx$.

8. Calculate the area bounded by $y = e^{-x}$ and the x axis between

   (a) $x = 0$ and $x = 1$ (b) $x = 1$ and $x = 2$
   (c) $x = 0$ and $x = 2$

   Deduce that

   $$\int_0^2 e^{-x}\,dx = \int_0^1 e^{-x}\,dx + \int_1^2 e^{-x}\,dx$$

9. Find the area under $y = 2e^{-t} + t$ from $t = 0$ to $t = 2$.

10. Evaluate

    $$\int_1^2 \dfrac{3}{x^3}\,dx$$

11. Calculate the area above the x axis and below $y = 4 - x^2$.

12. Calculate the area above the positive x axis and below $y = e^{-x}$.

13. Evaluate the following:

    (a) $\int_0^\pi \sin 3t + 2\cos 3t\,dt$

    (b) $\int_0^\pi \cos\left(\dfrac{t}{2}\right) - \sin\left(\dfrac{t}{3}\right)\,dt$

    (c) $\int_0^1 \dfrac{1}{2}\sin 2t - \cos 3t\,dt$

14. Calculate $\int_{-1}^2 x^2(3 + x)\,dx$.

15. By using a suitable trigonometrical identity, find $\int 2\sin t \cos t\,dt$.

16. By using appropriate trigonometrical identities, evaluate the following definite integrals:

    (a) $\int_0^1 \sin^2 t + \cos^2 t\,dt$

    (b) $\int_0^1 \cos^2 t\,dt$

    (c) $\int_0^1 \sin^2 t\,dt$

26 Tables and charts

Objectives

This chapter

- explains the distinction between discrete and continuous data
- shows how raw data can be organized using a tally chart
- explains what is meant by a frequency distribution and a relative frequency distribution
- shows how data can be represented in the form of bar charts, pie charts, pictograms and histograms

26.1 Introduction to data

In the modern world, information from a wide range of sources is gathered, presented and interpreted. Most newspapers, television news programmes and documentaries contain vast numbers of facts and figures concerning almost every aspect of life, including environmental issues, the lengths of hospital waiting lists, crime statistics and economic performance indicators. Information such as this, which is gathered by carrying out surveys and doing research, is known as **data**.

Sometimes data must take on a value from a specific set of numbers, and no other values are possible. For example, the number of children in a family must be 0, 1, 2, 3, 4 and so on, and no intermediate values are allowed. It is impossible to have 3.32 children, say. Usually one allowable value differs from the next by a fixed amount. Such data is said to be **discrete**. Other examples of discrete data include

- the number of car thefts in a city in one week – this must be 0, 1, 2, 3, and so on. You cannot have three-and-a-half thefts.

- shoe sizes – these can be $\ldots, 3\frac{1}{2}, 4, 4\frac{1}{2}, 5, 5\frac{1}{2}, 6, 6\frac{1}{2}, 7$ and so on. You cannot have a shoe size of 9.82.

Sometimes data can take on *any* value within a specified range. Such data is called **continuous**. The volume of liquid in a 1 litre jug can take any value from 0 to 1 litre. The lifespan of an electric light bulb could be any non-negative value.

Frequently data is presented in the form of tables and charts, the intention being to make the information readily understandable. When data is first collected, and before it is processed in any way, it is known as **raw data**. For example, in a test the mathematics marks out of 10 of a group of 30 students are

$$7 \quad 5 \quad 5 \quad 8 \quad 9 \quad 10 \quad 9 \quad 10 \quad 7 \quad 8 \quad 6 \quad 3 \quad 5 \quad 9 \quad 6$$

$$10 \quad 8 \quad 8 \quad 7 \quad 8 \quad 6 \quad 7 \quad 8 \quad 8 \quad 10 \quad 9 \quad 4 \quad 5 \quad 9 \quad 8$$

This is raw data. To try to make sense of this data it is helpful to find out how many students scored each particular mark. This can be done by means of a **tally chart.** To produce this we write down in a column all the possible marks, in order. We then go through the raw data and indicate the occurrence of each mark with a vertical line or tally like |. Every time a fifth tally is recorded this is shown by striking through the previous four, as in ⧸⧸⧸⧸. This makes counting up the tallies particularly easy. The tallies for all the marks are shown in Table 26.1. The number of occurrences of each mark is called its **frequency** and the frequencies can be found from the number of tallies. We see that the tally chart is a useful way of organizing the raw data into a form which will enable us to answer questions and obtain useful information about it.

Table 26.1
Tally chart for
mathematics marks

Mark	Tally	Frequency
0		0
1		0
2		0
3	|	1
4	|	1
5	||||	4
6	|||	3
7	||||	4
8	⧸⧸⧸⧸ |||	8
9	⧸⧸⧸⧸	5
10	||||	4
Total		30

Nowadays much of the tedium of producing tally charts for very large amounts of data can be avoided using computer programs available specifically for this purpose.

Self-assessment questions 26.1

1. Explain the distinction between discrete data and continuous data. Give a new example of each.
2. Explain the purpose of a tally chart.

Exercise 26.1

1. State whether each of the following is an example of discrete data or continuous data:

   (a) the number of matches found in a matchbox
   (b) the percentage of sulphur dioxide found in an air sample taken above a city
   (c) the number of peas in a packet of frozen peas
   (d) the radius of a ball bearing
   (e) the weight of a baby
   (f) the number of students in a particular type of accommodation
   (g) the weight of a soil sample
   (h) the temperature of a patient in hospital
   (i) the number of telephone calls received by an answering machine during one day

2. Twenty-five students were asked to state their year of birth. The information given was

1975	1975	1974	1974	1975	1976
1976	1974	1972	1973	1973	1970
1975	1975	1973	1970	1975	1974
1975	1974	1970	1973	1974	1974
1975					

   Produce a tally chart to organize this data and state the frequency of occurrence of each year of birth.

26.2	**Frequency tables and distributions**

Data is often presented in a **table**. For example, in a recent survey, 1000 students were asked to state the type of accommodation in which they lived during term time. The data gathered is given in Table 26.2. As we saw from the work in the previous section, the number of occurrences of each entry in the table is known as its **frequency.** So, for example, the frequency of students in private rented accommodation is 250. The table summarizes the various frequencies and is known as a **frequency distribution.**

A **relative frequency distribution** is found by expressing each frequency as a proportion of the total frequency.

Table 26.2
Frequency distribution

Type of accommodation	Number of students
Hall of residence	675
Parental home	50
Private rented	250
Other	25
Total	1000

KEY POINT

The relative frequency is found by dividing a frequency by the total frequency.

In this example, because the total frequency is 1000, the relative frequency is easy to calculate. The number of students living at their parental home is 50 out of a total of 1000. Therefore the relative frequency of this group of students is $\frac{50}{1000}$ or 0.050. Table 26.3 shows the relative frequency distribution of student accommodation. Note that the relative frequencies always sum to 1.

Table 26.3
Relative frequency
distribution

Type of accommodation	Number of students	Relative frequency
Hall of residence	675	0.675
Parental home	50	0.050
Private rented	250	0.250
Other	25	0.025
Total	1000	1.000

It is sometimes helpful to express the frequencies as percentages. This is done by multiplying each relative frequency by 100, which gives the results shown in Table 26.4. We see from this that 5% of students questioned live at their parental home.

KEY POINT

A frequency can be expressed as a percentage by multiplying its relative frequency by 100.

Table 26.4
Frequencies expressed as percentages

Type of accommodation	Number of students	Relative frequency × 100
Hall of residence	675	67.5%
Parental home	50	5.0%
Private rented	250	25.0%
Other	25	2.5%
Total	1000	100%

Worked example

26.1 In 1988 there were 4320 areas throughout the world which were designated as national parks. Table 26.5 is a frequency distribution showing how these parks are distributed in different parts of the world.

Table 26.5
The number of national parks throughout the world

Region	Number of national parks
Africa	486
N. America	587
S. America	315
Former Soviet Union	168
Asia	960
Europe	1032
Oceania	767
Antarctica	5
Total	4320

(a) Express this data as a relative frequency distribution and also in terms of percentages.

(b) What percentage of the world's national parks are in Africa?

Solution (a) The relative frequencies are found by expressing each frequency as a proportion of the total. For example, Europe has 1032 national parks out of a total of 4320, and so its relative frequency is $\frac{1032}{4320} = 0.2389$. Figures for the other regions have been calculated and are shown in Table 26.6. To express a relative frequency as a

Table 26.6
The number of national parks throughout the world

Region	Number	Relative frequency	Rel. freq. ×100
Africa	486	0.1125	11.25%
N. America	587	0.1359	13.59%
S. America	315	0.0729	7.29%
Former Soviet Union	168	0.0389	3.89%
Asia	960	0.2222	22.22%
Europe	1032	0.2389	23.89%
Oceania	767	0.1775	17.75%
Antarctica	5	0.0012	0.12%
Total	4320	1.000	100%

percentage it is multiplied by 100. Thus we find that Europe has 23.89% of the world's national parks. The percentages for the other regions are also shown in Table 26.6.

(b) We note from Table 26.6 that 11.25% of the world's national parks are in Africa.

When dealing with large amounts of data it is usual to group the data into classes. For example, in an environmental experiment the weights in grams of 30 soil samples were measured using a balance. We could list all 30 weights here but such a long list of data is cumbersome. Instead, we have already grouped the data into several weight ranges or **classes**. Table 26.7 shows the number of samples in each class. Such a table is called a **grouped frequency distribution.**

Table 26.7
Grouped frequency distribution

Weight range (grams)	Number of samples in that range
100–109	5
110–119	8
120–129	11
130–139	5
140–149	1
Total	30

A disadvantage of grouping the data in this way is that information about individual weights is lost. However, this disadvantage is usually outweighed by having a more compact set of data which is easier to work with. There are five samples which lie in the first class, eight samples in the second and so on. For the first class, the numbers 100 and 109 are called **class limits**, the smaller number being the **lower class limit** and the larger being the **upper class limit**. Theoretically, the class 100–109 includes all weights from 99.5 g up to but not including 109.5 g. These numbers are called the **class boundaries**. In practice the class boundaries are found by adding the upper class limit of one class to the lower limit of the next class and dividing by 2. The **class width** is the difference between the larger class boundary and the smaller. The class 100–109 has width $109.5 - 99.5 = 10$. We do not know the actual values of the weights in each class. Should we require an estimate, the best we can do is use the value in the middle of the class, known as the class **midpoint**. The midpoint of the first class is 104.5; the midpoint of the second class is 114.5 and so on. The midpoint can be found by adding half the class width to the lower class boundary.

Worked example

26.2   The blood pressure in millimetres of mercury of 20 workers was recorded to the nearest millimetre. The data collected was as follows:

121   123   124   129   130   119   129   124   119   121   122

124   124   128   129   136   120   119   121   136

(a) Group this data into classes 115–119, 120–124, and so on.

(b) State the class limits of the first three classes.

(c) State the class boundaries of the first three classes.

(d) What is the class width of the third class?

(e) What is the midpoint of the third class?

Solution   (a) To group the data a tally chart is used as shown in Table 26.8.

(b) The class limits of the first class are 115 and 119; those of the second class are 120 and 124; those of the third are 125 and 129.

(c) Theoretically, the class 115–119 will contain all blood pressures between 114.5 and 119.5 and so the class boundaries of the first class are 114.5 and 119.5. The class boundaries of the second class are 119.5 and 124.5, and those of the third class are 124.5 and 129.5.

Table 26.8
Tally chart for blood
pressures

Pressure (mm mercury)	Tally	Frequency
115–119	|||	3
120–124	JHT JHT	10
125–129	||||	4
130–134	|	1
135–139	||	2
Total		20

(d) The upper class boundary of the third class is 129.5. The lower class boundary is 124.5. The difference between these values gives the class width, that is, 129.5–124.5 = 5.

(e) The midpoint of the third class is found by adding half the class width to the lower class boundary, that is, 2.5 + 124.5 = 127.

Self-assessment questions 26.2

1. Why is it often useful to present data in the form of a *grouped* frequency distribution?
2. Explain the distinction between class boundaries and class limits.
3. How is the class width calculated?
4. How is the class midpoint calculated?

Exercise 26.2

1. The age of each patient over the age of 14 visiting a doctors' practice in one day was recorded as follows:

18	18	76	15	72	45	48	62
21	27	45	43	28	19	17	37
35	34	23	25	46	56	32	18
23	34	32	56	29	43		

(a) Use a tally chart to produce a grouped frequency distribution with age groupings 15–19, 20–24, 25–29, and so on.
(b) What is the relative frequency of patients in the age group 70–74?

(c) What is the relative frequency of patients aged 70 and over?
(d) Express the number of patients in the age group 25–29 as a percentage.

2. Consider the following table which shows the lifetimes, to the nearest hour, of 100 energy saving light bulbs.

(a) If a light bulb has a life of 5499.4 hours into which class will it be put?
(b) If a light bulb has a life of 5499.8 hours into which class will it be put?
(c) State the class boundaries of each class.

continued

(d) Find the class width.
(e) State the class midpoint of each class.

Lifetime (hours)	Frequency
5000–5499	6
5500–5999	32
6000–6499	58
6500–6999	4
Total	100

3. The percentages of 60 students in an Information Technology test are given as follows:

```
45  92  81  76  51  46  82  65  61
19  62  58  72  65  66  97  61  57
63  93  61  46  47  61  56  45  39
47  55  58  81  71  52  38  53  59
82  92  87  86  51  29  19  79  55
```

```
18  53  29  87  87  77  85  67  89
17  29  86  57  59  57
```

Form a frequency distribution by drawing up a tally chart using class intervals 0–9, 10–19 and so on.

4. The data in the table below gives the radius of 20 ball bearings. State the class boundaries of the second class and find the class width.

Radius (mm)	Frequency
20.56–20.58	3
20.59–20.61	6
20.62–20.64	8
20.65–20.67	3

26.3   Bar charts, pie charts, pictograms and histograms

In a **bar chart** information is represented by rectangles or bars. The bars may be drawn horizontally or vertically. The length of each bar corresponds to a frequency.

The data given earlier in Table 26.2 concerning student accommodation is presented in the form of a horizontal bar chart in Figure 26.1. Note that the scale on the horizontal axis must be uniform, that is, it must be evenly spaced, and that the type of accommodation is clearly identified on each bar. The bar chart must be given a title to explain the information that is being presented.

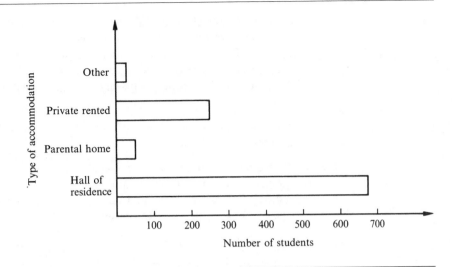

Figure 26.1
Horizontal bar chart
showing student
accommodation

Worked example

26.3   A reproduction antique furniture manufacturer has been producing
pine wardrobes for the past six years. The number of wardrobes sold
each year is given in Table 26.9. Produce a vertical bar chart to
illustrate this information.

Solution   The information is presented in the form of a vertical bar chart in
Figure 26.2.

Table 26.9
Wardrobes sold

Year	Number sold
1987	15
1988	16
1989	20
1990	20
1991	24
1992	30

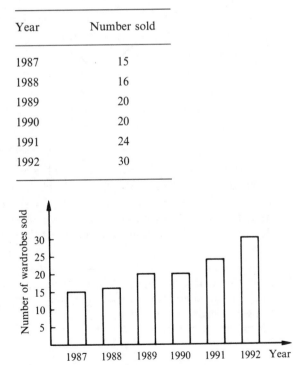

Figure 26.2
Vertical bar chart
showing the number of
wardrobes sold each year

In a **pie chart**, a circular 'pie' is divided into a number of portions, with each portion, or **sector**, of the pie representing a different category. The whole pie represents all categories together. The size of a particular portion must represent the number in its category and this is done by dividing the circle proportionately. Usually a protractor will be required.

Worked example

26.4 A sample of 360 people were asked to state their favourite flavour of potato crisp: 75 preferred 'plain' crisps, 120 preferred 'cheese and onion', 90 preferred 'salt and vinegar' and the remaining 75 preferred 'beef'. Draw a pie chart to present this data.

Solution All categories together, that is, all 360 people asked, make up the whole pie. Because there are 360° in a circle this makes the pie chart particularly easy to draw. A sector of angle 75° will represent 'plain crisps', a sector of 120° will represent 'cheese and onion' and so on. The pie chart is shown in Figure 26.3.

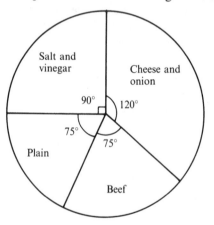

Figure 26.3
Pie chart showing preferred flavour of crisps of 360 people

The pie chart in the previous example was simple to draw because the total number of all categories was 360, the same as the number of degrees in a circle. Let us now see how to deal with a situation where this is not the case. First each relative frequency is found and then this is multiplied by 360. The result is the angle, in degrees, of the corresponding sector.

KEY POINT | The angle, in degrees, of each sector in a pie chart is found by multiplying the corresponding relative frequency by 360. This gives the angle of each sector.

Worked example

26.5 Five hundred students were asked to state how they usually travelled to and from their place of study. The results are given in Table 26.10. Present this information in a pie chart.

Table 26.10
Means of travel

Means of travel	Frequency f
Bus	50
Walk	180
Cycle	200
Car	40
Other	30
Total	500

Solution To find the angle of each sector of the pie chart we first find the corresponding relative frequencies. Recall that this is done by dividing each frequency by the total, 500. Then each relative frequency is multiplied by 360. The results are shown in Table 26.11. The table gives the required angles. Using a protractor the pie chart can then be drawn. This is shown in Figure 26.4.

Table 26.11
Means of travel

Means of travel	Frequency f	Rel. freq. $f/500$	Rel. freq. $\times 360$
Bus	50	$\frac{50}{500}$	$\frac{50}{500} \times 360 = 36.0°$
Walk	180	$\frac{180}{500}$	$\frac{180}{500} \times 360 = 129.6°$
Cycle	200	$\frac{200}{500}$	$\frac{200}{500} \times 360 = 144.0°$
Car	40	$\frac{40}{500}$	$\frac{40}{500} \times 360 = 28.8°$
Other	30	$\frac{30}{500}$	$\frac{30}{500} \times 360 = 21.6°$
Total	500		$360°$

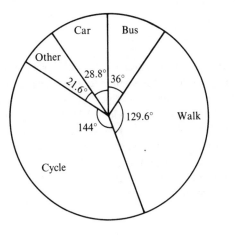

Figure 26.4
Pie chart showing how students travel to work

A **pictogram** uses a picture to represent data in an eye-catching way. There are many ways to produce pictograms but generally the number of objects drawn represents the number of items in a particular category. This is a common form of representation popular on television when the viewer only sees the information for a few seconds.

Worked example

26.6   The number of hours of sunshine in the month of August 1975 in four popular holiday resorts is given in Table 26.12. Show this information using a pictogram. Use one 'sun' to represent 100 hours of sunshine.

Table 26.12
Hours of sunshine in August 1975 in four resorts

Resort	Number of hours
Torbay	300
Scarborough	275
Blackpool	350
Brighton	325

Solution   The pictogram is shown in Figure 26.5.

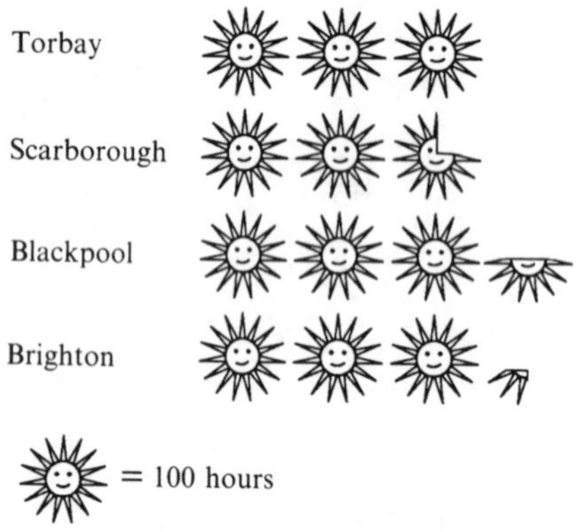

Figure 26.5
Pictogram showing hours of sunshine in August 1975

A **histogram** is often used to illustrate data which is continuous or has been grouped. It is similar to a vertical bar chart in that it is drawn by constructing vertical rectangles. However, in a histogram it is the *area* of each rectangle, and not its length, which is proportional to the frequency of each class. When all class widths are the same this point is not important and you can assume that the length of a rectangle represents the frequency. Class boundaries and not class limits must be used on the horizontal axis to distinguish one class from the next. Consider the following example.

Worked example

26.7 The heights, to the nearest centimetre, of 100 students are given in Table 26.13.

(a) State the class limits of the second class.

(b) Identify the class boundaries of the first class and the last class.

(c) In which class would a student whose actual height is 167.4 cm be placed?

(d) In which class would a student whose actual height is 167.5 cm be placed?

(e) Draw a histogram to depict this data.

Table 26.13
Height of 100 students

Height (cm)	Frequency
164–165	4
166–167	8
168–169	10
170–171	27
172–173	30
174–175	10
176–177	6
178–179	5
Total	100

Solution (a) The lower class limit of the second class is 166 cm. The upper class limit is 167 cm.

(b) The class boundaries of the first class are 163.5 cm and 165.5 cm. The class boundaries of the last class are 177.5 cm and 179.5 cm.

(c) A student with height 167.4 cm will have his height recorded to the nearest centimetre as 167. He will be placed in the class 166–167.

(d) A student with height 167.5 cm will have her height recorded to the nearest centimetre as 168. She will be placed in the class 168–169.

(e) The histogram is shown in Figure 26.6. Note that the class boundaries are used on the horizontal axis to distinguish one class from the next. The area of each rectangle is proportional to the frequency of each class.

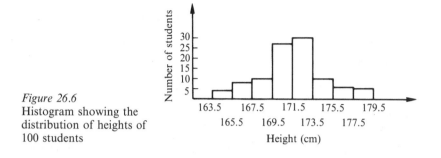

Figure 26.6
Histogram showing the
distribution of heights of
100 students

Self-assessment questions 26.3

1. There is an important distinction between a histogram and a vertical bar chart. Can you explain this?
2. Give one advantage and one disadvantage of using a pictogram to represent data.
3. On a histogram the class boundaries are used on the horizontal axis to distinguish one class from the next. Why do you think class boundaries as opposed to class limits are used?

Exercise 26.3

1. A local council wants to produce a leaflet explaining how the council tax is spent. For every one pound spent, 40 p went to Education, 15 p went to the Police, 15 p went to Cleansing and Refuse, 20 p went to Highways and the remaining 10 p went to a variety of other services. Show this information

    (a) using a pie chart, and (b) using a pictogram.

2. The composition of the atmosphere of the Earth is 78% nitrogen, 21% oxygen and 1% other gases. Draw a pie chart to show this information.

3. An insurance company keeps records on the distribution of the amount of small claims for theft it receives. Amounts are given to the nearest pound. This information is given in the following table.

Amount (£)	Frequency
0–99	5
100–199	8
200–299	16
300–399	24
400–499	26

(a) State the class boundaries.

(b) In which class would a claim for £199.75 be placed?

(c) Draw a histogram to show this information.

Test and assignment exercises 26

1. State whether the following are examples of discrete or continuous data:

   (a) the temperature of a furnace

   (b) the number of goals scored in a football match

   (c) the amount of money in a worker's wage packet at the end of the week.

2. A group of students were interviewed to obtain information on their termly expenditure. On average, for every pound spent, 60p was spent on accommodation, fuel, etc., 25p was spent on food, 10p was spent on entertainment and the remainder on a variety of other items. Draw a pie chart to show this expenditure.

3. Forty lecturers were asked to state which newspaper they bought most regularly. The results were: *Guardian*, 15; *Independent*, 8; *Times*, 2; *Daily Express*, 7; *Mail*, 2; None, 6. Produce a pie chart to show this data.

4. The percentage of total body weight formed by various parts of the body of a newborn baby is as follows: muscles, 26%; skeleton, 17%; skin, 20%; heart, lungs etc., 11%; remainder, 26%. Depict this information (a) on a pie chart, (b) using a bar chart.

5. The following marks were obtained by 27 students in a test. Draw up a frequency distribution by means of a tally chart.

   5 4 6 2 9 1 10 10 5 5 6
   5 5 9 8 7 9 6 7 6 8 7
   6 9 8 8 7

6. The number of cars sold by a major manufacturer of luxury cars in the last 10 years is given in the table below. Show this data on a horizontal bar chart.

Year	Number
1983	2000
1984	2320
1985	2380
1986	2600
1987	2650
1988	2750
1989	2750
1990	2800
1991	1500
1992	1425

7. The table below shows the frequency distribution of the lifetimes in hours of 100 light bulbs tested.

   (a) State the class limits of the first and second classes.

   (b) State the class boundaries of all classes.

Lifetime (hours)	Frequency
600–699	13
700–799	25
800–899	32
900–999	10
1000–1099	15
1100–1199	5

continued

(c) Determine each class width and midpoint.

(d) Draw a histogram for this frequency distribution.

8. In each year from 1988 to 1993 the numbers of visitors to a museum were respectively 12000, 13000, 14500, 18000, 18000 and 17500. Show this information in a pictogram.

27 Statistics

Objectives

This chapter

- introduces the three common averages – mean, median and mode – and shows how to calculate them
- explains what is meant by the variance and standard deviation and shows how to calculate them

27.1   Introduction

In the previous chapter we outlined how data gathered from a range of sources can be represented. In order to get the most out of such data and to be able to interpret it sensibly and reliably, techniques have been developed for its analysis. **Statistics** is the name given to this science. One important statistical quantity is an 'average', the purpose of which is to represent a whole set of data by a single number which in some way represents the set. This chapter explains three types of average and shows how to calculate them. Finally the 'variance' and 'standard deviation' are introduced. These are numbers which describe how widely the data is spread.

27.2   Averages: the mean, median and mode

When presented with a large amount of data it is often useful to ask 'is there a single number which typifies the data?' For example, following an examination taken by a large number of candidates, examiners may be dealing with hundreds of examination scripts. The marks on these will vary from those with very poor marks to those with very high or even top marks. In order to judge whether the group of students as a whole found the examination difficult, the examiner may be asked to provide a single value which gives a measure of how well the students have performed. Such a value is called an **average.** In statistics there are three important averages: the arithmetic mean, the median and the mode.

The arithmetic mean

The **arithmetic mean,** or simply the **mean,** of a set of values is found by adding up all the values and dividing the result by the total number of values in the set. It is given by the following formula.

KEY POINT

$$\text{mean} = \frac{\text{sum of the values}}{\text{total number of values}}$$

Worked examples

27.1   Eight students sit a mathematics test and their marks out of ten are 4, 6, 6, 7, 7, 7, 8, and 10. Find the mean mark.

Solution   The sum of the marks is $4 + 6 + 6 + 7 + 7 + 7 + 8 + 10 = 55$. The total number of values is 8. Therefore,

$$\text{mean} = \frac{\text{sum of the values}}{\text{total number of values}} = \frac{55}{8} = 6.875$$

The examiner can quote 6.875 out of 10 as the 'average mark' of the group of students.

27.2   In a hospital a patient's body temperature is recorded every hour for six hours. Find the mean temperature over the six-hour period if the six temperatures, in °C, were

36.5   36.8   36.9   36.9   36.9   37.0

Solution   To find the mean temperature the sum of all six values is found and the result is divided by 6. That is,

$$\text{mean} = \frac{36.5 + 36.8 + 36.9 + 36.9 + 36.9 + 37.0}{6}$$

$$= \frac{221.0}{6} = 36.83\,°\text{C}$$

In more advanced work we make use of a formula for the mean which requires knowledge of some special notation. Suppose we have n values and we call these x_1, x_2, x_3 and so on up to x_n. The mean of these values is given the symbol $\bar{x}$, pronounced 'x bar'. To calculate the mean we must add up these values and divide by n, that is,

$$\text{mean} = \bar{x} = \frac{x_1 + x_2 + x_3 + \cdots + x_n}{n}$$

A notation is often used to shorten this formula. In mathematics, the Greek letter sigma, written $\sum$, stands for a sum. The sum

$$x_1 + x_2 \text{ is written } \sum_{i=1}^{2} x_i$$

and the sum

$$x_1 + x_2 + x_3 + x_4 \text{ is written } \sum_{i=1}^{4} x_i$$

Note that i runs through all integer values from 1 to n

where the values above and below the sigma sign give the first and last values in the sum. Similarly $x_1 + x_2 + x_3 + \cdots + x_n$ is written $\sum_{i=1}^{n} x_i$. Using this notation, the formula for the mean can be written in the following way:

KEY POINT

$$\text{mean} = \bar{x} = \frac{\sum_{i=1}^{n} x_i}{n}$$

Worked examples

27.3   Express the following in sigma notation:

(a) $x_1 + x_2 + x_3 + \cdots + x_8 + x_9$ (b) $x_{10} + x_{11} + \cdots + x_{100}$

Solution   (a) $x_1 + x_2 + x_3 + \cdots + x_8 + x_9 = \sum_{i=1}^{9} x_i$

(b) $x_{10} + x_{11} + \cdots + x_{100} = \sum_{i=10}^{100} x_i$

27.4   Find the mean of the values $x_1 = 5$, $x_2 = 7$, $x_3 = 13$, $x_4 = 21$ and $x_5 = 29$.

Solution   The number of values equals 5, so we let $n = 5$. The sum of the values is

$$\sum_{i=1}^{5} x_i = x_1 + x_2 + x_3 + x_4 + x_5 = 5 + 7 + 13 + 21 + 29 = 75$$

The mean is

$$\text{mean} = \bar{x} = \frac{\sum_{i=1}^{5} x_i}{n} = \frac{75}{5} = 15$$

When the data is presented in the form of a frequency distribution the mean is found by first multiplying each data value by its frequency. The results are added. This is equivalent to adding up all the data values. The mean is found by dividing this sum by the sum of all the frequencies. Note that the sum of the frequencies is equal to the total number of values. Consider the following example.

Worked example

27.5 Thirty-eight students sit a mathematics test and their marks out of 10 are shown in Table 27.1. Find the mean mark.

Mark, m	Frequency, f
0	0
1	0
2	1
3	0
4	1
5	7
6	16
7	8
8	3
9	1
10	1
Total	38

Table 27.1
Marks of 38 students in a
test

Mark, m	Frequency, f	$m \times f$
0	0	0
1	0	0
2	1	2
3	0	0
4	1	4
5	7	35
6	16	96
7	8	56
8	3	24
9	1	9
10	1	10
Totals	38	236

Table 27.2
Marks of 38 students multiplied by
frequency

Solution Each data value, in this case the mark, is multiplied by its frequency, and the results are added. This is equivalent to adding up all the 38 individual marks. This is shown in Table 27.2.

Note that the sum of all the frequencies is equal to the number of students taking the test. The number 236 is equal to the sum of all the individual marks. The mean is found by dividing this sum by the sum of all the frequencies:

$$\text{mean} = \frac{236}{38} = 6.21$$

The mean mark is 6.21 out of 10.

Using the sigma notation the formula for the mean mark of a frequency distribution with N classes, where the frequency of value x_i is f_i, becomes:

KEY POINT

$$\text{mean} = \bar{x} = \frac{\sum_{i=1}^{N} f_i \times x_i}{\sum_{i=1}^{N} f_i}$$

Note that $\sum_{i=1}^{N} f_i = n$, that is, the sum of all the frequencies equals the total number of values.

When the data is in the form of a grouped distribution the class midpoint is used to calculate the mean. Consider the following example.

Worked example

27.6 The heights, to the nearest centimetre, of 100 students are given in Table 27.3. Find the mean height.

Table 27.3
Heights of 100 students

Height (cm)	Frequency
164–165	4
166–167	8
168–169	10
170–171	27
172–173	30
174–175	10
176–177	6
178–179	5
Total	100

Solution Because the actual heights of students in each class are not known we use the midpoint of the class as an estimate. The midpoint of the class 164–165 is 164.5. Other midpoints and the calculation of the mean are shown in Table 27.4.

Then

$$\text{mean} = \bar{x} = \frac{\sum_{i=1}^{N} f_i \times x_i}{\sum_{i=1}^{N} f_i} = \frac{17150}{100} = 171.5$$

The mean height is 171.5 cm.

Table 27.4
Heights of 100 students
with midpoints multiplied
by frequency

Height (cm)	Frequency f_i	Midpoint x_i	$f_i \times x_i$
164–165	4	164.5	658.0
166–167	8	166.5	1332.0
168–169	10	168.5	1685.0
170–171	27	170.5	4603.5
172–173	30	172.5	5175.0
174–175	10	174.5	1745.0
176–177	6	176.5	1059.0
178–179	5	178.5	892.5
Total	100		17150.0

The median

A second average which also typifies a set of data is the **median**.

KEY POINT

> The **median** of a set of numbers is found by listing the numbers in ascending order and then selecting the value that lies half-way along the list.

Worked example

27.7   Find the median of the numbers

1   2   6   7   9   11   11   11   14

Solution   The set of numbers is already given in order. The number half way along the list is 9, because there are four numbers before it and four numbers after it in the list. Hence the median is 9.

When there is an even number of values, the median is found by taking the mean of the two middle values.

Worked example

27.8   Find the median of the following salaries: £24000, £12000, £16000, £22000, £10000 and £25000.

Solution   The numbers are first arranged in order as £10000, £12000, £16000, £22000, £24000 and £25000. Because there is an even number of values there are two middle figures: £16000 and £22000. The mean of these is

$$\frac{16000 + 22000}{2} = 19000$$

The median salary is therefore £19000.

The mode

A third average is the **mode**.

KEY POINT

The **mode** of a set of values is that value which occurs most often.

Worked example

27.9   Find the mode of the set of numbers

1   1   4   4   5   6   8   8   8   9

Solution   The number which occurs most often is 8, which occurs three times. Therefore 8 is the mode. Usually a mode is quoted when we want to represent the most popular value in a set.

Sometimes a set of data may have more than one mode.

Worked example

27.10   Find the mode of the set of numbers

20   20   21   21   21   48   48   49   49   49

Solution   In this example there is no single value which occurs most frequently. The number 21 occurs three times, but so does the number 49. This set has two modes. The data is said to be **bimodal**.

Self-assessment questions 27.2

1. State the three different types of averages commonly used in statistical calculations.
2. In an annual report, an employer of a small firm claims that the median salary for his workforce is £18500. However, over discussions in the canteen it is apparent that no worker earns this amount. Explain how this might have arisen.

Exercise 27.2

1. Find the mean, median and mode of the following set of values: 2, 3, 5, 5, 5, 5, 8, 8, 9.

2. Find the mean of the set of numbers 1, 1, 1, 1, 1, 1, 256. Explain why the mean does not represent the data adequately. Which average might it have been more appropriate to use?

3. The marks of seven students in a test were 45%, 83%, 99%, 65%, 68%, 72% and 66%. Find the mean mark and the median mark.

4. Write out fully each of the following expressions:

   (a) $\sum\limits_{i=1}^{4} x_i$ (b) $\sum\limits_{i=1}^{7} x_i$

   (c) $\sum\limits_{i=1}^{3} (x_i - 3)^2$ (d) $\sum\limits_{i=1}^{4} (2 - x_i)^2$

5. Write the following concisely using sigma notation:

   (a) $x_3 + x_4 + x_5 + x_6$
   (b) $(x_1 - 1) + (x_2 - 1) + (x_3 - 1)$

6. Calculate the mean, median and mode of the following numbers: 1.00, 1.15, 1.25, 1.38, 1.39 and 1.40.

7. The prices of the eight terrace houses advertised by a local estate agent are

   £29000   £37500   £32500   £29995

   £31995   £32750   £29950   £32950

   Find the mean price of these houses.

8. The data in the table gives the radius of 20 ball bearings. Find the class midpoints and hence calculate the average radius.

Radius (mm)	Frequency
20.56–20.58	3
20.59–20.61	6
20.62–20.64	8
20.65–20.67	3

| 27.3 | **The variance and standard deviation** |

Suppose we consider the test marks of two groups of three students. Suppose that the marks out of 10 for the first group are

$$4 \quad 7 \quad \text{and} \quad 10$$

while those of the second group are

$$7 \quad 7 \quad \text{and} \quad 7$$

It is easy to calculate the mean mark of each group: the first group has mean mark

$$\frac{4+7+10}{3} = \frac{21}{3} = 7$$

The second group has mean mark

$$\frac{7+7+7}{3} = \frac{21}{3} = 7$$

We see that both groups have the same mean mark even though the marks in the first group are widely spread, whilst the marks in the second group are all the same. If the teacher quotes just the mean mark of each group this gives no information about how widely the marks are spread. The **variance** and **standard deviation** are important and widely used statistical quantities which contain this information. Most calculators are pre-programmed to calculate these quantities. Once you understand the processes involved, check to see if your calculator can be used to find the variance and standard deviation of a set of data.

Suppose we have n values x_1, x_2, x_3 up to x_n. Their mean is $\bar{x}$ given by

$$\bar{x} = \frac{\sum_{i=1}^{n} x_i}{n}$$

The variance is found from the following formula.

KEY POINT

$$\text{variance} = \frac{\sum_{i=1}^{n} (x_i - \bar{x})^2}{n}$$

If you study this carefully you will see that to calculate the variance we must

- calculate the mean value $\bar{x}$
- subtract the mean from each value in turn, that is, find $x_i - \bar{x}$
- square each answer to get $(x_i - \bar{x})^2$

- add up all these squared quantities to get $\sum_{i=1}^{n}(x_i - \bar{x})^2$

- divide the result by n to get

$$\frac{\sum_{i=1}^{n}(x_i - \bar{x})^2}{n}; \text{ this is the variance.}$$

The standard deviation is found by taking the square root of the variance.

KEY POINT

$$\text{standard deviation} = \sqrt{\frac{\sum_{i=1}^{n}(x_i - \bar{x})^2}{n}}$$

Let us calculate the variance and standard deviation of each of the two sets of marks 4, 7, 10 and 7, 7, 7. We have already noted that the mean of each set is 7. Each stage of the calculation of the variance of the set 4, 7, 10 is shown in Table 27.5. The mean is subtracted from each number and

Table 27.5

x_i	$x_i - \bar{x}$	$(x_i - \bar{x})^2$
4	$4 - 7 = -3$	$(-3)^2 = 9$
7	$7 - 7 = 0$	$0^2 = 0$
10	$10 - 7 = 3$	$3^2 = 9$
Total		18

the results are squared and then added. Note that when a negative number is squared the result is positive. The squares are added to give 18. In this example, the number of values equals 3. So, taking $n = 3$,

$$\text{variance} = \frac{\sum_{i=1}^{n}(x_i - \bar{x})^2}{n} = \frac{18}{3} = 6$$

The standard deviation is the square root of the variance, that is, $\sqrt{6} = 2.449$. Similarly, calculation of the variance and standard deviation of the set 7, 7, 7 is shown in Table 27.6. Again, n equals 3. So

$$\text{variance} = \frac{\sum_{i=1}^{n}(x_i - \bar{x})^2}{n} = \frac{0}{3} = 0$$

The standard deviation is the square root of the variance and so it also equals zero. The fact that the standard deviation is zero reflects that there is no spread of values. By comparison it is easy to check that the standard

Table 27.6

x_i	$x_i - \bar{x}$	$(x_i - \bar{x})^2$
7	$7 - 7 = 0$	$0^2 = 0$
7	$7 - 7 = 0$	$0^2 = 0$
7	$7 - 7 = 0$	$0^2 = 0$
Total		0

deviation of the set 6, 7 and 8, which also has mean 7, is equal to 0.816. The fact that the set 4, 7 and 10 has standard deviation 2.449 shows that the values are more widely spread than in the set 6, 7 and 8.

Worked example

27.11   Find the variance and standard deviation of 10, 15.8, 19.2 and 8.7.

Solution   First the mean is found:

$$\bar{x} = \frac{10 + 15.8 + 19.2 + 8.7}{4} = \frac{53.7}{4} = 13.425$$

The calculation to find the variance is given in Table 27.7:

Table 27.7

x_i	$x_i - \bar{x}$	$(x_i - \bar{x})^2$
10	$10 - 13.425 = -3.425$	$(-3.425)^2 = 11.731$
15.8	$15.8 - 13.425 = 2.375$	$2.375^2 = 5.641$
19.2	$19.2 - 13.425 = 5.775$	$5.775^2 = 33.351$
8.7	$8.7 - 13.425 = -4.725$	$(-4.725)^2 = 22.326$
Total		73.049

$$\text{variance} = \frac{\sum_{i=1}^{n}(x_i - \bar{x})^2}{n} = \frac{73.049}{4} = 18.262$$

The standard deviation is the square root of the variance:

$$\sqrt{18.262} = 4.273$$

When dealing with a grouped frequency distribution with N classes the formula for the variance becomes:

KEY POINT

$$\text{variance} = \frac{\sum_{i=1}^{N} f_i(x_i - \bar{x})^2}{\sum_{i=1}^{N} f_i}$$

As before the standard deviation is the square root of the variance.

Worked example

27.12   In a period of 30 consecutive days in July the temperature in °C was recorded as follows:

18   19   20   23   24   24   21   18   17   16

16   17   17   17   18   19   20   20   22   23

24   24   25   23   21   21   20   19   19   18

(a) Produce a grouped frequency distribution showing data grouped from 16–17 °C, 18–19 °C and so on.

(b) Find the mean temperature of the grouped data.

(c) Find the standard deviation of the grouped data.

Solution   (a) The data is grouped using a tally chart as in Table 27.8.

Table 27.8

Temperature range (°C)	Tally	Frequency
16–17	ⅢⅡ I	6
18–19	ⅢⅡ III	8
20–21	ⅢⅡ II	7
22–23	IIII	4
24–25	ⅢⅡ	5
Total		30

(b) When calculating the mean temperature from the grouped data we do not know the actual temperatures. We do know the frequency of each class. The best we can do is use the midpoint of each class as an estimate of the values in that class. The class midpoints and the calculation to determine the mean are shown in Table 27.9. The mean is then

$$\bar{x} = \frac{\sum_{i=1}^{N} f_i \times x_i}{\sum_{i=1}^{N} f_i} = \frac{603}{30} = 20.1 \,°C$$

Table 27.9

Temperature range (°C)	Frequency f_i	Class midpoint x_i	$f_i \times x_i$
16–17	6	16.5	99.0
18–19	8	18.5	148.0
20–21	7	20.5	143.5
22–23	4	22.5	90.0
24–25	5	24.5	122.5
Total	30		603

(c) To find the variance and hence the standard deviation we must subtract the mean, 20.1, from each value, square and then add the results. Finally this sum is divided by 30. The complete calculation is shown in Table 27.10.

Table 27.10

Temperature range (°C)	Frequency f_i	Class midpoint x_i	$x_i - \bar{x}$	$(x_i - \bar{x})^2$	$f_i(x_i - \bar{x})^2$
16–17	6	16.5	−3.60	12.96	77.76
18–19	8	18.5	−1.60	2.56	20.48
20–21	7	20.5	0.40	0.16	1.12
22–23	4	22.5	2.40	5.76	23.04
24-25	5	24.5	4.40	19.36	96.80
Total	30				219.20

Finally the variance is given by

$$\text{variance} = \frac{\sum_{i=1}^{N} f_i(x_i - \bar{x})^2}{\sum_{i=1}^{N} f_i} = \frac{219.20}{30} = 7.307$$

The standard deviation is the square root of the variance, that is, $\sqrt{7.307} = 2.703$.

Self-assessment questions 27.3

1. Why is an average often an insufficient way of describing a set of values?
2. Describe the stages involved in calculating the variance and standard deviation of a set of values $x_1, x_2, \cdots, x_n$.

Exercise 27.3

1. Find the standard deviation of each of the following sets of numbers and comment upon your answers:

   (a) 20, 20, 20, 20 (b) 16, 17, 23, 24 (c) 0, 20, 20, 40

2. Calculate the variance and standard deviation of the following set of numbers: 1, 2, 3, 3, 3, 4, 7, 8, 9, 10.

3. The examination results of two students, Jane and Tony, are shown in the table. Find the mean and standard deviation of each of them. Comment upon the results.

	Jane	Tony
Maths	50	29
Physics	42	41
Chemistry	69	60
French	34	48
Spanish	62	80

4. The marks of 50 students in a mathematics examination are given as

45	50	62	62	68	47	45	44	48
73	62	63	62	67	80	45	41	40
23	55	21	83	67	49	48	48	62
63	79	58	71	37	32	58	54	50
62	66	68	62	81	92	62	45	49
71	72	70	49	51				

   (a) Produce a grouped frequency distribution using classes 25–29, 30–34, 35–39 and so on.

   (b) The modal class is the class with the highest frequency. State the modal class.

   (c) Calculate the mean, variance and standard deviation of the grouped frequency distribution.

Test and assignment exercises 27

1. Calculate the mean, median and mode of the following numbers: 2.7, 2.8, 3.1, 3.1, 3.1, 3.4, 3.8, 4.1.

2. Calculate the mean of

   (a) the first ten whole numbers, $1, 2, \cdots, 10$

   (b) the first 12 whole numbers, $1, 2, \cdots, 12$

   (c) the first n whole numbers, $1, 2, \cdots, n$.

3. A manufacturer of breakfast cereals uses two machines which pack 250 g packets of cereal automatically. In order to check the average weight, sample packets are taken and weighed to the nearest gram. The results of checking eight packets from each of the two machines are given in the table.

Machine								
1	250	250	251	257	253	250	268	259
2	249	251	250	251	252	258	254	250

continued

(a) Find the mean weight of each sample of eight packets.

(b) Find the standard deviation of each sample.

(c) Comment upon the performance of the two machines.

4. What is the median of the five numbers 18, 28, 39, 42, 43? If 50 is added as the sixth number what would the median become?

5. Express the following sums in sigma notation:

(a) $y_1 + y_2 + y_3 + \cdots + y_7 + y_8$

(b) $y_1^2 + y_2^2 + y_3^2 + \cdots + y_7^2 + y_8^2$

(c) $(y_1 - \bar{y})^2 + (y_2 - \bar{y})^2 + \cdots + (y_7 - \bar{y})^2 + (y_8 - \bar{y})^2$

28 Probability

Objectives	This chapter
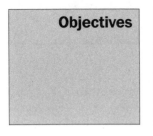	• introduces theoretical and experimental probability and how to calculate them
	• explains the meaning of the term 'complementary events'
	• explains the meaning of the term 'independent events'

28.1 Introduction

Some events in life are impossible; other events are quite certain to happen. For example, it is impossible for a human being to live for 1000 years. We say that the probability of a human being living for 1000 years is zero. On the other hand it is quite certain that all of us will die someday. We say that the probability that all of us will die is 1. Using the letter P to stand for probability we can write

$$P(\text{a human will live for 1000 years}) = 0$$

and

$$P(\text{all of us will die someday}) = 1$$

Many other events in life are neither impossible nor certain. These have varying degrees of likelihood, or chance. For example, it is quite unlikely but not impossible that the British weather throughout a particular summer will be dominated by snowfall. It is quite likely but not certain that the life expectancy of the UK population will continue to rise in the foreseeable future. Events such as these can be assigned probabilities varying from 0 up to 1. Those having probabilities close to 1 are quite likely to happen. Those having probabilities close to 0 are almost impossible. Several events and their probabilities are shown on the probability scale in Figure 28.1. It is important to note that no probability can lie outside the range 0 to 1.

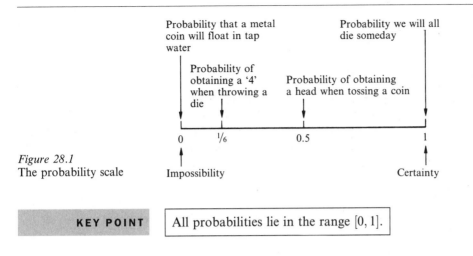

Figure 28.1
The probability scale

All probabilities lie in the range $[0, 1]$.

Complementary events

Consider the following situation. A light bulb is tested. Clearly it either works or it does not work. Here we have two events: the first is that the light bulb works and the second is that the light bulb does not work. When the bulb is tested, one or other of these events must occur. Furthermore, each event excludes the other. In such a situation we say the two events are **complementary**. In general, two events are complementary if one of them must happen and, when it does, the other event cannot. The sum of the probabilities of the two complementary events must always equal 1. This is known as **total probability**. We shall see how this result can be used to calculate probabilities shortly.

Self-assessment questions 28.1

1. All probabilities must lie in a certain range. What is this range?
2. Explain the meaning of the term 'complementary events'.

28.2 ## Calculating theoretical probabilities

Sometimes we have sufficient information about a set of circumstances to calculate the probability of an event occurring. For example, suppose we roll a die and ask what is the probability of obtaining a score of '5'. If the die were fair, or **unbiased**, the chance of getting a '5' is the same as the chance of getting any other score. You would expect to get a '5' one time in every six. That is,

the probability of throwing a '5' is $\dfrac{1}{6}$ or 0.167

In other words there is a one in six chance of scoring a '5'. The fact that the probability is closer to 0 than to 1 means that it is quite unlikely that a '5' will be thrown although it is not impossible. Such a probability is known as a **theoretical probability** and, when all events are equally likely, it is calculated from the following formula.

KEY POINT

$$P\begin{pmatrix} \text{obtaining our} \\ \text{chosen event} \end{pmatrix} = \frac{\text{number of ways the chosen event can occur}}{\text{total number of possibilities}}$$

For example, suppose we ask what is the probability of obtaining a score more than 4. A score more than 4 can occur in two ways, by scoring a '5' or a '6'. If the die is fair, all possible events are equally likely. The total number of possibilities is 6, therefore,

$$P(\text{score more than 4}) = \frac{2}{6} = \frac{1}{3}$$

Worked examples

28.1   A fair die is thrown. What is the probability of obtaining an even score?

Solution   The chosen event is throwing an even score, that is, a '2', '4' or '6'. There are therefore three ways that this chosen event can occur out of a total of six, equally likely, ways. So

$$P(\text{even score}) = \frac{3}{6} = \frac{1}{2}$$

28.2   A fair coin is tossed. What is the probability that it will land with its head uppermost?

Solution   There are two equally likely ways the coin can land: head uppermost or tail uppermost. The chosen event, that the coin lands with its head uppermost, is just one of these ways. The chance of getting a head is the same as the chance of getting a tail. Therefore,

$$P(\text{head}) = \frac{1}{2}$$

Note from the previous example that $P(\text{tail}) = \frac{1}{2}$ and that also

$$P(\text{tail}) + P(\text{head}) = 1$$

Note that the two events, getting a head and getting a tail, are complementary because one of them must happen and either one excludes the other. Therefore the sum of the two probabilities, the total probability, equals 1.

Worked example

28.3   Two coins are tossed.

(a) Write down all the possible outcomes.

(b) What is the probability of obtaining two tails?

Solution   (a) Letting H stand for head and T for tail, the possible outcomes are:

$$H, H \qquad H, T \qquad T, H \qquad T, T$$

There are four possible outcomes, each one equally likely to occur.

(b) Obtaining two tails is just one of the four possible outcomes. Therefore, the probability of obtaining two tails is $\frac{1}{4}$.

Exercise 28.2

1. A die is thrown. Find

   (a) the probability of obtaining a score less than 6,
   (b) the probability of obtaining a score more than 6,
   (c) the probability of obtaining an even score less than 5,
   (d) the probability of obtaining an even score less than 2.

2. There are four Aces in a pack of 52 playing cards. What is the probability that a card selected at random is not an Ace?

3. A drawer contains six red socks, six black socks and eight blue socks. Find the probability that a sock selected at random from the drawer is

   (a) black (b) red (c) red or blue

4. Two dice are thrown together and their scores are added together. By considering all the possible outcomes, find the probability that the total score will be

   (a) 12 (b) 0 (c) 1 (d) 2 (e) more than 5

5. A basket contains 87 good apples and three bad ones. What is the probability that an apple chosen at random is bad?

6. A box contains 16 red blocks, 20 blue blocks, 24 orange blocks and 10 black blocks. A block is picked at random. Calculate the probability that the block is

   (a) black (b) orange (c) blue
   (d) red or blue (e) red or blue or orange
   (f) not orange

7. Three coins are tossed. By considering all possible outcomes calculate the probability of obtaining

   (a) two heads and one tail
   (b) at least two heads
   (c) no heads

28.3   Calculating experimental probabilities

In some circumstances we do not have sufficient information to calculate a theoretical probability. We know that if a coin is unbiased the probability of obtaining a head is $\frac{1}{2}$. But suppose the coin is biased so that it is more likely to land with its tail uppermost. We can experiment by tossing the coin a large number of times and counting the number of tails obtained. Suppose we toss the coin 100 times and obtain 65 tails. We can then estimate the probability of obtaining a tail as $\frac{65}{100}$ or 0.65. Such a probability is known as an **experimental probability** and is only accurate if a very large number of experiments are performed. Generally, we can calculate an experimental probability from the following formula.

KEY POINT

$$P\left(\begin{array}{c}\text{chosen event}\\ \text{occurs}\end{array}\right) = \frac{\text{number of ways the chosen event occurs}}{\text{total number of times the experiment is repeated}}$$

Worked examples

28.4   A biased die is thrown 1000 times and a score of '6' is obtained on 200 occasions. If the die is now thrown again what is the probability of obtaining a score of '6'?

Solution   Using the formula for the experimental probability we find

$$P(\text{throwing a six}) = \frac{200}{1000} = 0.2$$

If the die were unbiased the theoretical probability of throwing a six would be $\frac{1}{6} = 0.167$, so the die has been biased in favour of throwing a '6'.

28.5   A manufacturer produces microwave ovens. It is known from experience that the probability that a microwave oven is of an acceptable standard is 0.92. Find the probability that an oven selected at random is not of an acceptable standard.

Solution   When an oven is tested either it is of an acceptable standard or it is not. An oven cannot be both acceptable and unacceptable. The two events, that the oven is acceptable or that the oven is unacceptable, are therefore complementary. Recall that the sum of the probabilities of complementary events is 1 and so

$$P(\text{oven is not acceptable}) = 1 - 0.92 = 0.08$$

Self-assessment questions 28.3

1. In what circumstances is it appropriate to use an experimental probability?
2. A series of experiments is performed three times. In the first of the series an experiment is carried out 100 times. In the second it is carried out 1000 times and in the third 10 000 times. Which of the series of experiments is likely to lead to the best estimate of probability? Why?

Exercise 28.3

1. A new component is fitted to a washing machine. In a sample of 150 machines tested, seven failed to function correctly. Calculate the probability that a machine fitted with the new component (a) works correctly, (b) does not work correctly.

2. In a sample containing 5000 nails manufactured by a company, 5% are too short or too long. A nail is picked at random from the production line; estimate the probability that it is of the right length.

3. The probability that the post office delivers first class mail on the following working day after posting is 0.96. If 93500 first class letters are posted on Wednesday, how many are likely to be delivered on Thursday?

4. The probability a car rescue service reaches a car in less than one hour is 0.87. If the rescue service is called out 17 300 times in one day, calculate the number of cars reached in less than one hour.

5. Out of 50 000 components tested, 48 700 were found to be working well. A batch of 3000 components is delivered to a depot. How many are likely to be not working well?

28.4   Independent events

Two events are **independent** if the occurrence of either one in no way affects the occurrence of the other. For example, if an unbiased die is thrown twice the score on the second throw is in no way affected by the score on the first. The two scores are independent. The **multiplication law** for independent events states the following:

KEY POINT

If events A and B are independent, then the probability of obtaining A and B is given by

$$P(A \text{ and } B) = P(A) \times P(B)$$

Worked example

28.6 A die is thrown and a coin is tossed. What is the probability of obtaining a six and a head?

Solution These events are independent since the score on the die in no way affects the result of tossing the coin, and vice versa. Therefore

$$P(\text{throwing a '6' and tossing a head}) = P(\text{throwing a '6'})$$
$$\times P(\text{tossing a head})$$
$$= \frac{1}{6} \times \frac{1}{2}$$
$$= \frac{1}{12}$$

When several events are independent of each other the multiplication law becomes

$$P(A \text{ and } B \text{ and } C \text{ and } D \ldots) = P(A) \times P(B) \times P(C) \times P(D) \ldots$$

Worked example

28.7 A coin is tossed three times. What is the probability of obtaining three heads?

Solution The three tosses are all independent events since the result of any one has no effect on the others. Therefore

$$P(3 \text{ heads}) = \frac{1}{2} \times \frac{1}{2} \times \frac{1}{2} = \frac{1}{8}$$

Self-assessment questions 28.4

1. Explain what is meant by saying two events are independent.
2. Suppose we have two packs each of 52 playing cards. A card is selected from each pack. Event A is that the card from the first pack is the Ace of Spades. Event B is that the card from the second pack is a Spade. Are these two events dependent or independent?
3. From a single pack of cards, two are removed. The first is examined. Suppose event A is that the first card is the Ace of Spades. The second card is examined. Event B is that the second card is a Spade. Are the two events dependent or independent?

Exercise 28.4

1. A die is thrown and a coin is tossed. What is the probability of getting an even score on the die and a tail?

2. Suppose you have two packs each of 52 playing cards. A card is drawn from the first and a card is drawn from the second. What is the probability that both cards are the Ace of Spades?

3. A coin is tossed eight times. What is the probability of obtaining eight tails?

4. A die is thrown four times. What is the probability of obtaining four 1s?

5. Suppose that there is an equal chance of a mother giving birth to a boy or a girl.

   (a) If one child is born find the probability that it is a boy.

   (b) If two children are born find the probability that they are both boys, assuming that the sex of neither one can influence the sex of the other.

6. The probability a component is working well is 0.96. If 4 components are picked at random calculate the probability that

   (a) they all work well
   (b) none of them work well.

7. The probability a student passes a module is 0.91. If three modules are studied calculate the probability that the student passes

   (a) all three modules (b) two modules
   (c) one module (d) no modules.

Test and assignment exercises 28

1. The following numbers are probabilities of a certain event happening. One of them is an error. Which one?

   (a) 0.5 (b) $\frac{3}{4}$ (c) 0.001 (d) $\frac{13}{4}$

2. Which of the following cannot be a probability?

   (a) 0.125 (b) −0.2 (c) 1 (d) 0

3. Events A, B and C are defined as follows. In each case state the complementary event.

   A: the lifespan of a light bulb is greater than 1000 hours
   B: it will rain on Christmas Day 2005
   C: your car will be stolen sometime in the next 100 days

4. The probability of a component manufactured in a factory being defective is 0.01. If three components are selected at random what is the probability that they all work as required?

5. A parcel delivery company guarantees 'next-day' delivery for 99% of its parcels. Out of a sample of 1800 deliveries 15 were delivered later than the following day. Is the company living up to its promises?

6. At a certain bus stop the probability of a bus arriving late is $\frac{3}{20}$. The probability of it arriving on time is $\frac{4}{5}$. Find

   (a) the probability that the bus arrives early,
   (b) the probability that the bus does not arrive late.

7. Suppose you have two packs each of 52 playing cards. A card is drawn from the first and a card is drawn from the second. What is the probability that

   (a) both cards are diamonds?
   (b) both cards are black?
   (c) both cards are Kings?

continued

8. Find the probability of obtaining an odd number when throwing a fair die once.

9. A bag contains eight red beads, four white beads and five blue beads. A bead is drawn at random. Determine the probability that it is

   (a) red (b) white (c) black (d) blue (e) red or white (f) not blue.

10. Two cards are selected from a pack of 52. Find the probability that they are both Aces if the first card is (a) replaced and (b) not replaced.

11. A pack contains twenty cards, numbered one to twenty. A card is picked at random. Calculate the probability that the number on the card is

   (a) even (b) 16 or more (c) divisible by 3.

12. Out of 42 300 components, 846 were defective. How many defective components would you expect in a batch of 1500?

13. A biased die has the following probabilities

   $p(1) = 0.1$, $p(2) = 0.15$, $p(3) = 0.1$, $p(4) = 0.2$, $p(5) = 0.2$, $p(6) = 0.25$

   The die is thrown twice. Calculate the probability that

   (a) both scores are 6s
   (b) both scores are 1s
   (c) the first score is odd and the second is even
   (d) the total score is 10.

Solutions to exercises

Solutions to Chapter 1

Exercise 1.1

1. (a) 3 (b) 9 (c) 11 (d) 21 (e) 30 (f) 56 (g) -13
   (h) -19 (i) -19 (j) -13 (k) 29 (l) -75
   (m) -75 (n) 29
2. (a) -24 (b) -32 (c) -30 (d) 16 (e) -42
3. (a) -5 (b) 3 (c) -3 (d) 3 (e) -3 (f) -6
   (g) 6 (h) -6
4. (a) Sum $= 3 + 6 = 9$
   Product $= 3 \times 6 = 18$
   (b) Sum $= 17$, Product $= 70$
   (c) Sum $= 2 + 3 + 6 = 11$
   Product $= 2 \times 3 \times 6 = 36$
5. (a) Difference $= 18 - 9 = 9$
   Quotient $= \frac{18}{9} = 2$
   (b) Difference $= 20 - 5 = 15$
   Quotient $= \frac{20}{5} = 4$
   (c) Difference $= 100 - 20 = 80$
   Quotient $= \frac{100}{20} = 5$

Exercise 1.2

1. (a) $6 - 2 \times 2 = 6 - 4 = 2$
   (b) $(6 - 2) \times 2 = 4 \times 2 = 8$
   (c) $6 \div 2 - 2 = 3 - 2 = 1$
   (d) $(6 \div 2) - 2 = 3 - 2 = 1$
   (e) $6 - 2 + 3 \times 2 = 6 - 2 + 6 = 10$
   (f) $6 - (2 + 3) \times 2 = 6 - 5 \times 2$
   $= 6 - 10 = -4$
   (g) $(6 - 2) + 3 \times 2 = 4 + 3 \times 2 = 4 + 6 = 10$
   (h) $\frac{16}{-2} = -8$
   (i) $\frac{-24}{-3} = 8$
   (j) $(-6) \times (-2) = 12$
   (k) $(-2)(-3)(-4) = -24$

Exercise 1.3

1. 13, 2 and 29 are prime.
2. (a) 2×13 (b) $2 \times 2 \times 5 \times 5$ (c) $3 \times 3 \times 3$
   (d) 71 (e) $2 \times 2 \times 2 \times 2 \times 2 \times 2$ (f) 3×29
   (g) 19×23 (h) 29×31
3. $30 = 2 \times 3 \times 5$
   $42 = 2 \times 3 \times 7$
   2 and 3 are common prime factors.

Exercise 1.4

1. (a) $12 = 2 \times 2 \times 3$ $15 = 3 \times 5$ $21 = 3 \times 7$
   Hence h.c.f. $= 3$
   (b) $16 = 2 \times 2 \times 2 \times 2$ $24 = 2 \times 2 \times 2 \times 3$
   $40 = 2 \times 2 \times 2 \times 5$
   So h.c.f. $= 2 \times 2 \times 2 = 8$
   (c) $28 = 2 \times 2 \times 7$ $70 = 2 \times 5 \times 7$
   $120 = 2 \times 2 \times 2 \times 3 \times 5$
   $160 = 2 \times 2 \times 2 \times 2 \times 2 \times 5$
   So h.c.f. $= 2$
   (d) $35 = 5 \times 7$ $38 = 2 \times 19$

2. (a) $6 \times (12 - 3) + 1 = 6 \times 9 + 1$
   $= 54 + 1 = 55$
   (b) $6 \times 12 - (3 + 1) = 72 - 4 = 68$
   (c) $6 \times (12 - 3 + 1) = 6 \times 10 = 60$
   (d) $5 \times (4 - 3) + 2 = 5 \times 1 + 2 = 5 + 2 = 7$
   (e) $5 \times 4 - (3 + 2) = 20 - 5 = 15$
   or $5 \times (4 - 3 + 2) = 5 \times 3 = 15$
   (f) $5 \times (4 - (3 + 2)) = 5 \times (4 - 5)$
   $= 5 \times (-1) = -5$

$42 = 2 \times 3 \times 7$

So h.c.f. $= 1$

(e) $96 = 2 \times 2 \times 2 \times 2 \times 2 \times 3$

$120 = 2 \times 2 \times 2 \times 3 \times 5$

$144 = 2 \times 2 \times 2 \times 2 \times 3 \times 3$

So h.c.f. $= 2 \times 2 \times 2 \times 3 = 24$

2. (a) 5 $6 = 2 \times 3$ $8 = 2 \times 2 \times 2$

So l.c.m. $= 2 \times 2 \times 2 \times 3 \times 5 = 120$

(b) $20 = 2 \times 2 \times 5$ $30 = 2 \times 3 \times 5$

So l.c.m. $= 2 \times 2 \times 3 \times 5 = 60$

(c) 7 $9 = 3 \times 3$ $12 = 2 \times 2 \times 3$

So l.c.m. $= 2 \times 2 \times 3 \times 3 \times 7 = 252$

(d) $100 = 2 \times 2 \times 5 \times 5$

$150 = 2 \times 3 \times 5 \times 5$

$235 = 5 \times 47$

So l.c.m. $=$

$2 \times 2 \times 3 \times 5 \times 5 \times 47 = 14100$

(e) $96 = 2 \times 2 \times 2 \times 2 \times 2 \times 3$

$120 = 2 \times 2 \times 2 \times 3 \times 5$

$144 = 2 \times 2 \times 2 \times 2 \times 3 \times 3$

So l.c.m. $=$

$2 \times 2 \times 2 \times 2 \times 2 \times 3 \times 3 \times 5 = 1440$

Solutions to Chapter 2

Exercise 2.1

1. (a) Proper (b) Proper (c) Improper
   (d) Proper (e) Improper

Exercise 2.2

1. (a) $\dfrac{18}{27} = \dfrac{2}{3}$ (b) $\dfrac{12}{20} = \dfrac{3}{5}$ (c) $\dfrac{15}{45} = \dfrac{1}{3}$

   (d) $\dfrac{25}{80} = \dfrac{5}{16}$ (e) $\dfrac{15}{60} = \dfrac{1}{4}$ (f) $\dfrac{90}{200} = \dfrac{9}{20}$

   (g) $\dfrac{15}{20} = \dfrac{3}{4}$ (h) $\dfrac{2}{18} = \dfrac{1}{9}$ (i) $\dfrac{16}{24} = \dfrac{2}{3}$

   (j) $\dfrac{30}{65} = \dfrac{6}{13}$ (k) $\dfrac{12}{21} = \dfrac{4}{7}$ (l) $\dfrac{100}{45} = \dfrac{20}{9}$

   (m) $\dfrac{6}{9} = \dfrac{2}{3}$ (n) $\dfrac{12}{16} = \dfrac{3}{4}$ (o) $\dfrac{13}{42}$

   (p) $\dfrac{13}{39} = \dfrac{1}{3}$ (q) $\dfrac{11}{33} = \dfrac{1}{3}$ (r) $\dfrac{14}{30} = \dfrac{7}{15}$

   (s) $-\dfrac{12}{16} = -\dfrac{3}{4}$ (t) $\dfrac{11}{-33} = -\dfrac{1}{3}$ (u) $\dfrac{-14}{-30} = \dfrac{7}{15}$

2. $\dfrac{3}{4} = \dfrac{21}{28}$

3. $4 = \dfrac{20}{5}$

4. $\dfrac{5}{12} = \dfrac{15}{36}$

5. $2 = \dfrac{8}{4}$

6. $6 = \dfrac{18}{3}$

7. $\dfrac{2}{3} = \dfrac{8}{12}, \dfrac{5}{4} = \dfrac{15}{12}, \dfrac{5}{6} = \dfrac{10}{12}$

8. $\dfrac{4}{9} = \dfrac{8}{18}, \dfrac{1}{2} = \dfrac{9}{18}, \dfrac{5}{6} = \dfrac{15}{18}$

9. (a) $\dfrac{1}{2} = \dfrac{6}{12}$ (b) $\dfrac{3}{4} = \dfrac{9}{12}$ (c) $\dfrac{5}{2} = \dfrac{30}{12}$

   (d) $5 = \dfrac{60}{12}$ (e) $4 = \dfrac{48}{12}$ (f) $12 = \dfrac{144}{12}$

Exercise 2.3

1. (a) $\dfrac{1}{4} + \dfrac{2}{3} = \dfrac{3}{12} + \dfrac{8}{12} = \dfrac{11}{12}$

   (b) $\dfrac{3}{5} + \dfrac{5}{3} = \dfrac{9}{15} + \dfrac{25}{15} = \dfrac{34}{15}$

   (c) $\dfrac{12}{14} - \dfrac{2}{7} = \dfrac{6}{7} - \dfrac{2}{7} = \dfrac{4}{7}$

   (d) $\dfrac{3}{7} - \dfrac{1}{2} + \dfrac{2}{21} = \dfrac{18}{42} - \dfrac{21}{42} + \dfrac{4}{42} = \dfrac{1}{42}$

   (e) $1\dfrac{1}{2} + \dfrac{4}{9} = \dfrac{3}{2} + \dfrac{4}{9} = \dfrac{27}{18} + \dfrac{8}{18} = \dfrac{35}{18}$

(f) $2\frac{1}{4} - 1\frac{1}{3} + \frac{1}{2} = \frac{9}{4} - \frac{4}{3} + \frac{1}{2}$

$= \frac{27}{12} - \frac{16}{12} + \frac{6}{12} = \frac{17}{12}$

(g) $\frac{10}{15} - 1\frac{2}{5} + \frac{8}{3} = \frac{10}{15} - \frac{7}{5} + \frac{8}{3}$

$= \frac{10}{15} - \frac{21}{15} + \frac{40}{15} = \frac{29}{15}$

(h) $\frac{9}{10} - \frac{7}{16} + \frac{1}{2} - \frac{2}{5} = \frac{72}{80} - \frac{35}{80} + \frac{40}{80} - \frac{32}{80}$

$= \frac{45}{80} = \frac{9}{16}$

2. (a) $\frac{7}{8} + \frac{1}{3} = \frac{21}{24} + \frac{8}{24} = \frac{29}{24}$

(b) $\frac{1}{2} - \frac{3}{4} = \frac{2}{4} - \frac{3}{4} = -\frac{1}{4}$

(c) $\frac{3}{5} + \frac{2}{3} + \frac{1}{2} = \frac{18}{30} + \frac{20}{30} + \frac{15}{30} = \frac{53}{30} = 1\frac{23}{30}$

(d) $\frac{3}{8} + \frac{1}{3} + \frac{1}{4} = \frac{9}{24} + \frac{8}{24} + \frac{6}{24} = \frac{23}{24}$

(e) $\frac{2}{3} - \frac{4}{7} = \frac{14}{21} - \frac{12}{21} = \frac{2}{21}$

(f) $\frac{1}{11} - \frac{1}{2} = \frac{2}{22} - \frac{11}{22} = -\frac{9}{22}$

(g) $\frac{3}{11} - \frac{5}{8} = \frac{24}{88} - \frac{55}{88} = \frac{-31}{88}$

3. (a) $\frac{5}{2}$ (b) $\frac{11}{3}$ (c) $\frac{41}{4}$ (d) $\frac{37}{7}$ (e) $\frac{56}{9}$ (f) $\frac{34}{3}$ (g) $\frac{31}{2}$

(h) $\frac{55}{4}$ (i) $\frac{133}{11}$ (j) $\frac{41}{3}$ (k) $\frac{113}{2}$

4. (a) $3\frac{1}{3}$ (b) $3\frac{1}{2}$ (c) $3\frac{3}{4}$ (d) $4\frac{1}{6}$

Exercise 2.4

1. (a) $\frac{2}{3} \times \frac{6}{7} = \frac{2}{1} \times \frac{2}{7} = \frac{4}{7}$

(b) $\frac{8}{15} \times \frac{25}{32} = \frac{1}{15} \times \frac{25}{4} = \frac{1}{3} \times \frac{5}{4} = \frac{5}{12}$

(c) $\frac{1}{4} \times \frac{8}{9} = \frac{1}{1} \times \frac{2}{9} = \frac{2}{9}$

(d) $\frac{16}{17} \times \frac{34}{48} = \frac{1}{17} \times \frac{34}{3} = \frac{1}{1} \times \frac{2}{3} = \frac{2}{3}$

(e) $2 \times \frac{3}{5} \times \frac{5}{12} = \frac{3}{5} \times \frac{5}{6} = \frac{3}{1} \times \frac{1}{6} = \frac{1}{2}$

(f) $2\frac{1}{3} \times 1\frac{1}{4} = \frac{7}{3} \times \frac{5}{4} = \frac{35}{12}$

(g) $1\frac{3}{4} \times 2\frac{1}{2} = \frac{7}{4} \times \frac{5}{2} = \frac{35}{8}$

(h) $\frac{3}{4} \times 1\frac{1}{2} \times 3\frac{1}{2} = \frac{3}{4} \times \frac{3}{2} \times \frac{7}{2} = \frac{63}{16}$

2. (a) $\frac{2}{3} \times \frac{3}{4} = \frac{1}{2}$

(b) $\frac{4}{7} \times \frac{21}{30} = \frac{6}{15} = \frac{2}{5}$

(c) $\frac{9}{10} \times 80 = 72$

(d) $\frac{6}{7} \times 42 = 36$

3. Yes, because $\frac{3}{4} \times \frac{12}{15} = \frac{12}{15} \times \frac{3}{4}$

4. (a) $-\frac{5}{21}$ (b) $-\frac{3}{8}$ (c) $-\frac{5}{11}$ (d) $\frac{10}{7}$

5. (a) $5\frac{1}{2} \times \frac{1}{2} = \frac{11}{2} \times \frac{1}{2} = \frac{11}{4}$

(b) $3\frac{3}{4} \times \frac{1}{3} = \frac{15}{4} \times \frac{1}{3} = \frac{5}{4}$

(c) $\frac{2}{3} \times 5\frac{1}{9} = \frac{2}{3} \times \frac{46}{9} = \frac{92}{27}$

(d) $\frac{3}{4} \times 11\frac{1}{2} = \frac{3}{4} \times \frac{23}{2} = \frac{69}{8}$

6. (a) $\frac{3}{5} \times 11\frac{1}{4} = \frac{3}{5} \times \frac{45}{4} = \frac{27}{4}$

(b) $\frac{2}{3} \times 15\frac{1}{2} = \frac{2}{3} \times \frac{31}{2} = \frac{31}{3}$

(c) $\frac{1}{4} \times \left(-8\frac{1}{3}\right) = \frac{1}{4} \times \left(-\frac{25}{3}\right) = -\frac{25}{12}$

Exercise 2.5

1. (a) $\frac{3}{4} \div \frac{1}{8} = \frac{3}{4} \times \frac{8}{1} = \frac{3}{1} \times \frac{2}{1} = 6$

(b) $\dfrac{8}{9} \div \dfrac{4}{3} = \dfrac{8}{9} \times \dfrac{3}{4} = \dfrac{2}{9} \times \dfrac{3}{1} = \dfrac{2}{3}$

(c) $-\dfrac{2}{7} \div \dfrac{4}{21} = -\dfrac{2}{7} \times \dfrac{21}{4} = -\dfrac{2}{1} \times \dfrac{3}{4} = -\dfrac{3}{2}$

(d) $\dfrac{9}{4} \div 1\dfrac{1}{2} = \dfrac{9}{4} \div \dfrac{3}{2} = \dfrac{9}{4} \times \dfrac{2}{3} = \dfrac{3}{4} \times \dfrac{2}{1} = \dfrac{3}{2}$

(e) $\dfrac{5}{6} \div \dfrac{5}{12} = \dfrac{5}{6} \times \dfrac{12}{5} = \dfrac{1}{6} \times \dfrac{12}{1} = 2$

(f) $\dfrac{99}{100} \div 1\dfrac{4}{5} = \dfrac{99}{100} \div \dfrac{9}{5} = \dfrac{99}{100} \times \dfrac{5}{9}$

$= \dfrac{11}{100} \times \dfrac{5}{1} = \dfrac{11}{20}$

(g) $3\dfrac{1}{4} \div 1\dfrac{1}{8} = \dfrac{13}{4} \div \dfrac{9}{8} = \dfrac{13}{4} \times \dfrac{8}{9}$

$= \dfrac{13}{1} \times \dfrac{2}{9} = \dfrac{26}{9}$

(h) $\left(2\dfrac{1}{4} \div \dfrac{3}{4}\right) \times 2 = \left(\dfrac{9}{4} \times \dfrac{4}{3}\right) \times 2$

$= \left(\dfrac{3}{4} \times 4\right) \times 2 = 3 \times 2 = 6$

(i) $2\dfrac{1}{4} \div \left(\dfrac{3}{4} \times 2\right) = \dfrac{9}{4} \div \left(\dfrac{3}{4} \times \dfrac{2}{1}\right)$

$= \dfrac{9}{4} \div \dfrac{3}{2} = \dfrac{9}{4} \times \dfrac{2}{3} = \dfrac{3}{4} \times \dfrac{2}{1} = \dfrac{3}{2}$

(j) $6\dfrac{1}{4} \div 2\dfrac{1}{2} + 5 = \dfrac{25}{4} \div \dfrac{5}{2} + 5$

$= \dfrac{25}{4} \times \dfrac{2}{5} + 5 = \dfrac{5}{2} + 5 = \dfrac{15}{2}$

(k) $6\dfrac{1}{4} \div \left(2\dfrac{1}{2} + 5\right) = \dfrac{25}{4} \div \left(\dfrac{5}{2} + 5\right)$

$= \dfrac{25}{4} \div \dfrac{15}{2} = \dfrac{25}{4} \times \dfrac{2}{15} = \dfrac{5}{4} \times \dfrac{2}{3} = \dfrac{5}{6}$

Solutions to Chapter 3

Exercise 3.1

1. (a) $\dfrac{7}{10}$ (b) $\dfrac{4}{5}$ (c) $\dfrac{9}{10}$

2. (a) $\dfrac{11}{20}$ (b) $\dfrac{79}{500}$ (c) $\dfrac{49}{50}$ (d) $\dfrac{99}{1000}$

3. (a) $4\dfrac{3}{5}$ (b) $5\dfrac{1}{5}$ (c) $8\dfrac{1}{20}$ (d) $11\dfrac{59}{100}$ (e) $121\dfrac{9}{100}$

4. (a) 0.697 (b) 0.083 (c) 0.517

Exercise 3.2

1. (a) 6960 (b) 70.4 (c) 0.0123 (d) 0.0110
   (e) 45.6 (f) 2350
2. (a) 66.00 (b) 66.0 (c) 66 (d) 70 (e) 66.00
   (f) 66.0
3. (a) 10 (b) 10.0
4. (a) 65.456 (b) 65.46 (c) 65.5 (d) 65.456
   (e) 65.46 (f) 65.5 (g) 65 (h) 70

Solutions to Chapter 4

Exercise 4.1

1. 23% of $124 = \dfrac{23}{100} \times 124 = 28.52$

2. (a) $\dfrac{9}{11} = \dfrac{9}{11} \times 100\% = \dfrac{900}{11}\% = 81.82\%$

(b) $\dfrac{15}{20} = \dfrac{15}{20} \times 100\% = 75\%$

(c) $\dfrac{9}{10} = \dfrac{9}{10} \times 100\% = 90\%$

(d) $\dfrac{45}{50} = \dfrac{45}{50} \times 100\% = 90\%$

(e) $\dfrac{75}{90} = \dfrac{75}{90} \times 100\% = 83.33\%$

3. $\dfrac{13}{12} = \dfrac{13}{12} \times 100\% = 108.33\%$

4. 217% of $500 = \dfrac{217}{100} \times 500 = 1085$

5. New weekly wage is 106% of £400,

$$106\% \text{ of } 400 = \dfrac{106}{100} \times 400 = 424$$

The new weekly wage is £424.

6. 17% of $1200 = \dfrac{17}{100} \times 1200 = 204$

The debt is decreased by £204 to
£1200 − 204 = £996.

7. (a) $50\% = \dfrac{50}{100} = 0.5$

(b) $36\% = \dfrac{36}{100} = 0.36$

(c) $75\% = \dfrac{75}{100} = 0.75$

(d) $100\% = \dfrac{100}{100} = 1$

(e) $12.5\% = \dfrac{12.5}{100} = 0.125$

8. £204.80
9. £1125

Exercise 4.2
1. $8 + 1 + 3 = 12$. The first number is $\frac{8}{12}$ of
180, that is, 120; the second number is $\frac{1}{12}$ of
180, that is, 15; and the third number is $\frac{3}{12}$
of 180, that is, 45. Hence 180 is divided into
120, 15 and 45.
2. $1 + 1 + 3 = 5$. We calculate $\frac{1}{5}$ of 930 to
be 186 and $\frac{3}{5}$ of 930 to be 558. The length is
divided into 186 cm, 186 cm and 558 cm.
3. $2 + 3 + 4 = 9$. The first piece is $\frac{2}{9}$ of 6 m,
that is, 1.33 m; the second piece is $\frac{3}{9}$ of 6 m,
that is, 2 m; the third piece is $\frac{4}{9}$ of 6 m, that
is, 2.67 m.
4. $1 + 2 + 3 + 4 = 10$: $\frac{1}{10}$ of 1200 = 120; $\frac{2}{10}$
of 1200 = 240; $\frac{3}{10}$ of 1200 = 360; $\frac{4}{10}$ of 1200
= 480. The number 1200 is divided into
120, 240, 360 and 480.
5. $2\frac{3}{4} : 1\frac{1}{2} : 2\frac{1}{4} = \dfrac{11}{4} : \dfrac{3}{2} : \dfrac{9}{4} = 11 : 6 : 9$
Now, $11 + 6 + 9 = 26$, so

$$\dfrac{11}{26} \times 2600 = 1100 \qquad \dfrac{6}{26} \times 2600 = 600$$

$$\dfrac{9}{26} \times 2600 = 900$$

Alan receives £1100, Bill receives £600 and
Claire receives £900.

6. 8 kg, 10.67 kg, 21.33 kg
7. (a) 1:2 (b) 1:2 (c) 1:2:4 (d) 1:21
8. 24, 84

Solutions to Chapter 5

Exercise 5.2
1. $2^4 = 16$; $\left(\frac{1}{2}\right)^2 = \frac{1}{4}$; $1^8 = 1$; $3^5 = 243$; $0^3 = 0$.
2. $10^4 = 10000$; $10^5 = 100000$; $10^6 = 1000000$.
3. $11^4 = 14641$; $16^8 = 4294967296$;
   $39^4 = 2313441$; $1.5^7 = 17.0859375$.
4. (a) $a^4 b^2 c = a \times a \times a \times a \times b \times b \times c$
   (b) $xy^2 z^4 = x \times y \times y \times z \times z \times z \times z$
5. (a) $x^4 y^2$ (b) $x^2 y^2 z^3$ (c) $x^2 y^2 z^2$ (d) $a^2 b^2 c^2$

6. (a) $7^4 = 2401$ (b) $7^5 = 16807$
   (c) $7^4 \times 7^5 = 40353607$
   (d) $7^9 = 40353607$ (e) $8^3 = 512$
   (f) $8^7 = 2097152$
   (g) $8^3 \times 8^7 = 1073741824$
   (h) $8^{10} = 1073741824$
   The rule states that $a^m \times a^n = a^{m+n}$; the
   powers are added.

7. $(-3)^3 = -27$; $(-2)^2 = 4$; $(-1)^7 = -1$;
$(-1)^4 = 1$.
8. -4492.125; 324; -0.03125.
9. (a) 36 (b) 9 (c) -64 (d) -8
$-6^2 = -36$, $-3^2 = -9$,
$-4^3 = -64$, $-2^3 = -8$

Exercise 5.3

1. 60
2. 69
3. (a) 314.2 cm^2 (b) 28.28 cm^2 (c) 0.126 cm^2
4. $3x^2 = 3 \times 4^2 = 3 \times 16 = 48$;
$(3x)^2 = 12^2 = 144$
5. $5x^2 = 5(-2)^2 = 20$; $(5x)^2 = (-10)^2 = 100$
6. (a) 33.95 (b) 23.5225 (c) 26.75
(d) 109.234125
7. (a) $a + b + c = 25.5$ (b) $ab = 46.08$
(c) $bc = 32.76$ (d) $abc = 419.328$
8. $C = \frac{5}{9}(100 - 32) = \frac{5}{9}(68) = 37.78$
9. (a) $x^2 = 49$ (b) $-x^2 = -49$
(c) $(-x)^2 = (-7)^2 = 49$

10. (a) 4 (b) 4 (c) -4 (d) 12 (e) -12 (f) 36
11. (a) 3 (b) 9 (c) -1 (d) 36 (e) -36 (f) 144
12. $x^2 - 7x + 2 = (-9)^2 - 7(-9) + 2$
$= 81 + 63 + 2 = 146$
13. $2x^2 + 3x - 11 = 2(-3)^2 + 3(-3) - 11$
$= 18 - 9 - 11 = -2$
14. $-x^2 + 3x - 5 = -(-1)^2 + 3(-1) - 5$
$= -1 - 3 - 5 = -9$
15. 0
16. (a) 49 (b) 43 (c) 1 (d) 5
17. (a) 21 (b) 27 (c) 0 (d) $\frac{1}{6}$
18. (a) 3 (b) $3\frac{4}{5}$ (c) 23 (d) 23
19. (a) $-\frac{1}{2}$ (b) 4 (c) 32
20. (a) 0 (b) 16 (c) $40\frac{1}{2}$ (d) $2\frac{1}{2}$
21. (a) 17 (b) -0.5 (c) 7
22. (a) 6000 (b) 2812.5
23. (a) 17151 (b) 276951

Solutions to Chapter 6

Exercise 6.1

1. (a) $5^7 \times 5^{13} = 5^{20}$ (b) $9^8 \times 9^5 = 9^{13}$
(c) $11^2 \times 11^3 \times 11^4 = 11^9$
2. (a) $15^3/15^2 = 15^1 = 15$ (b) $4^{18}/4^9 = 4^9$
(c) $5^{20}/5^{19} = 5^1 = 5$
3. (a) a^{10} (b) a^9 (c) b^{22}
4. (a) x^{15} (b) y^{21}
5. $19^8 \times 17^8$ cannot be simplified using the laws of indices because the two bases are not the same.
6. (a) $(7^3)^2 = 7^{3 \times 2} = 7^6$ (b) $(4^2)^8 = 4^{16}$
(c) $(7^9)^2 = 7^{18}$
7. $1/(5^3)^8 = 1/5^{24}$
8. (a) $x^5 y^5$ (b) $a^3 b^3 c^3$
9. (a) $x^{10} y^{20}$ (b) $81x^6$ (c) $-27x^3$ (d) $x^8 y^{12}$
10. (a) z^3 (b) y^2 (c) 1

Exercise 6.2

1. (a) $\frac{1}{4}$ (b) $\frac{1}{8}$ (c) $\frac{1}{9}$ (d) $\frac{1}{27}$ (e) $\frac{1}{25}$
(f) $\frac{1}{16}$ (g) $\frac{1}{9}$ (h) $\frac{1}{121}$ (i) $\frac{1}{7}$

2. (a) 0.1 (b) 0.01 (c) 0.000001
(d) 0.01 (e) 0.001 (f) 0.0001

3. (a) $\frac{1}{x^4}$ (b) x^5 (c) $\frac{1}{x^7}$ (d) $\frac{1}{y^2}$ (e) $y^1 = y$
(f) $\frac{1}{y^1} = \frac{1}{y}$ (g) $\frac{1}{y^2}$ (h) $\frac{1}{z^1} = \frac{1}{z}$ (i) $z^1 = z$

4. (a) $x^{-3} = \frac{1}{x^3}$ (b) $x^{-5} = \frac{1}{x^5}$ (c) $x^{-1} = \frac{1}{x}$

(d) x^5 (e) $x^{-13} = \dfrac{1}{x^{13}}$ (f) $x^{-8} = \dfrac{1}{x^8}$

(g) $x^{-9} = \dfrac{1}{x^9}$ (h) $x^{-4} = \dfrac{1}{x^4}$

5. (a) a^{11} (b) x^{-16} (c) x^{-18} (d) 4^{-6}
6. (a) 0.001 (b) 0.0001 (c) 0.00001
7. $4^{-8}/4^{-6} = 4^{-2} = 1/4^2 = \frac{1}{16}$;
$3^{-5}/3^{-8} = 3^3 = 27$

Exercise 6.3

1. (a) $64^{1/3} = \sqrt[3]{64} = 4$ since $4^3 = 64$
   (b) $144^{1/2} = \sqrt{144} = \pm 12$
   (c) $16^{-1/4} = 1/16^{1/4} = 1/\sqrt[4]{16} = \pm\frac{1}{2}$
   (d) $25^{-1/2} = 1/25^{1/2} = 1/\sqrt{25} = \pm\frac{1}{5}$
   (e) $1/32^{-1/5} = 32^{1/5} = \sqrt[5]{32} = 2$ since
   $2^5 = 32$
2. (a) $(3^{-1/2})^4 = 3^{-2} = 1/3^2 = \frac{1}{9}$
   (b) $(8^{1/3})^{-1} = 8^{-1/3} = 1/8^{1/3} = 1/\sqrt[3]{8} = \frac{1}{2}$
3. (a) $8^{1/2}$ (b) $12^{1/3}$ (c) $16^{1/4}$ (d) $13^{3/2}$ (e) $4^{7/3}$
4. (a) $x^{1/2}$ (b) $y^{1/3}$ (c) $x^{5/2}$ (d) $5^{7/3}$

Exercise 6.4

1. (a) 743 (b) 74300 (c) 70 (d) 0.0007
2. (a) 3×10^2 (b) 3.56×10^2 (c) 0.32×10^2
   (d) 0.0057×10^2

Exercise 6.5

1. (a) $45 = 4.5 \times 10^1$ (b) $45000 = 4.5 \times 10^4$
   (c) $-450 = -4.5 \times 10^2$
   (d) $90000000 = 9.0 \times 10^7$
   (e) $0.15 = 1.5 \times 10^{-1}$
   (f) $0.00036 = 3.6 \times 10^{-4}$ (g) 3.5 is already in
   standard form.
   (h) $-13.2 = -1.32 \times 10^1$
   (i) $1000000 = 1 \times 10^6$
   (j) $0.0975 = 9.75 \times 10^{-2}$
   (k) $45.34 = 4.534 \times 10^1$
2. (a) $3.75 \times 10^2 = 375$ (b) $3.97 \times 10^1 = 39.7$
   (c) $1.875 \times 10^{-1} = 0.1875$
   (d) $-8.75 \times 10^{-3} = -0.00875$

Solutions to Chapter 7

Exercise 7.1

1. (a) $-5p + 19q$ (b) $-5r - 13s + z$ (c) not
   possible to simplify
   (d) $8x^2 + 3y^2 - 2y$ (e) $4x^2 - x + 9$
2. (a) $-12y + 8p + 9q$ (b) $21x^2 - 11x^3 + y^3$
   (c) $7xy + y^2$ (d) $2xy$ (e) 0

Exercise 7.2

1. (a) 84 (b) 84 (c) 84
2. (a) 40 (b) 40
3. (a) $14z$ (b) $30y$ (c) $6x$ (d) $27a$ (e) $55a$ (f) $6x$
4. (a) $20x^2$ (b) $6y^3$ (c) $22u^2$ (d) $8u^2$ (e) $26z^2$
5. (a) $21x^2$ (b) $21a^2$ (c) $14a^2$
6. (a) $15y^2$ (b) $8y$
7. (a) $a^3b^2c^2$ (b) x^3y^2 (c) x^2y^4
8. No difference; both equal x^2y^4.
9. $(xy^2)(xy^2) = x^2y^4$; $xy^2 + xy^2 = 2xy^2$
10. (a) $-21z^2$ (b) $-4z$
11. (a) $-3x^2$ (b) $2x$
12. (a) $2x^2$ (b) $-3x$

Exercise 7.3

1. (a) $4x + 4$ (b) $-4x - 4$ (c) $4x - 4$
   (d) $-4x + 4$
2. (a) $5x - 5y$ (b) $19x + 57y$ (c) $8a + 8b$
   (d) $5y + xy$ (e) $12x + 48$
   (f) $17x - 153$ (g) $-a + 2b$ (h) $x + \frac{1}{2}$
   (i) $6m - 12m^2 - 9mn$
3. (a) $18 - 13x - 26 = -8 - 13x$ (b) $x^2 + xy$
   will not simplify any further.
4. (a) $x^2 + 7x + 6$ (b) $x^2 + 9x + 20$
   (c) $x^2 + x - 6$ (d) $x^2 + 5x - 6$
   (e) $xm + ym + nx + yn$
   (f) $12 + 3y + 4x + yx$ (g) $25 - x^2$
   (h) $51x^2 - 79x - 10$
5. (a) $x^2 - 4x - 21$ (b) $6x^2 + 11x - 7$
   (c) $16x^2 - 1$ (d) $x^2 - 9$ (e) $6 + x - 2x^2$
6. (a) $\frac{29}{2}x - \frac{5}{2}y$ (b) $\frac{5}{4}x + \frac{5}{4}$
7. (a) $-x + y$ (b) $-a - 2b$ (c) $-\frac{3}{2}p - \frac{1}{2}q$
8. $(x + 1)(x + 2) = x^2 + 3x + 2$. So
   $(x + 1)(x + 2)(x + 3) = (x^2 + 3x + 2)(x + 3)$
   $= (x^2 + 3x + 2)(x) + (x^2 + 3x + 2)(3)$
   $= x^3 + 3x^2 + 2x + 3x^2 + 9x + 6$
   $= x^3 + 6x^2 + 11x + 6$

Solutions to Chapter 8

Exercise 8.1

1. (a) $9x + 27$ (b) $-5x + 10$ (c) $\frac{1}{2}x + \frac{1}{2}$
   (d) $-a + 3b$ (e) $1/(2x + 2y)$ (f) $x/(yx - y^2)$
2. (a) $4x^2$ has factors $1, 2, 4, x, 2x, 4x, x^2, 2x^2,$
   $4x^2$; (b) $6x^3$ has factors $1, 2, 3, 6, x, 2x, 3x,$
   $6x, x^2, 2x^2, 3x^2, 6x^2, x^3, 2x^3, 3x^3, 6x^3$
3. (a) $3(x + 6)$ (b) $3(y - 3)$ (c) $-3(y + 3)$
   (d) $-3(1 + 3y)$ (e) $5(4 + t)$ (f) $5(4 - t)$
   (g) $-5(t + 4)$ (h) $3(x + 4)$ (i) $17(t + 2)$
   (j) $4(t - 9)$
4. (a) $x(x^3 + 2)$ (b) $x(x^3 - 2)$ (c) $x(3x^3 - 2)$
   (d) $x(3x^3 + 2)$ (e) $x^2(3x^2 + 2)$ (f) $x^3(3x + 2)$
   (g) $z(17 - z)$ (h) $x(3 - y)$ (i) $y(3 - x)$
   (j) $x(1 + 2y + 3yz)$
5. (a) $10(x + 2y)$ (b) $3(4a + b)$ (c) $2x(2 - 3y)$
   (d) $7(a + 2)$
   (e) $5(2m - 3)$ (f) $1/[5(a + 7b)]$
   (g) $1/[5a(a + 7b)]$
6. (a) $3x(5x + 1)$ (b) $x(4x - 3)$ (c) $4x(x - 2)$
   (d) $3(5 - x^2)$
   (e) $5x^2(2x + 1 + 3y)$ (f) $6ab(a - 2b)$
   (g) $8b(2ac - ba + 3c)$

Exercise 8.2

1. (a) $(x + 2)(x + 1)$ (b) $(x + 7)(x + 6)$
   (c) $(x + 5)(x - 3)$
   (d) $(x + 10)(x - 1)$ (e) $(x - 8)(x - 3)$
   (f) $(x - 10)(x + 10)$
   (g) $(x + 2)(x + 2)$ or $(x + 2)^2$
   (h) $(x + 6)(x - 6)$ (i) $(x + 5)(x - 5)$
   (j) $(x + 1)(x + 9)$ (k) $(x + 9)(x - 1)$

(l) $(x + 1)(x - 9)$ (m) $(x - 1)(x - 9)$
(n) $x(x - 5)$

2. (a) $(2x + 1)(x - 3)$ (b) $(3x + 1)(x - 2)$
   (c) $(5x + 3)(2x + 1)$
   (d) $2x^2 + 12x + 16$ has a common factor of
   2 which should be written outside a bracket
   to give $2(x^2 + 6x + 8)$. The bracket can then
   be factorized to give $2(x + 4)(x + 2)$
   (e) $(2x + 3)(x + 1)$ (f) $(3s + 2)(s + 1)$
   (g) $(3z + 2)(z + 5)$
   (h) $9(x^2 - 4) = 9(x + 2)(x - 2)$
   (i) $(2x + 5)(2x - 5)$
3. (a) $(x + y)(x - y) = x^2 + yx - yx - y^2$
   $= x^2 - y^2$
   (b) (i) $(4x + 1)(4x - 1)$ (ii) $(4x + 3)(4x - 3)$
   (iii) $(5t - 4r)(5t + 4r)$
4. (a) $(x - 2)(x + 5)$ (b) $(2x + 5)(x - 4)$
   (c) $(3x + 1)(3x - 1)$ (d) $10x^2 + 14x - 12$ has
   a common factor of 2 which is written
   outside a bracket to give $2(5x^2 + 7x - 6)$,
   the bracket can then be factorized to give
   $2(5x - 3)(x + 2)$; (e) $(x + 13)(x + 2)$
   (f) $(-x + 1)(x + 3)$
5. (a) $(10 + 7x)(10 - 7x)$
   (b) $(6x + 5y)(6x - 5y)$

   (c) $\left(\dfrac{1}{2} + 3v\right)\left(\dfrac{1}{2} - 3v\right)$

   (d) $\left(\dfrac{x}{y} + 2\right)\left(\dfrac{x}{y} - 2\right)$

Solutions to Chapter 9

Exercise 9.2

1. (a) $\dfrac{3x}{y}$ (b) $\dfrac{9}{x}$ (c) $3y$ (d) $3x$ (e) 9 (f) 3
2. (a) $\dfrac{5x}{y}$ (b) $\dfrac{3x}{y}$ (c) $15y$ (d) 15

   (e) $-x^2$ (f) $-\dfrac{1}{y^4} = -y^{-4}$ (g) $\dfrac{1}{y} = y^{-1}$ (h) y^{-7}

3. (a) $\dfrac{1}{3 + 2x}$ (b) $1 + 2x$ (c) $\dfrac{1}{2 + 7x}$ (d) $\dfrac{x}{2 + 7x}$
   (e) $\dfrac{x}{1 + 7x}$ (f) $\dfrac{1}{7x + y}$ (g) $\dfrac{y}{7x + y}$ (h) $\dfrac{x}{7x + y}$

4. (a) $5x + 1$ (b) $\dfrac{3(5x + 1)}{3(x + 2y)} = \dfrac{5x + 1}{x + 2y}$

(c) $\dfrac{3}{x + 2}$ (d) $\dfrac{3}{y + 2}$ (e) $\dfrac{13}{x + 5}$ (f) $\dfrac{17}{9y + 4}$

5. (a) $\dfrac{1}{3 + 2x}$ (b) $\dfrac{2}{x + 7}$

(c) $\dfrac{2(x + 4)}{(x - 2)(x + 4)} = \dfrac{2}{x - 2}$

(d) $\dfrac{7}{ab + 9}$ (e) $\dfrac{y}{y + 1}$

6. (a) $\dfrac{1}{x - 2}$ (b) $\dfrac{2(x - 2)}{(x - 2)(x + 3)} = \dfrac{2}{x + 3}$

(c) $\dfrac{1}{x + 2}$ (d) $\dfrac{(x + 1)(x + 1)}{(x + 1)(x - 3)} = \dfrac{x + 1}{x - 3}$

(e) $\dfrac{2}{x - 3}$ (f) $\dfrac{1}{x - 3}$ (g) $\dfrac{1}{2(x - 3)}$ (h) $\dfrac{2}{x - 3}$

(i) $\dfrac{1}{2(x + 4)}$ (j) $\dfrac{1}{2}$ (k) 2 (l) $x + 4$

(m) $\dfrac{1}{x - 3}$ (n) $\dfrac{1}{x + 4}$ (o) $\dfrac{1}{2}$ (p) $\dfrac{x + 4}{2x + 9}$

Exercise 9.3

1. (a) $\dfrac{y}{6}$ (b) $\dfrac{z}{6}$ (c) $\dfrac{2}{5y}$ (d) $\dfrac{2}{5x}$ (e) $\dfrac{3x}{4y}$ (f) $\dfrac{3x^2}{5y}$ (g) $\dfrac{3x}{5y}$

(h) $\dfrac{7x}{16y}$ (i) $\dfrac{1}{4x}$ (j) $\dfrac{x}{4}$ (k) $\dfrac{1}{x}$ (l) $\dfrac{x}{9}$ (m) $\dfrac{1}{x}$

(n) $\dfrac{1}{9x}$ (o) $\dfrac{x}{2}$

2. (a) $\dfrac{1}{x}$ (b) $\dfrac{x}{4}$ (c) 1 (d) x (e) $\dfrac{1}{x}$ (f) $\dfrac{4}{x}$ (g) $\dfrac{6}{x}$

3. (a) $\dfrac{a}{20}$ (b) $\dfrac{5a}{4b}$ (c) $\dfrac{1}{2ab}$ (d) $\dfrac{6x^2}{y^3}$ (e) $\dfrac{3b}{5a^2}$ (f) $\dfrac{x}{4y}$

(g) $\dfrac{x}{3(x + y)}$ (h) $\dfrac{1}{3(x + 4)}$

4. (a) $\dfrac{3y}{z^3}$ (b) $\dfrac{3 + x}{y}$ (c) $\dfrac{x}{12}$ (d) $\dfrac{b}{c}$

5. (a) $\dfrac{1}{x + 4}$ (b) $\dfrac{4(x - 2)}{x}$ (c) $\dfrac{12ab}{5ef} \times \dfrac{f}{4ab^2} = \dfrac{3}{5eb}$

(d) $\dfrac{x + 3y}{2x} \times \dfrac{4x^2}{y} = \dfrac{2x(x + 3y)}{y}$ (e) $\dfrac{9}{xyz}$

6. $\dfrac{2}{x + 3}$

7. $\dfrac{x + 4}{x + 3}$

Exercise 9.4

1. (a) $\dfrac{5z}{6}$ (b) $\dfrac{7x}{12}$ (c) $\dfrac{6y}{25}$

2. (a) $\dfrac{x + 2}{2x}$ (b) $\dfrac{1 + 2x}{2}$ (c) $\dfrac{1 + 3y}{3}$

(d) $\dfrac{y + 3}{3y}$ (e) $\dfrac{8y + 1}{y}$

3. (a) $\dfrac{10 - x}{2x}$ (b) $\dfrac{5 + 2x}{x}$ (c) $\dfrac{9 - x}{3x}$

(d) $\dfrac{2x - 3}{6}$ (e) $\dfrac{9 + x}{3x}$

4. (a) $\dfrac{3}{x} + \dfrac{4}{y} = \dfrac{3y}{xy} + \dfrac{4x}{xy} = \dfrac{3y + 4x}{xy}$

(b) $\dfrac{3}{x^2} + \dfrac{4y}{x} = \dfrac{3}{x^2} + \dfrac{4xy}{x^2} = \dfrac{3 + 4xy}{x^2}$

(c) $\dfrac{4ab}{x} + \dfrac{3ab}{2y} = \dfrac{8aby}{2xy} + \dfrac{3abx}{2xy} = \dfrac{8aby + 3abx}{2xy}$

(d) $\dfrac{4xy}{a} + \dfrac{3xy}{2b} = \dfrac{8xyb}{2ab} + \dfrac{3xya}{2ab} = \dfrac{8xyb + 3xya}{2ab}$

(e) $\dfrac{3}{x} - \dfrac{6}{2x} = \dfrac{3}{x} - \dfrac{3}{x} = 0$

(f) $\dfrac{3x}{2y} - \dfrac{7y}{4x} = \dfrac{6x^2}{4xy} - \dfrac{7y^2}{4xy} = \dfrac{6x^2 - 7y^2}{4xy}$

(g) $\dfrac{3}{x + y} - \dfrac{2}{y} = \dfrac{3y}{(x + y)y} - \dfrac{2(x + y)}{(x + y)y}$

$= \dfrac{3y - 2x - 2y}{(x + y)y} = \dfrac{y - 2x}{(x + y)y}$

(h) $\dfrac{1}{a + b} - \dfrac{1}{a - b}$

$= \dfrac{a - b}{(a + b)(a - b)} - \dfrac{a + b}{(a + b)(a - b)}$

$= \dfrac{a - b - a - b}{(a + b)(a - b)} = \dfrac{-2b}{(a + b)(a - b)}$

(i) $2x + \dfrac{1}{2x} = \dfrac{4x^2}{2x} + \dfrac{1}{2x} = \dfrac{4x^2 + 1}{2x}$

(j) $2x - \dfrac{1}{2x} = \dfrac{4x^2}{2x} - \dfrac{1}{2x} = \dfrac{4x^2 - 1}{2x}$

5. (a) $\dfrac{x}{y} + \dfrac{3x^2}{z} = \dfrac{xz}{yz} + \dfrac{3x^2y}{yz} = \dfrac{xz + 3x^2y}{yz}$

(b) $\dfrac{4}{a} + \dfrac{5}{b} = \dfrac{4b + 5a}{ab}$

(c) $\dfrac{6x}{y} - \dfrac{2y}{x} = \dfrac{6x^2 - 2y^2}{xy}$

(d) $3x - \dfrac{3x+1}{4} = \dfrac{12x}{4} - \dfrac{3x+1}{4}$

$= \dfrac{12x - 3x - 1}{4} = \dfrac{9x-1}{4}$

(e) $\dfrac{5a}{12} + \dfrac{9a}{18} = \dfrac{15a + 18a}{36} = \dfrac{33a}{36} = \dfrac{11a}{12}$

(f) $\dfrac{x-3}{4} + \dfrac{3}{5} = \dfrac{5(x-3) + 12}{20}$

$= \dfrac{5x - 15 + 12}{20} = \dfrac{5x-3}{20}$

6. (a) $\dfrac{2x+3}{(x+1)(x+2)}$ (b) $\dfrac{3x+1}{(x-1)(x+3)}$

(c) $\dfrac{4x+17}{(x+5)(x+4)}$ (d) $\dfrac{4x-10}{(x-2)(x-4)}$

(e) $\dfrac{5x+4}{(2x+1)(x+1)}$ (f) $\dfrac{x+1}{x(1-2x)}$

(g) $\dfrac{3x+7}{(x+1)^2}$ (h) $\dfrac{x}{(x-1)^2}$

Solutions to Chapter 10

Exercise 10.1

1. (a) $x = \dfrac{y}{3}$ (b) $x = \dfrac{1}{y}$ (c) $x = \dfrac{5+y}{7}$

(d) $x = 2y + 14$ (e) $x = \dfrac{1}{2y}$

(f) $2x + 1 = \dfrac{1}{y}$ so $2x = \dfrac{1}{y} - 1 = \dfrac{1-y}{y}$

finally, $x = \dfrac{1-y}{2y}$

(g) $y - 1 = \dfrac{1}{2x}$ so $2x = \dfrac{1}{y-1}$

and then $x = \dfrac{1}{2(y-1)}$

(h) $x = \dfrac{y+21}{18}$ (i) $x = \dfrac{19-y}{8}$

2. (a) $m = \dfrac{y-c}{x}$ (b) $x = \dfrac{y-c}{m}$ (c) $c = y - mx$

3. (a) $y = 13x - 26$ and so $x = \dfrac{y+26}{13}$

(b) $y = x + 1$ and so $x = y - 1$

(c) $y = a + tx - 3t$ and so $tx = y + 3t - a$

and then $x = \dfrac{y + 3t - a}{t}$

4. (a) $x = \dfrac{y - 11}{7}$ (b) $I = \dfrac{V}{R}$ (c) $r^3 = \dfrac{3V}{4\pi}$

and so $r = \sqrt[3]{\dfrac{3V}{4\pi}}$ (d) $m = \dfrac{F}{a}$

5. $n = (l - a)/d + 1$
6. $\sqrt{x} = (m - n)/t$ and so $x = [(m-n)/t]^2$
7. (a) $x^2 = 1 - y$ and so $x = \pm\sqrt{1-y}$

(b) $1 - x^2 = 1/y$ and so $x^2 = 1 - 1/y$,

finally, $x = \pm\sqrt{1 - 1/y}$

(c) Write $(1 + x^2)y = 1 - x^2$, remove

brackets to get $y + x^2 y = 1 - x^2$,

rearrange and factorize to give

$x^2(y + 1) = 1 - y$ from which

$x^2 = (1 - y)/(1 + y)$, finally

$x = \pm\sqrt{(1 - y)/(1 + y)}$

Solutions to Chapter 11

Exercise 11.1

2. (a) $x = \frac{9}{3} = 3$ (b) $x = 3 \times 9 = 27$

(c) $3t = -6$, so $t = \frac{-6}{3} = -2$ (d) $x = 22$

(e) $3x = 4$ and so $x = \frac{4}{3}$ (f) $x = 36$

(g) $5x = 3$ and so $x = \frac{3}{5}$ (h) $x + 3 = 6$

and so $x = 3$

(i) adding the two terms on the left we obtain

$$\frac{3x+2}{2} + 3x = \frac{3x+2}{2} + \frac{6x}{2}$$

$$= \frac{3x+2+6x}{2} = \frac{9x+2}{2}$$

and the equation becomes $(9x+2)/2 = 1$, therefore $9x + 2 = 2$, that is, $9x = 0$ and so $x = 0$

3. (a) $5x + 10 = 13$, so $5x = 3$ and so $x = \frac{3}{5}$
   (b) $3x - 21 = 2x + 2$ and so $x = 23$
   (c) $5 - 10x = 8 - 4x$ so that $-3 = 6x$
   finally $x = \frac{-3}{6} = -\frac{1}{2}$
4. (a) $t = 9$ (b) $v = \frac{17}{7}$ (c) $3s + 2 = 14s - 14$, so that $11s = 16$ and so $s = \frac{16}{11}$
5. (a) $t = 3$ (b) $t = 5$ (c) $t = -5$ (d) $t = 3$

   (e) $x = \dfrac{12}{5}$ (f) $x = -9$ (g) $x = 24$

   (h) $x = 15$ (i) $x = 23$ (j) $x = -\dfrac{43}{19}$

6. (a) $x = \dfrac{1}{5}$ (b) $x = \dfrac{2}{5}$ (c) $x = -\dfrac{3}{5}$ (d) $x = 1\dfrac{2}{5}$

   (e) $x = -1$ (f) $x = -3/5$ (g) $x = -1/2$
   (h) $x = -1/10$

Exercise 11.2

2. (a) Adding the two equations gives

$$\begin{array}{rcl} 3x + y &=& 1 \\ 2x - y &=& 2 \\ \hline 5x &=& 3 \end{array}$$

from which $x = \frac{3}{5}$. Substitution into either equation gives $y = -\frac{4}{5}$.
   (b) Subtracting the given equations eliminates y to give $x = 4$. From either equation $y = 1$.
   (c) Multiplying the second equation by 2 and subtracting the first gives

$$\begin{array}{rcl} 2x - y &=& 17 \\ 2x + 6y &=& 24 \\ \hline 7y &=& 7 \end{array} -$$

from which $y = 1$. Substitute into either equation to get $x = 9$.
   (d) Multiplying the second equation by -2 and subtracting from the first gives

$$\begin{array}{rcl} -2x + y &=& -21 \\ -2x - 6y &=& 28 \\ \hline 7y &=& -49 \end{array} -$$

from which $y = -7$. Substitution into either equation gives $x = 7$.
   (e) Multiplying the first equation by 3 and adding the second gives

$$\begin{array}{rcl} -3x + 3y &=& -30 \\ 3x + 7y &=& 20 \\ \hline 10y &=& -10 \end{array} +$$

from which $y = -1$. Substitution into either equation gives $x = 9$.
   (f) Multiplying the second equation by 2 and subtracting this from the first gives

$$\begin{array}{rcl} 4x - 2y &=& 2 \\ 6x - 2y &=& 8 \\ \hline -2x &=& -6 \end{array} -$$

from which $x = 3$. Substitution into either equation gives $y = 5$.
3. (a) $x = 7$, $y = 1$ (b) $x = -7$, $y = 2$
   (c) $x = 10$, $y = 0$ (d) $x = 0$, $y = 5$
   (e) $x = -1$, $y = -1$

Exercise 11.3

1. (a) $(x+2)(x-1) = 0$ so that $x = -2$ and 1
   (b) $(x-3)(x-5) = 0$ so that $x = 3$ and 5
   (c) $2(2x+1)(x+1) = 0$ so that $x = -1$ and $x = -\frac{1}{2}$ (d) $(x-3)(x-3) = 0$ so that $x = 3$ twice
   (e) $(x+9)(x-9) = 0$ so that $x = -9$ and 9
   (f) $-1, -3$ (g) $1, -3$ (h) $1, -4$ (i) $-1, -5$

(j) 5, 7 (k) -5, -7 (l) 1, $-\frac{3}{2}$ (m) $-\frac{3}{2}$, 2
(n) $-\frac{3}{2}$, 5 (o) 1, $-1/3$ (p) $-1/3$, $5/3$
(q) 0, $-\frac{1}{7}$ (r) $-\frac{3}{2}$ twice

2. (a) $x = \dfrac{-(-6) \pm \sqrt{(-6)^2 - 4(3)(-5)}}{6}$

$= \dfrac{6 \pm \sqrt{96}}{6}$

$= 2.633$ and -0.633

(b) $x = \dfrac{-3 \pm \sqrt{9 - 4(1)(-77)}}{2}$

$= \dfrac{-3 \pm \sqrt{317}}{2} = 7.402$ and -10.402

(c) $x = \dfrac{-(-9) \pm \sqrt{(-9)^2 - 4(2)(2)}}{4}$

$= \dfrac{9 \pm \sqrt{65}}{4} = 4.266$ and 0.234

(d) 1, -4 (e) 1.758, -0.758
(f) 0.390, -0.640 (g) 7.405, -0.405
(h) 0.405, -7.405 (i) no solutions
(j) 2.766, -1.266

3. (a) $(3x + 2)(2x + 3) = 0$
from which $x = -2/3$ and $-3/2$
(b) $(t + 3)(3t + 4) = 0$
from which $t = -3$ and $t = -4/3$
(c) $t = (7 \pm \sqrt{37})/2 = 6.541$ and 0.459

Solutions to Chapter 12

Exercises 12.2

1. (a) Multiply the input by 10 (b) multiply the input by -1 and then add 2, alternatively we could say subtract the input from 2 (c) raise the input to the power 4 and then multiply the result by 3 (d) divide 4 by the square of the input (e) take three times the square of the input, subtract twice the input from the result and finally add 9 (f) the output is always 5 whatever the value of the input (g) the output is always zero whatever the value of the input.

2. (a) Take three times the square of the input and add to twice the input (b) take three times the square of the input and add to twice the input. The functions in parts (a) and (b) are the same. Both instruct us to do the same thing.

3. (a) $f(x) = x^3/12$ – letters other than f and x are also valid.
(b) $f(x) = (x + 3)^2$
(c) $f(x) = x^2 + 4x - 10$
(d) $f(x) = x/(x^2 + 5)$ (e) $f(x) = x^3 - 1$
(f) $f(x) = (x - 1)^2$ (g) $f(x) = (7 - 2x)/4$
(h) $f(x) = -13$

4. (a) $A(2) = 3$ (b) $A(3) = 7$ (c) $A(0) = 1$
(d) $A(-1) = (-1)^2 - (-1) + 1 = 3$
5. (a) $y(1) = 1$
(b) $y(-1) = [2(-1) - 1]^2 = (-3)^2 = 9$
(c) $y(-3) = 49$ (d) $y(0.5) = 0$
(e) $y(-0.5) = 4$
6. (a) $f(t + 1) = 4(t + 1) + 6 = 4t + 10$
(b) $f(t + 2) = 4(t + 2) + 6 = 4t + 14$
(c) $f(t + 1) - f(t) = (4t + 10) - (4t + 6) = 4$
(d) $f(t + 2) - f(t) = (4t + 14) - (4t + 6) = 8$
7. (a) $f(n) = 2n^2 - 3$ (b) $f(z) = 2z^2 - 3$
(c) $f(t) = 2t^2 - 3$
(d) $f(2t) = 2(2t)^2 - 3 = 8t^2 - 3$
(e) $f(1/z) = 2(1/z)^2 - 3 = 2/z^2 - 3$
(f) $f(3/n) = 2(3/n)^2 - 3 = 18/n^2 - 3$
(g) $f(-x) = 2(-x)^2 - 3 = 2x^2 - 3$
(h) $f(-4x) = 2(-4x)^2 - 3 = 32x^2 - 3$
(i) $f(x + 1) = 2(x + 1)^2 - 3 = 2(x^2 + 2x + 1)$
$-3 = 2x^2 + 4x - 1$
(j) $f(2x - 1) = 2(2x - 1)^2 - 3$
$= 8x^2 - 8x - 1$
8. $a(p + 1) = (p + 1)^2 + 3(p + 1) + 1$
$= p^2 + 2p + 1 + 3p + 3 + 1 = p^2 + 5p + 5$
$a(p + 1) - a(p) = (p^2 + 5p + 5)$
$-(p^2 + 3p + 1) = 2p + 4$

9. (a) $f(3) = 6$ (b) $h(2) = 3$
   (c) $f[h(2)] = f(3) = 6$
   (d) $h[f(3)] = h(6) = 7$
10. (a) $f[h(t)] = f(t+1) = 2(t+1) = 2t+2$
    (b) $h[f(t)] = h(2t) = 2t+1$. Note that $h[f(t)]$ is not equal to $f[h(t)]$
11. (a) $f(0.5) = 0.5$ (first part) (b) $f(1.1) = 1$ (third part) (c) $f(1) = 2$ (second part)

Exercise 12.3

1. (a) $f(g(x)) = f(3x-2) = 4(3x-2) = 12x-8$
   (b) $g(f(x)) = g(4x) = 3(4x) - 2 = 12x - 2$
2. (a) $y(x(t)) = y(t^3) = 2t^3$
   (b) $x(y(t)) = x(2t) = (2t)^3 = 8t^3$
3. (a) $r(s(x)) = r(3x) = \frac{1}{6x}$
   (b) $t(s(x)) = t(3x) = 3x - 2$
   (c) $t(r(s(x))) = t(\frac{1}{6x}) = \frac{1}{6x} - 2$
   (d) $r(t(s(x))) = r(3x-2) = \frac{1}{2(3x-2)}$
   (e) $r(s(t(x))) = r(s(x-2)) = r(3(x-2)) = \frac{1}{6(x-2)}$
4. (a) $v(v(t)) = v(2t+1) = 2(2t+1) + 1 = 4t+3$
   (b) $v(v(v(t))) = v(4t+3) = 2(4t+3) + 1 = 8t+7$
5. (a) $m(n(t)) = m(t^2-1) = (t^2)^3 = t^6$
   (b) $n(m(t)) = n((t+1)^3) = (t+1)^6 - 1$
   (c) $m(p(t)) = m(t^2) = (t^2+1)^3$
   (d) $p(m(t)) = p((t+1)^3) = (t+1)^6$
   (e) $n(p(t)) = n(t^2) = t^4 - 1$
   (f) $p(n(t)) = p(t^2-1) = (t^2-1)^2$
   (g) $m(n(p(t))) = m(t^4-1) = (t^4)^3 = t^{12}$
   (h) $p(p(t)) = p(t^2) = t^4$
   (i) $n(n(t)) = n(t^2-1) = (t^2-1)^2 - 1$
   (j) $m(m(t)) = m((t+1)^3) = [(t+1)^3 + 1]^3$

Exercise 12.4

1. In each case let the inverse be denoted by g
   (a) $g(x) = x/3$
   (b) $g(x) = 4x$ (c) $g(x) = x - 1$
   (d) $g(x) = x + 3$ (e) $g(x) = 3 - x$
   (f) $g(x) = (x-6)/2$ (g) $g(x) = (7-x)/3$
   (h) $g(x) = 1/x$
   (i) $g(x) = 3/x$ (j) $g(x) = -3/4x$

2. (a) $f^{-1}(x) = \frac{x}{6}$
   (b) $f^{-1}(x) = \frac{x-1}{6}$
   (c) $f^{-1}(x) = x - 6$
   (d) $f^{-1}(x) = 6x$
   (e) $f^{-1}(x) = \frac{6}{x}$
3. (a) $g^{-1}(t) = \frac{t-1}{3}$
   (b) We require $g^{-1}\left(\frac{1}{3t+1}\right) = t$

   Let $z = \dfrac{1}{3t+1}$

   so that $t = \dfrac{1}{3}\left(\dfrac{1}{z} - 1\right)$

   Then $g^{-1}(z) = \dfrac{1}{3}\left(\dfrac{1}{z} - 1\right)$

   i.e. $g^{-1}(t) = \dfrac{1}{3}\left(\dfrac{1}{t} - 1\right)$

   (c) $g^{-1}(t) = t^{\frac{1}{3}}$
   (d) $g^{-1}(t) = (\frac{t}{3})^{\frac{1}{3}}$
   (e) $g^{-1}(t) = (\frac{t-1}{3})^{\frac{1}{3}}$
   (f) We require $g^{-1}\left(\dfrac{3}{t^3+1}\right) = t$

   Let $z = \dfrac{3}{t^3+1}$

   so that $t = \left(\dfrac{3}{z} - 1\right)^{\frac{1}{3}}$

   Then $g^{-1}(z) = \left(\dfrac{3}{z} - 1\right)^{\frac{1}{3}}$

   i.e. $g^{-1}(t) = \left(\dfrac{3}{t} - 1\right)^{\frac{1}{3}}$

4. (a) $h^{-1}(t) = \frac{t-3}{4}$
   (b) $g^{-1}(t) = \frac{t+1}{2}$
   (c)
   $$g^{-1}(h^{-1}(t)) = g^{-1}\left(\frac{t-3}{4}\right)$$
   $$= \frac{\frac{t-3}{4} + 1}{2}$$
   $$= \frac{t+1}{8}$$

   (d) $h(g(t)) = h(2t-1) = 4(2t-1) + 3 = 8t - 1$
   (e) $h(g(t))^{-1} = \frac{t+1}{8}$. We note that $[h(g(t))]^{-1}$ is the same as $g^{-1}(h^{-1}(t))$

Solutions to Chapter 13

Exercise 13.1
1.

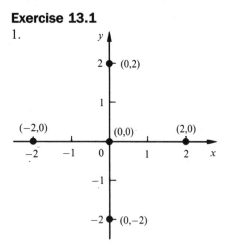

2. All y coordinates must be the same.
3. All x coordinates must be the same.

Exercise 13.2
1. (a) $\{x : x \in \mathbb{R}, 2 \le x \le 6\}$

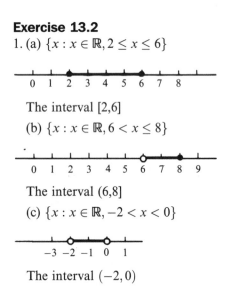

The interval [2,6]

(b) $\{x : x \in \mathbb{R}, 6 < x \le 8\}$

The interval (6,8]

(c) $\{x : x \in \mathbb{R}, -2 < x < 0\}$

The interval $(-2, 0)$

(d) $\{x : x \in \mathbb{R}, -3 \le x < -1.5\}$

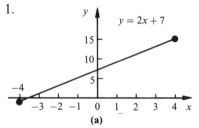

The interval $[-3, -1.5)$

2. (a) T (b) T (c) T (d) T (e) F (f) T (g) F
(h) T

Exercise 13.3
1.

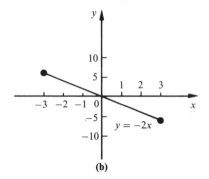

(a)

(b)

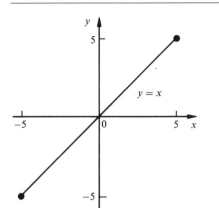

(c)

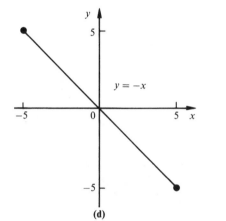

(d)

2.

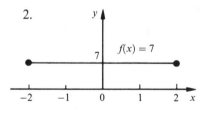

3. (a)

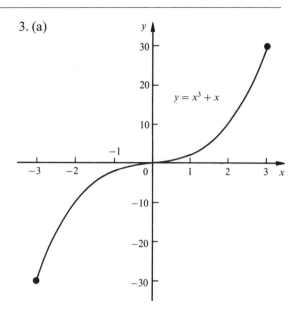

(b) (1,2) and (−1,−2) lie on the curve.

4.

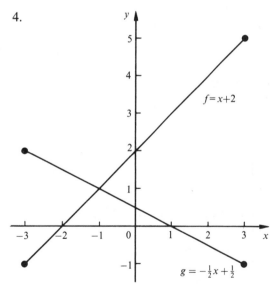

Graphs intersect at (−1,1)

5.

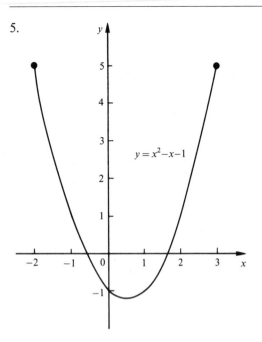

$y = x^2 - x - 1$

(a) The curve cuts the horizontal axis at $x = -0.6$ and $x = 1.6$.
(b) the curve cuts the vertical axis at $y = -1$.

6.

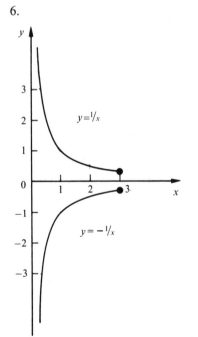

$y = 1/x$

$y = -1/x$

7.

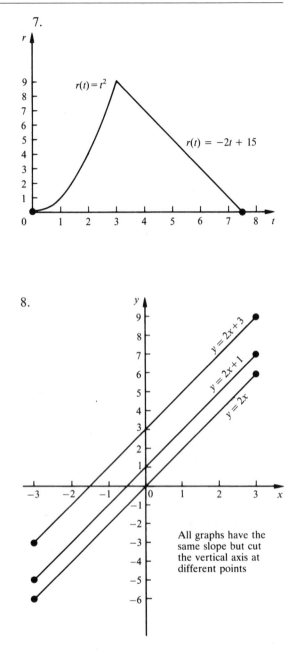

$r(t) = t^2$

$r(t) = -2t + 15$

8.

$y = 2x + 3$

$y = 2x + 1$

$y = 2x$

All graphs have the same slope but cut the vertical axis at different points

Exercise 13.4

1.

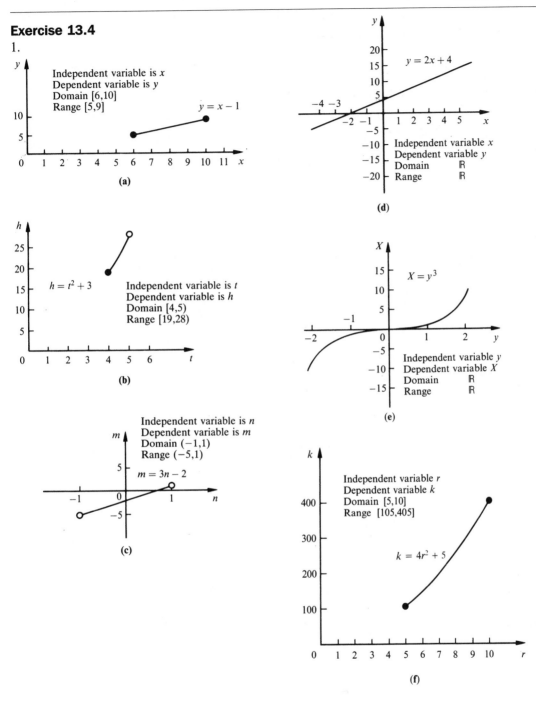

(a)

Independent variable is x
Dependent variable is y
Domain [6,10]
Range [5,9]

$y = x - 1$

(b)

$h = t^2 + 3$

Independent variable is t
Dependent variable is h
Domain [4,5)
Range [19,28)

(c)

Independent variable is n
Dependent variable is m
Domain (−1,1)
Range (−5,1)

$m = 3n - 2$

(d)

$y = 2x + 4$

Independent variable x
Dependent variable y
Domain   ℝ
Range   ℝ

(e)

$X = y^3$

Independent variable y
Dependent variable X
Domain   ℝ
Range   ℝ

(f)

Independent variable r
Dependent variable k
Domain [5,10]
Range [105,405]

$k = 4r^2 + 5$

2.

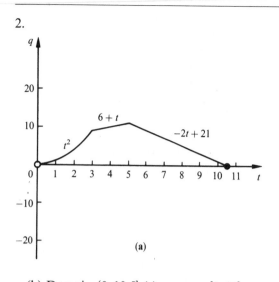

(a)

(b) Domain $(0, 10.5]$ (c) range $= [0, 11]$
(d) $q(1) = 1$, $q(3) = 9$, $q(5) = 11$, $q(7) = 7$

2.

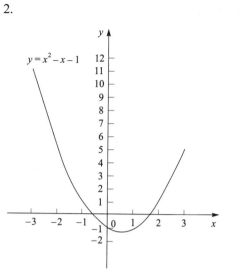

$y = x^2 - x - 1$

From the graph the roots of $x^2 - x - 1 = 0$ are $x = -0.6$ and $x = 1.6$.

Exercise 13.5
1.

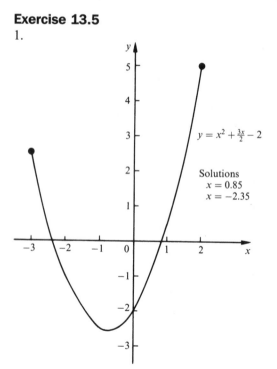

$y = x^2 + \frac{3x}{2} - 2$

Solutions
$x = 0.85$
$x = -2.35$

3.

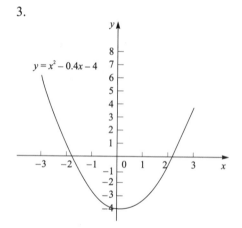

$y = x^2 - 0.4x - 4$

The required roots are $x = -1.8$ and $x = 2.2$.

4. (a)

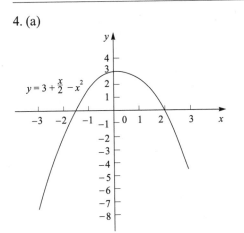

(b) From the graph the roots of
$x^2 - \frac{x}{2} - 3 = 0$ are $x = -1.5$ and $x = 2$.

Exercise 13.6

1. (a) We write the equations in the form
$$y = -\frac{3}{2}x + 2, \qquad y = x - 3$$

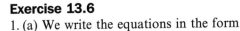

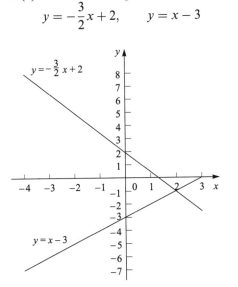

The point of intersection is
$x = 2$, $y = -1$ which is the solution to
the simultaneous equations.

(b) The equations are written as
$$y = -2x - 2, \qquad y = \frac{x}{4} + \frac{5}{2}$$

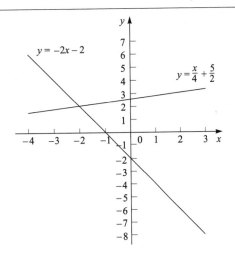

From the graph the solution is
$x = -2$, $y = 2$.

(c) The equations are written in the form
$$y = -2x + 4, \qquad y = \frac{4}{3}x + \frac{7}{3}$$

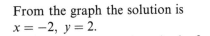

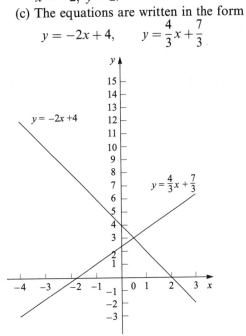

The solution is $x = \frac{1}{2}$, $y = 3$.

(d) The equations are written

$$y = \frac{x}{4} + \frac{7}{4}, \qquad y = 2x + 7$$

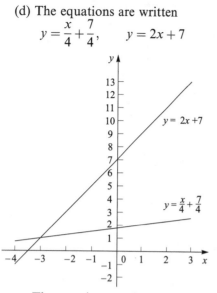

The root is $x = -3$, $y = 1$.

(e) The equations are written

$$y = -\frac{x}{2} + \frac{1}{2}, \qquad y = -\frac{x}{10} - \frac{1}{10}$$

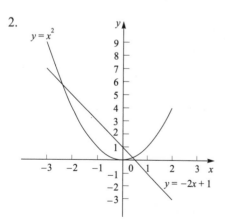

The root is $x = 1.5$, $y = -0.25$.

2.

From the graph the required roots are approximately $x = -2.4$, $y = 5.8$ and $x = 0.4$, $y = 0.2$.

3.

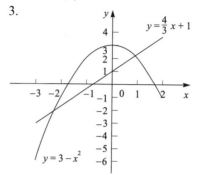

The roots of the simultaneous equations are approximately $x = -2.2$, $y = -2$ and $x = 0.9$, $y = 2.2$.

4.

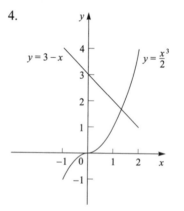

From the point of intersection we see the required root is approximately $x = 1.4$, $y = 1.5$.

5.

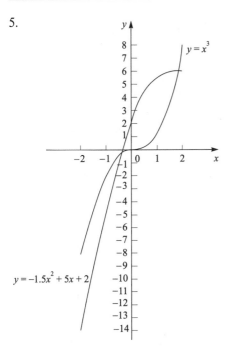

The points of intersection, and hence the
required roots are approximately
$x = -0.4$, $y = -0.2$ and $x = 1.8$, $y = 6.1$.

Solutions to Chapter 14

Exercise 14.1
1. (a), (b), (c) and (e) are straight lines.
2. (a) $m = 9$, $c = -11$ (b) $m = 8$, $c = 1.4$
   (c) $m = \frac{1}{2}$, $c = -11$
   (d) $m = -2$, $c = 17$ (e) $m = \frac{2}{3}$, $c = \frac{1}{3}$
   (f) $m = -\frac{2}{5}$, $c = \frac{4}{5}$
   (g) $m = 3$, $c = -3$ (h) $m = 0$, $c = 4$
3. (a) (i)-1 (ii) 6
   (b) (i) 2 (ii) -1
   (c) (i) 2 (ii) -1.5
   (d) (i) $\frac{3}{4}$ (ii) 3
   (e) (i) -6 (ii) 18

Exercise 14.2
1. $y = 2x + 5$

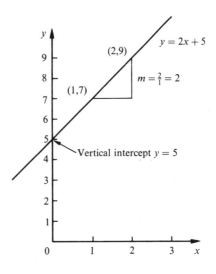

2. $y = 6x - 10$

3. $y = 2$

4. $y = x$

5. $y = -x$

6. $y = 2x + 8$

7. (b), (c) and (d) lie on the line.

8. The equation of the line is $y = mx + c$.
   The gradient is $m = -1$ and so

   $$y = -x + c$$

   When $x = -3$, $y = 7$ and so $c = 4$. Hence
   the required equation is $y = -x + 4$.

9. The gradient of $y = 3x + 17$ is 3. Let the
   required equation be $y = mx + c$.
   Since the lines are parallel then $m = 3$ and
   so $y = 3x + c$. When $x = -1$, $y = -6$ and
   so $c = -3$. Hence the equation is
   $y = 3x - 3$.

10. The equation is $y = mx + c$.
    The vertical intercept is $c = -2$ and so

    $$y = mx - 2$$

    When $x = 3, y = 10$ and so $m = 4$. Hence
    the required equation is $y = 4x - 2$.

2.

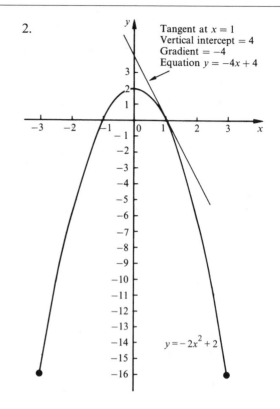

Tangent at $x = 1$
Vertical intercept $= 4$
Gradient $= -4$
Equation $y = -4x + 4$

$y = -2x^2 + 2$

Exercise 14.3

1.

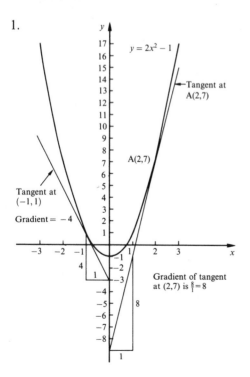

$y = 2x^2 - 1$

Tangent at A(2,7)

A(2,7)

Tangent at (−1, 1)

Gradient $= -4$

Gradient of tangent at (2,7) is $\frac{8}{1} = 8$

Solutions to Chapter 15

Exercise 15.1

1. (a) 9.9742; (b) 6.6859; (c) 0.5488; (d) 0.8808; (e) −8.4901.

2. (a) $e^2 \cdot e^7 = e^{2+7} = e^9$ (b) $\frac{e^7}{e^4} = e^{7-4} = e^3$

   (c) $\frac{e^{-2}}{e^3} = e^{-2-3} = e^{-5}$ or $\frac{1}{e^5}$

   (d) $\frac{e^3}{e^{-1}} = e^{3-(-1)} = e^4$;

   (e) $\frac{(4e)^2}{(2e)^3} = \frac{4^2 e^2}{2^3 e^3} = \frac{16 e^{2-3}}{8} = 2e^{-1} = \frac{2}{e}$

   (f) $e^2(e + \frac{1}{e}) - e = e^2 e + \frac{e^2}{e} - e$
       $= e^3 + e - e = e^3$

   (g) $e^{1.5} e^{2.7} = e^{1.5+2.7} = e^{4.2}$

   (h) $\sqrt{e}\sqrt{4e} = e^{0.5}(4e)^{0.5} = e^{0.5} 4^{0.5} e^{0.5}$
       $= 2e^1 = 2e$

3. (a) $e^x e^{-3x} = e^{x-3x} = e^{-2x}$

   (b) $e^{3x} e^{-x} = e^{3x-x} = e^{2x}$

   (c) $e^{-3x} e^{-x} = e^{-3x-x} = e^{-4x}$

   (d) $(e^{-x})^2 e^{2x} = e^{-2x} e^{2x} = e^0 = 1$

   (e) $\frac{e^{3x}}{e^x} = e^{3x-x} = e^{2x}$

   (f) $\frac{e^{-3x}}{e^{-x}} = e^{-3x-(-x)} = e^{-2x}$

4. (a) $\frac{e^t}{e^3} = e^{t-3}$

   (b) $\frac{2e^{3x}}{4e^x} = \frac{1}{2} e^{3x-x} = \frac{e^{2x}}{2}$

   (c) $\frac{(e^x)^3}{3e^x} = \frac{e^{3x}}{3e^x} = \frac{e^{3x-x}}{3} = \frac{e^{2x}}{3}$

   (d) $e^x + 2e^{-x} + e^x = 2e^x + 2e^{-x}$
       $= 2(e^x + e^{-x})$

   (e) $\frac{e^x e^y e^z}{e^{(x/2)-y}} = e^{x+y+z-((\frac{x}{2})-y)} = e^{(\frac{x}{2})+2y+z}$

   (f) $\frac{(e^{\frac{t}{3}})^6}{(e^t + e^t)} = \frac{e^{(\frac{t}{3})\times 6}}{2e^t} = \frac{e^{2t}}{2e^t} = \frac{e^t}{2}$

   (g) $\frac{e^t}{e^{-t}} = e^{t-(-t)} = e^{2t}$

   (h) $\frac{1+e^{-t}}{e^{-t}} = (1 + e^{-t})e^t = e^t + e^{-t}e^t$
       $= e^t + e^0 = e^t + 1$

   (i) $(e^z e^{-\frac{z}{2}})^2 = (e^{z-\frac{z}{2}})^2 = (e^{\frac{z}{2}})^2 = e^z$

   (j) $\frac{e^{-3+t}}{2e^t} = \frac{e^{-3+t-t}}{2} = \frac{e^{-3}}{2} = \frac{1}{2e^3}$

   (k) $(\frac{e^{-1}}{e^{-x}})^{-1} = (e^{-1-(-x)})^{-1} = (e^{x-1})^{-1}$
       $= e^{(x-1)(-1)} = e^{1-x}$

Exercise 15.2

1.

x	−1	−0.8	−0.6	−0.4	−0.2
e^{3x}	0.0498	0.0907	0.1653	0.3012	0.5488

x	0	0.2	0.4	0.6	0.8	1
e^{3x}	1	1.8221	3.3201	6.0496	11.0232	20.0855

The graph has similar shape and properties to that of $y = e^x$.

2. (a)

t	0	1	2	3	4	5
P	5	8.1606	9.3233	9.7511	9.9084	9.9663

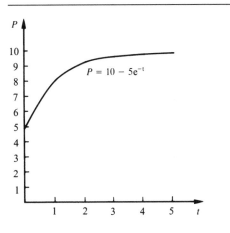

(b) Maximum population is 10, obtained for large values of t.

3. (a)

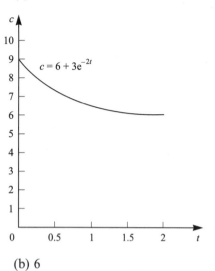

$$c = 6 + 3e^{-2t}$$

(b) 6

4. (a) $\cosh x + \sinh x$

$$= \frac{e^x + e^{-x}}{2} + \frac{e^x - e^{-x}}{2}$$

$$= \frac{e^x + e^{-x} + e^x - e^{-x}}{2} = e^x$$

(b) $\cosh x - \sinh x$

$$= \frac{e^x + e^{-x}}{2} - \frac{(e^x - e^{-x})}{2}$$

$$= \frac{e^x + e^{-x} - e^x + e^{-x}}{2} = e^{-x}$$

5. $3 \sinh x + 7 \cosh x$

$$= 3\left(\frac{e^x - e^{-x}}{2}\right) + 7\left(\frac{e^x + e^{-x}}{2}\right)$$

$$= 5e^x + 2e^{-x}$$

6. $6e^x - 9e^{-x} = 6(\cosh x + \sinh x)$

$$- 9(\cosh x - \sinh x)$$

$$= 15 \sinh x - 3 \cosh x$$

Exercise 15.3

1. (a)

x	0	0.5	1	1.5	2	2.5	3
$15 - x^2$	15	14.75	14	12.75	11	8.75	6
e^x	1	1.65	2.72	4.48	7.39	12.18	20.09

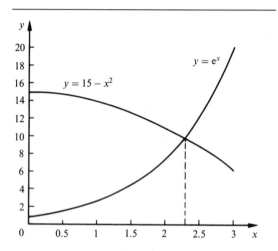

(b) Solving $e^x + x^2 = 15$ is equivalent to solving $e^x = 15 - x^2$. Approximate solution is $x = 2.3$.

2. (a)

x	-1	-0.75	-0.5	-0.25
$12x^2 - 1$	11	5.75	2	-0.25
$3e^{-x}$	8.15	6.35	4.95	3.85

x	0	0.25	0.5	0.75	1
$12x^2 - 1$	-1	-0.25	2	5.75	11
$3e^{-x}$	3	2.34	1.82	1.42	1.10

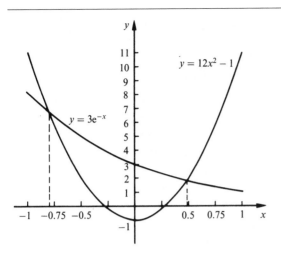

(b) Approximate solutions are $x = -0.80$ and $x = 0.49$.

3.

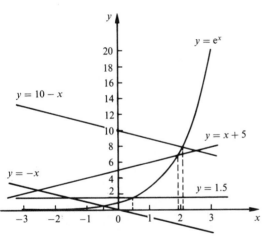

(a) The equation $e^x + x = 0$ is equivalent to $e^x = -x$. Hence draw $y = -x$. Intersection of $y = e^x$ and $y = -x$ will give solution to $e^x + x = 0$. Approximate solution: $x = -0.57$.
(b) The equation $e^x - 1.5 = 0$ is equivalent to $e^x = 1.5$. Hence also draw $y = 1.5$. This is a horizontal straight line. Intersection point gives solution to $e^x - 1.5 = 0$.

Approximate solution: $x = 0.41$.
(c) The equation $e^x - x - 5 = 0$ can be written as $e^x = x + 5$. Hence draw $y = x + 5$. Approximate solution: $x = 1.94$.
(d) The equation

$$\frac{e^x}{2} + \frac{x}{2} - 5 = 0$$

can be written as

$$e^x = 10 - x$$

Hence also draw $y = 10 - x$. Approximate solution: $x = 2.07$.

4. (a)

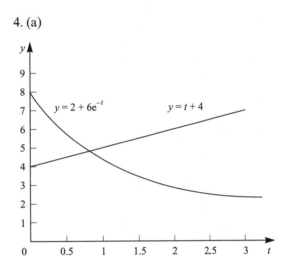

(b) The t value corresponding to $y = 5$ on the graph is $t = 0.7$. Hence the root of $2 + 6e^{-t} = 5$ is $t = 0.7$.
(c) $y = t + 4$ is also shown. The point of intersection of $y = 2 + 6e^{-t}$ and $y = t + 4$ is a solution of
$$2 + 6e^{-t} = t + 4$$
$$6e^{-t} = t + 2$$

The solution is $t = 0.77$.

Solutions to Chapter 16

Exercise 16.1

1. (a) 2.1761 (b) 5.0106 (c) −0.5003
   (d) −2.3026
2. (a) $\log_3 6561 = 8$ (b) $\log_6 7776 = 5$
   (c) $\log_2 1024 = 10$
   (d) $\log_{10} 100000 = 5$ (e) $\log_4 16384 = 7$
   (f) $\log_{0.5} 0.03125 = 5$
   (g) $\log_{12} 1728 = 3$ (h) $\log_9 6561 = 4$
3. (a) $1296 = 6^4$ (b) $225 = 15^2$ (c) $512 = 8^3$
   (d) $2401 = 7^4$
   (e) $243 = 3^5$ (f) $216 = 6^3$ (g) $8000 = 20^3$
   (h) $4096 = 16^3$
   (i) $4096 = 2^{12}$
4. (a) 0.4771
   (b) $\qquad 3 = 10^{0.4771}$

   $$3 \times 10^2 = 10^2 \times 10^{0.4771}$$
   $$300 = 10^{2.4771}$$
   $$\log 300 = 2.4771$$

   (c) $\qquad 3 = 10^{0.4771}$

   $$3 \times 10^{-2} = 10^{-2} \times 10^{0.4771}$$
   $$0.03 = 10^{-1.5229}$$
   $$\log 0.03 = -1.5229$$

5. (a) 7
   (b) $\qquad \log 7 = 0.8451$

   $$7 = 10^{0.8451}$$
   $$7 \times 10^2 = 10^2 \times 10^{0.8451}$$
   $$700 = 10^{2.8451}$$
   $$\log 700 = 2.8451$$

   (c) $\qquad 7 = 10^{0.8451}$

   $$10^{-2} \times 7 = 10^{-2} \times 10^{0.8451}$$
   $$0.07 = 10^{-1.1549}$$
   $$\log 0.07 = -1.1549$$

Exercise 16.2

1. (a) $\log_4 6 = \log 6 / \log 4 = 1.2925$
   (b) $\log_3 10 = \log 10 / \log 3 = 2.0959$
   (c) $\log_{20} 270 = \log 270 / \log 20 = 1.8688$
   (d) $\log_5 0.65 = \log 0.65 / \log 5 = -0.2677$
   (e) $\log_2 100 = \log 100 / \log 2 = 6.6439$
   (f) $\log_2 0.03 = \log 0.03 / \log 2 = -5.0589$
   (g) $\log_{100} 10 = \log 10 / \log 100 = \frac{1}{2}$
   (h) $\log_7 7 = 1$

2. $\quad \ln X = \dfrac{\log_{10} X}{\log_{10} e} = \dfrac{\log_{10} X}{0.4343} = 2.3026 \log_{10} X$

3. (a) $\quad \log_3 7 + \log_4 7 + \log_5 7$

   $$= \frac{\log 7}{\log 3} + \frac{\log 7}{\log 4} + \frac{\log 7}{\log 5} = 4.3840$$

   (b) $\quad \log_8 4 + \log_8 0.25 = \dfrac{\log 4}{\log 8} + \dfrac{\log 0.25}{\log 8}$

   $$= 0.6667 + (-0.6667)$$
   $$= 0$$

   (c) $\quad \log_{0.7} 2 = \dfrac{\log 2}{\log 0.7} = -1.9434$

   (d) $\quad \log_2 0.7 = \dfrac{\log 0.7}{\log 2} = -0.5146$

Exercise 16.3

1. (a) $\log 5 + \log 9 = \log(5 \times 9) = \log 45$
   (b) $\log 9 - \log 5 = \log(\frac{9}{5}) = \log 1.8$
   (c) $\log 5 - \log 9 = \log(\frac{5}{9})$
   (d) $2\log 5 + \log 1 = 2\log 5 = \log 5^2 = \log 25$
   (e) $2\log 4 - 3\log 2 = \log 4^2 - \log 2^3$
      $= \log 16 - \log 8 = \log(\frac{16}{8}) = \log 2$
   (f) $\log 64 - 2\log 2 = \log(64/2^2) = \log 16$

(g) $3\log 4 + 2\log 1 + \log 27 - 3\log 12$

$\quad = \log 4^3 + 2(0) + \log 27 - \log 12^3$

$\quad = \log\left(\dfrac{4^3 \times 27}{12^3}\right)$

$\quad = \log 1 = 0$

2. (a) $\log 3x$ (b) $\log 8x$ (c) $\log 1.5$
   (d) $\log T^2$ (e) $\log 10X^2$

3. (a) $3\log X - \log X^2 = 3\log X - 2\log X = \log X$
   (b) $\log y - 2\log \sqrt{y} = \log y - \log(\sqrt{y})^2$
   $\qquad\qquad\qquad\quad = \log y - \log y = 0$

(c) $5\log x^2 + 3\log\dfrac{1}{x} = 10\log x - 3\log x$

$\qquad\qquad\qquad = 7\log x$

(d) $4\log X - 3\log X^2 + \log X^3$
   $\quad = 4\log X - 6\log X + 3\log X$
   $\quad = \log X$

(e) $3\log y^{1.4} + 2\log y^{0.4} - \log y^{1.2}$
   $\quad = 4.2\log y + 0.8\log y - 1.2\log y$
   $\quad = 3.8\log y$

4. (a) $\log 4x - \log x = \log(4x/x) = \log 4$
   (b) $\log t^3 + \log t^4 = \log(t^3 \times t^4) = \log t^7$

(c) $\log 2t - \log\left(\dfrac{t}{4}\right) = \log\left(2t \div \dfrac{t}{4}\right)$

$\quad = \log\left(2t \times \dfrac{4}{t}\right) = \log 8$

(d) $\log 2 + \log\left(\dfrac{3}{x}\right) - \log\left(\dfrac{x}{2}\right)$

$\quad = \log\left(2 \times \dfrac{3}{x}\right) - \log\left(\dfrac{x}{2}\right)$

$\quad = \log\left(\dfrac{6}{x} \div \dfrac{x}{2}\right)$

$\quad = \log\left(\dfrac{6}{x} \times \dfrac{2}{x}\right)$

$\quad = \log\left(\dfrac{12}{x^2}\right)$

(e) $\log\left(\dfrac{t^2}{3}\right) + \log\left(\dfrac{6}{t}\right) - \log\left(\dfrac{1}{t}\right)$

$\quad = \log\left(\dfrac{t^2}{3} \times \dfrac{6}{t}\right) - \log\left(\dfrac{1}{t}\right)$

$\quad = \log(2t) - \log\left(\dfrac{1}{t}\right)$

$\quad = \log\left(\dfrac{2t}{1/t}\right)$

$\quad = \log 2t^2$

(f) $2\log y - \log y^2 = \log y^2 - \log y^2 = 0$

(g) $3\log\left(\dfrac{1}{t}\right) + \log t^2 = \log\left(\dfrac{1}{t}\right)^3 + \log t^2$

$\quad = \log\left(\dfrac{1}{t^3}\right) + \log t^2$

$\quad = \log\left(\dfrac{t^2}{t^3}\right)$

$\quad = \log\left(\dfrac{1}{t}\right) = -\log t$

(h) $4\log\sqrt{x} + 2\log\left(\dfrac{1}{x}\right) = 4\log x^{0.5} + \log\left(\dfrac{1}{x}\right)^2$

$\quad = \log x^2 + \log\left(\dfrac{1}{x^2}\right)$

$\quad = \log\left(x^2\dfrac{1}{x^2}\right)$

$\quad = \log 1 = 0$

(i) $2\log x + 3\log t = \log x^2 + \log t^3 = \log(x^2 t^3)$

(j) $\log A - \dfrac{1}{2}\log 4A = \log A - \log(4A)^{1/2}$

$\quad = \log\left[\dfrac{A}{(4A)^{1/2}}\right]$

$\quad = \log\left(\dfrac{A^{1/2}}{2}\right)$

(k) $\dfrac{\log 9x + \log 3x^2}{3} = \dfrac{\log(9x.3x^2)}{3}$

$= \dfrac{\log(27x^3)}{3}$

$= \dfrac{\log(3x)^3}{3}$

$= \dfrac{3\log(3x)}{3}$

$= \log(3x)$

(l) $\log xy + 2\log\left(\dfrac{x}{y}\right) + 3\log\left(\dfrac{y}{x}\right)$

$= \log xy + \log\left(\dfrac{x^2}{y^2}\right) + \log\left(\dfrac{y^3}{x^3}\right)$

$= \log\left(xy\dfrac{x^2}{y^2}\dfrac{y^3}{x^3}\right)$

$= \log\left(\dfrac{x^3y^4}{x^3y^2}\right)$

$= \log y^2$

(m) $\log\left(\dfrac{A}{B}\right) - \log\left(\dfrac{B}{A}\right) = \log\left(\dfrac{A}{B} \div \dfrac{B}{A}\right)$

$= \log\left(\dfrac{A}{B} \times \dfrac{A}{B}\right) = \log\left(\dfrac{A^2}{B^2}\right)$

(n) $\log\left(\dfrac{2t}{3}\right) + \dfrac{1}{2}\log 9t - \log\left(\dfrac{1}{t}\right)$

$= \log\left(\dfrac{2t}{3}\right) + \log(9t)^{1/2} - \log t^{-1}$

$= \log\left[\dfrac{2t}{3}(9t)^{1/2}\right] + \log t$

$= \log\left[\dfrac{2t}{3}(9t)^{1/2}t\right]$

$= \log\left(\dfrac{2t}{3}3t^{1/2}t\right)$

$= \log(2t^{2.5})$

5. $\log_{10} X + \ln X = \log_{10} X + \dfrac{\log_{10} X}{\log_{10} e}$

$= \log_{10} X + 2.3026\log_{10} X$

$= 3.3026\log_{10} X$

$= \log X^{3.3026}$

6. (a) $\log(9x-3) - \log(3x-1) = \log\left(\dfrac{9x-3}{3x-1}\right)$

$= \log 3$

(b) $\log(x^2-1) - \log(x+1) = \log\left(\dfrac{x^2-1}{x+1}\right)$

$= \log(x-1)$

(c) $\log(x^2+3x) - \log(x+3) = \log\left(\dfrac{x^2+3x}{x+3}\right)$

$= \log x$

Exercise 16.4

1 (a) $\log x = 1.6000$

$x = 10^{1.6000} = 39.8107$

(b) $10^x = 75$

$x = \log 75 = 1.8751$

(c) $\ln x = 1.2350$

$x = e^{1.2350} = 3.4384$

(d) $e^x = 36$

$x = \ln 36 = 3.5835$

2. (a) $\log(3t) = 1.8$

$3t = 10^{1.8}$

$t = \dfrac{10^{1.8}}{3} = 21.0319$

(b) $10^{2t} = 150$

$2t = \log 150$

$t = \dfrac{1}{2}\log 150 = 1.0880$

(c) $\ln(4t) = 2.8$

$4t = e^{2.8}$

$t = \dfrac{1}{4}e^{2.8} = 4.1112$

(d) $e^{3t} = 90$

$\quad 3t = \ln 90$

$\quad t = \dfrac{1}{3} \ln 90 = 1.4999$

3. (a) $\log x = 0.3940$

$\quad x = 10^{0.3940} = 2.4774$

(b) $\ln x = 0.3940$

$\quad x = e^{0.3940} = 1.4829$

(c) $10^y = 5.5$

$\quad y = \log 5.5 = 0.7404$

(d) $e^z = 500$

$\quad z = \ln 500 = 6.2146$

(e) $\log(3v) = 1.6512$

$\quad 3v = 10^{1.6512}$

$\quad v = \dfrac{10^{1.6512}}{3} = 14.9307$

(f) $\ln\left(\dfrac{t}{6}\right) = 1$

$\quad \dfrac{t}{6} = e^1 = e$

$\quad t = 6e = 16.3097$

(g) $10^{2r+1} = 25$

$\quad 2r + 1 = \log 25$

$\quad 2r = \log 25 - 1$

$\quad r = \dfrac{\log 25 - 1}{2} = 0.1990$

(h) $e^{(2t-1)/3} = 7.6700$

$\quad \dfrac{2t - 1}{3} = \ln 7.6700$

$\quad 2t - 1 = 3\ln 7.6700$

$\quad 2t = 3\ln 7.6700 + 1$

$\quad t = \dfrac{3\ln 7.6700 + 1}{2} = 3.5560$

(i) $\log(4b^2) = 2.6987$

$\quad 4b^2 = 10^{2.6987}$

$\quad b^2 = \dfrac{10^{2.6987}}{4} = 124.9223$

$\quad b = \sqrt{124.9223} = \pm 11.1769$

(j) $\log\left(\dfrac{6}{2+t}\right) = 1.5$

$\quad \dfrac{6}{2+t} = 10^{1.5}$

$\quad 6 = 10^{1.5}(2 + t)$

$\quad \dfrac{6}{10^{1.5}} = 2 + t$

$\quad t = \dfrac{6}{10^{1.5}} - 2 = -1.8103$

(k) $\ln(2r^3 + 1) = 3.0572$

$\quad 2r^3 + 1 = e^{3.0572}$

$\quad 2r^3 = e^{3.0572} - 1$

$\quad r^3 = \dfrac{e^{3.0572} - 1}{2} = 10.1340$

$\quad r = 2.1640$

(l) $\ln(\log t) = -0.3$

$\quad \log t = e^{-0.3} = 0.7408$

$\quad t = 10^{0.7408} = 5.5058$

(m) $10^{t^2-1} = 180$

$\quad t^2 - 1 = \log 180$

$\quad t^2 = \log 180 + 1 = 3.2553$

$\quad t = \sqrt{3.2553} = \pm 1.8042$

(n) $10^{3r^4} = 170000$

$\quad 3r^4 = \log 170000$

$\quad r^4 = \dfrac{\log 170000}{3} = 1.7435$

$\quad r = \pm 1.1491$

(o) $\log 10^t = 1.6$

$t \log 10 = 1.6$

$t = 1.6$

(p) $\ln(e^x) = 20000$

$x \ln e = 20000$

$x = 20000$

4. (a) $e^{3x}e^{2x} = 59$

$e^{5x} = 59$

$5x = \ln 59$

$x = \dfrac{\ln 59}{5} = 0.8155$

(b) $10^{3t}.10^{4-t} = 27$

$10^{4+2t} = 27$

$4 + 2t = \log 27$

$2t = \log 27 - 4$

$t = \dfrac{\log 27 - 4}{2} = -1.2843$

(c) $\log(5 - t) + \log(5 + t) = 1.2$

$\log(5 - t)(5 + t) = 1.2$

$\log(25 - t^2) = 1.2$

$25 - t^2 = 10^{1.2}$

$t^2 = 25 - 10^{1.2}$

$= 9.1511$

$t = \pm 3.0251$

(d)
$$\log x + \ln x = 4$$
$$\log x + \frac{\log x}{\log e} = 4$$
$$\log x \left(1 + \frac{1}{\log e} \right) = 4$$
$$3.3026 \log x = 4$$
$$\log x = \frac{4}{3.3026} = 1.2112$$
$$x = 10^{1.2112} = 16.2619$$

Exercise 16.5

1. (a)

x	0.5	1	2	3	4
$\log x^2$	−0.6021	0	0.6021	0.9542	1.2041

x	5	6	7	8	9	10
$\log x^2$	1.3979	1.5563	1.6902	1.8062	1.9085	2

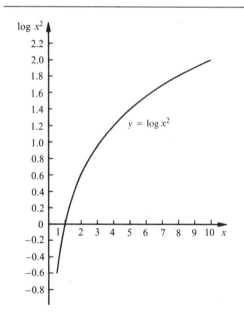

(b)

x	0.1	0.2	0.5	0.75	1
$\log(1/x)$	1	0.6990	0.3010	0.1249	0

x	2	3	4	5	6
$\log(1/x)$	−0.3010	−0.4771	−0.6021	−0.6990	−0.7782

x	7	8	9	10
$\log(1/x)$	−0.8451	−0.9031	−0.9542	−1

2. (a)

x	0.1	0.5	1	1.5	2	2.5	3
$\log x$	−1	−0.3010	0	0.1761	0.3010	0.3979	0.4771

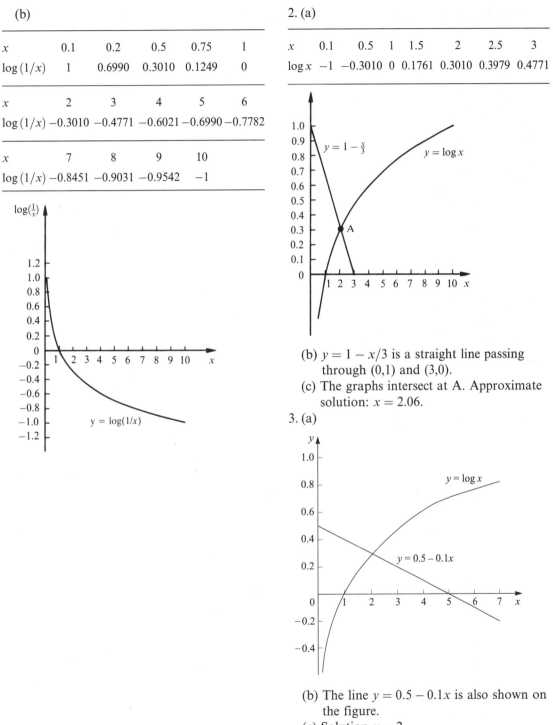

(b) $y = 1 - x/3$ is a straight line passing through (0,1) and (3,0).

(c) The graphs intersect at A. Approximate solution: $x = 2.06$.

3. (a)

(b) The line $y = 0.5 - 0.1x$ is also shown on the figure.

(c) Solution $x = 2$.

Solutions to Chapter 17

Exercise 17.1

1. (a) $240° = \pi/180 \times 240 = 4.19$
   (b) $300° = \pi/180 \times 300 = 5.24$
   (c) $400° = \pi/180 \times 400 = 6.98$
   (d) $37° = \pi/180 \times 37 = 0.65$
   (e) $1000° = \pi/180 \times 1000 = 17.45$
2. (a) $\pi/10 = 180/10 = 18°$
   (b) $2\pi/9 = \frac{2}{9} \times 180 = 40°$
   (c) $2 = 180/\pi \times 2 = 114.59°$
   (d) $3.46 = 180/\pi \times 3.46 = 198.24°$
   (e) $1.75 = 180/\pi \times 1.75 = 100.27°$
3. (a) $720°$ (b) $630°$ (c) $286.5°$ (d) $89.4°$
4. (a) 2.8π (b) 1.2π (c) $\frac{7\pi}{3}$ (d) 0.7π (e) 1.8π

Exercise 17.2

1. (a) length of arc $= r\theta = 15(2) = 30$ cm
   (b) length of arc $= r\theta = 15(3) = 45$ cm
   (c) length of arc $= r\theta = 15(1.2) = 18$ cm
   (d) length of arc $= r\theta = 15(100 \times \pi/180)$
       $= 26.18$ cm
   (e) length of arc $= r\theta = 15(217 \times \pi/180)$
       $= 56.81$ cm
2. (a) angle subtended

$$= \theta = \frac{\text{arc length}}{\text{radius}} = \frac{12}{12} = 1$$

   (b) $\theta = \dfrac{\text{arc length}}{\text{radius}} = \dfrac{6}{12} = 0.5$

   (c) $\theta = \dfrac{\text{arc length}}{\text{radius}} = \dfrac{24}{12} = 2$

   (d) $\theta = \dfrac{\text{arc length}}{\text{radius}} = \dfrac{15}{12} = 1.25$

   (e) $\theta = \dfrac{\text{arc length}}{\text{radius}} = \dfrac{2}{12} = \dfrac{1}{6}$

3. (a) Arc length $= r\theta = 18 \times 1.5 = 27$ cm
   (b) 19.8 cm (c) 18.85 cm (͡ cm

4. (a) Angle $= \dfrac{\text{arc length}}{\text{radius}} =$

   (b) 1.25 (c) 0.8333 (d) 2.5

Exercise 17.3

1. (a)
   $$\text{area of sector} = \frac{r^2\theta}{2} = \frac{9^2(1.5)}{2} = 60.75 \text{ cm}^2$$

   (b) area of sector $= \dfrac{r^2\theta}{2} = \dfrac{9^2(2)}{2} = 81 \text{ cm}^2$

   (c) area of sector $= \dfrac{r^2\theta}{2} = \dfrac{9^2}{2}\left(100 \times \dfrac{\pi}{180}\right)$

   $$= 70.69 \text{ cm}^2$$

   (d) area of sector $= \dfrac{r^2\theta}{2} = \dfrac{9^2}{2}\left(215 \times \dfrac{\pi}{180}\right)$

   $$= 151.97 \text{ cm}^2$$

2. area of sector $= \dfrac{r^2\theta}{2}$

   $$\theta = \frac{2 \times \text{area}}{r^2}$$

   When $r = 16$ we have

   $$\theta = \frac{\text{area}}{128}$$

   (a) Angle $= \theta = \frac{100}{128} = 0.78$
   (b) angle $= \frac{5}{128} = 0.039$
   (c) angle $= \frac{520}{128} = 4.06$

3. (a) Area $= \dfrac{r^2\theta}{2} = \dfrac{(18)^2(1.5)}{2} = 243 \text{ cm}^2$
   (b) 356.4 cm^2
   (c) $120° = \frac{2\pi}{3}$ and so

   $$\text{Area} = \frac{(18)^2(2\pi/3)}{2} = 339.3 \text{ cm}^2$$

   (d) $217° = \dfrac{217}{180}\pi = 3.7874$

   $$\text{Area} = \frac{(18)^2(3.7874)}{2} = 613.6 \text{ cm}^2$$

4. Angle subtended by arc AB $= \frac{17}{25} = 0.68$

   $$\text{Area} = \frac{r^2\theta}{2} = \frac{(25)^2(0.68)}{2} = 212.5 \text{ cm}^2$$

5. Area $= \dfrac{r^2\theta}{2}$

$$370 = \dfrac{(12)^2\theta}{2}$$

$$\theta = \dfrac{2(370)}{12^2} = 5.1389$$

Arc length $= r\theta = 12(5.1389) = 61.67\,\text{cm}$

6. Arc length $= r\theta$, so $r = \dfrac{16}{1.2} = 13.33\,\text{cm}$

$$\text{Area} = \dfrac{r^2\theta}{2} = \dfrac{(13.33)^2(1.2)}{2} = 107\,\text{cm}^2$$

Solutions to Chapter 18

Exercise 18.1
1. (a) 0.5774 (b) 0.3420 (c) 0.2588
   (d) 0.9320 (e) 0.6294 (f) 1

2. (a) $\sin 70° = 0.9397, \quad \cos 20° = 0.9397$
   (b) $\sin 25° = 0.4226, \quad \cos 65° = 0.4226$
   (c) in Figure 18.1, $A = \theta$ and $C = 90° - \theta$, hence
      $\sin\theta = BC/AC$ and $\cos(90° - \theta) = BC/AC$,
      so $\sin\theta = \cos(90° - \theta)$
   (d) In Figure 18.1
      $\cos A = \cos\theta = AB/AC$
      $\sin C = \sin(90° - \theta) = AB/AC$
      So
      $\cos\theta = \sin(90° - \theta)$

3. (a) $\sin\theta = BC/AC = 3/\sqrt{18}$
   (b) $\cos\theta = AB/AC = 3/\sqrt{18}$
   (c) $\tan\theta = BC/AB = \frac{3}{3} = 1$

4. (a) $\sin\theta = \frac{4}{\sqrt{116}} = 0.3714$
   (b) $\cos\theta = \frac{10}{\sqrt{116}} = 0.9285$
   (c) $\tan\theta = \frac{4}{10} = 0.4$
   (d) $\sin\alpha = \frac{10}{\sqrt{116}} = 0.9285$
   (e) $\cos\alpha = \frac{4}{\sqrt{116}} = 0.3714$
   (f) $\tan\alpha = \frac{10}{4} = 2.5$

Exercise 18.2
1. (a) 53.13° (b) 78.46° (c) 52.43°
2. (a) 0.6816 (b) 1.3181 (c) 1.1607
3. (a) 26.57° (b) 66.80° (c) 53.13°
4. (a) 0.1799 (b) 0.9818 (c) 0.7956

Solutions to Chapter 19

Exercise 19.1
1. When $\sin\alpha > 0$ then α must be in the first or second quadrant. When $\cos\alpha < 0$ then α must be in the second or third quadrant. To satisfy both conditions, α must be in the second quadrant.
2. β can be in the third or fourth quadrant.

3. 2nd and 3rd quadrants.

4. 3rd and 4th quadrants.

5. (a) $\sin\theta = \dfrac{1}{\sqrt{10}}$, $\cos\theta = -\dfrac{3}{\sqrt{10}}$, $\tan\theta = -\dfrac{1}{3}$

(b) $-\dfrac{5}{\sqrt{29}}$, $-\dfrac{2}{\sqrt{29}}$, $\dfrac{5}{2}$

(c) $-\dfrac{3}{\sqrt{18}} = -\dfrac{1}{\sqrt{2}}$, $\dfrac{3}{\sqrt{18}} = \dfrac{1}{\sqrt{2}}$, -1

Exercise 19.2

1. (a) $\sin 0° = 0$, $\cos 0° = 1$

(b) 1, 0

(c) 0, −1

(d) −1, 0

(e) 0, 1

2. (a)

θ	0	30	60	90	120	150
$\sin(\theta + 45°)$	0.71	0.97	0.97	0.71	0.26	−0.26

θ	180	210	240	270	300	330	360
$\sin(\theta + 45°)$	−0.71	−0.97	−0.97	−0.71	−0.26	0.26	0.71

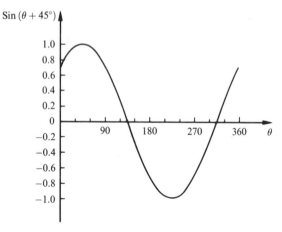

(b)

θ	0	60	120	180	240	300	360
$3\cos(\theta/2)$	3	2.60	1.5	0	−1.5	−2.60	−3

θ	420	480	540	600	660	720
$3\cos(\theta/2)$	−2.60	−1.5	0	1.5	2.60	3

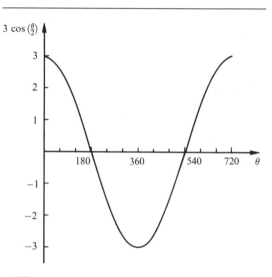

(c)

θ	0	30	60	90	120	150	180
$2\tan(\theta + 60°)$	3.46	–	−3.46	−1.15	0	1.15	3.46

θ	210	240	270	300	330	360
$2\tan(\theta + 60°)$	–	−3.46	−1.15	0	1.15	3.46

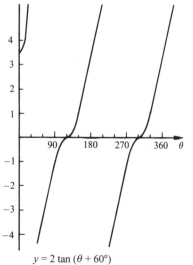

$y = 2\tan(\theta + 60°)$

Solutions to Chapter 20

Exercise 20.1

1. (a) $\theta = 50°$; $\theta = 180 - 50 = 130°$

   (b) $\theta = 180 - 40 = 140°$ and also
   $\theta = 180 + 40 = 220°$

   (c) $\theta = 20°$; $\theta = 180 + 20 = 200°$

   (d) $\theta = 180 + 70 = 250°$;
   $\theta = 360 - 70 = 290°$

   (e) $\theta = 10°$; $\theta = 360 - 10 = 350°$

   (f) $\theta = 180 - 80 = 100°$;
   $\theta = 360 - 80 = 280°$

   (g) $\theta = 0°$; $\theta = 180°, \ 360°$

   (h) $\theta = 90°$; $\theta = 360 - 90 = 270°$

2. (a) $20°, 340°$ (b) $190°, 350°$ (c) $140°, 320°$

3. $\cos 2A = \cos^2 A - \sin^2 A$
   $$= \cos^2 A - (1 - \cos^2 A)$$
   $$= 2\cos^2 A - 1$$

4. $\sin^2 A + \cos^2 A = 1$. Dividing this equation by $\cos^2 A$ gives:
   $$\frac{\sin^2 A}{\cos^2 A} + \frac{\cos^2 A}{\cos^2 A} = \frac{1}{\cos^2 A}$$
   $$\tan^2 A + 1 = \frac{1}{\cos^2 A}$$

5. $\cos 2\theta + \cos 8\theta$
   $$= 2\cos\left(\frac{2\theta + 8\theta}{2}\right)\cos\left(\frac{2\theta - 8\theta}{2}\right)$$
   $$= 2\cos 5\theta \cos(-3\theta)$$

   Recall that $\cos(-x) = \cos x$ and so
   $$\cos 2\theta + \cos 8\theta = 2\cos 5\theta \cos 3\theta$$

   and so
   $$\frac{\cos 2\theta + \cos 8\theta}{2\cos 3\theta} = \cos 5\theta$$

6. (a) $\cos(270° - \theta)$
   $$= \cos 270° \cos\theta + \sin 270° \sin\theta = -\sin\theta$$

   (b) $\cos(270° + \theta)$
   $$= \cos 270° \cos\theta - \sin 270° \sin\theta = \sin\theta$$

   (c) $\sin(270° + \theta)$
   $$= \sin 270° \cos\theta + \sin\theta \cos 270° = -\cos\theta$$

   (d)
   $$\tan(135° + \theta) = \frac{\tan 135° + \tan\theta}{1 - \tan 135° \tan\theta}$$
   $$= \frac{-1 + \tan\theta}{1 + \tan\theta}$$
   $$= \frac{\tan\theta - 1}{\tan\theta + 1}$$

   (e) $\sin(270° - \theta)$
   $$= \sin 270° \cos\theta - \cos 270° \sin\theta$$
   $$= -\cos\theta$$

7. $\dfrac{1 - \tan A}{1 + \tan A} = \dfrac{\tan 45° - \tan A}{1 + \tan 45° \tan A} = \tan(45° - A)$

8. Note that
   $$\cos 2\theta = \cos^2\theta - \sin^2\theta = 1 - 2\sin^2\theta$$
   using $\cos^2\theta = 1 - \sin^2\theta$
   and so $1 - \cos 2\theta = 2\sin^2\theta$.
   Also note that
   $$\cos 2\theta = \cos^2\theta - \sin^2\theta = 2\cos^2\theta - 1$$
   using $\sin^2\theta = 1 - \cos^2\theta$
   Finally we note that $\sin 2\theta = 2\sin\theta \cos\theta$.
   Hence

   $$\frac{1 - \cos 2\theta + \sin 2\theta}{1 + \cos 2\theta + \sin 2\theta} = \frac{2\sin^2\theta + 2\sin\theta\cos\theta}{2\cos^2\theta + 2\sin\theta\cos\theta}$$
   $$= \frac{2\sin\theta(\sin\theta + \cos\theta)}{2\cos\theta(\cos\theta + \sin\theta)}$$
   $$= \tan\theta$$

9. $\dfrac{\sin 4\theta + \sin 2\theta}{\cos 4\theta + \cos 2\theta} = \dfrac{2\sin 3\theta \cos\theta}{2\cos 3\theta \cos\theta} = \tan 3\theta$

10. $\dfrac{\tan A}{\tan^2 A + 1} = \dfrac{\sin A / \cos A}{\left(\frac{\sin A}{\cos A}\right)^2 + 1}$

    Multiplying numerator and denominator by $\cos^2 A$ gives:

    $$\frac{\sin A \cos A}{\sin^2 A + \cos^2 A} = \sin A \cos A = \frac{1}{2}\sin 2A$$

11. (a) $\dfrac{1}{\cos A} - \cos A = \dfrac{1 - \cos^2 A}{\cos A}$

$$= \frac{\sin^2 A}{\cos A} = \sin A \tan A$$

(b) $\dfrac{1}{\sin A} - \sin A = \dfrac{1 - \sin^2 A}{\sin A}$

$$= \frac{\cos^2 A}{\sin A} = \frac{\cos A}{\tan A}$$

12. $\sin 3\theta = \sin(2\theta + \theta)$

$$= \sin 2\theta \cos \theta + \sin \theta \cos 2\theta$$
$$= (2 \sin \theta \cos \theta) \cos \theta$$
$$\quad + \sin \theta(\cos^2 \theta - \sin^2 \theta)$$
$$= 2 \sin \theta \cos^2 \theta + \sin \theta \cos^2 \theta - \sin^3 \theta$$
$$= 3 \sin \theta \cos^2 \theta - \sin^3 \theta$$
$$= 3 \sin \theta(1 - \sin^2 \theta) - \sin^3 \theta$$
$$= 3 \sin \theta - 4 \sin^3 \theta$$

Exercises 20.2

1. (a) Using a calculator, we see
$\sin^{-1}(0.9) = 64.16°$.
Also $\sin^{-1}(0.9) = 180 - 64.16 = 115.84°$.
(b) Using a calculator we see
$\cos^{-1}(0.45) = 63.26°$.
Also $\cos^{-1}(0.45) = 360 - 63.26 = 296.74°$.
(c) Using a calculator we see
$\tan^{-1} 1.3 = 52.43°$.
Also $\tan^{-1} 1.3 = 180 + 52.43 = 232.43°$.
(d) Using a calculator we see
$\sin^{-1} 0.6 = 36.87$. Hence
$\sin^{-1}(-0.6) = 180 + 36.87 = 216.87°$ and
$360 - 36.87 = 323.13°$.
(e) Using a calculator we see
$\cos^{-1} 0.75 = 41.41°$. So
$\cos^{-1}(-0.75) = 180 - 41.41 = 138.59°$
and $180 + 41.41 = 221.41°$.
(f) We have $\tan^{-1} 0.3 = 16.70°$ and so
$\tan^{-1}(-0.3) = 180 - 16.70 = 163.30°$
and $360 - 16.70 = 343.30°$.
(g) $\sin^{-1} 1 = 90°$
(h) $\cos^{-1}(-1) = 180°$

2. (a) $\sin(2\theta + 50°) = 0.5$
$$2\theta + 50° = (30),\ 150,\ 390,\ 510,$$
$$750, \ ...$$
$$2\theta = (-20),\ 100,\ 340,\ 460,$$
$$700, \ ...$$
$$\theta = 50,\ 170,\ 230,\ 350,\ ...$$

Solutions in the range $0°$ to $360°$ are
$$\theta = 50°,\ 170°,\ 230°,\ 350°$$

(b) $\cos(\theta + 110°) = 0.3$
$$\theta + 110° = (72.54),\ 287.46,\ 432.54$$
$$\theta = 177.46°,\ 322.54°$$

(c) $\tan\left(\dfrac{\theta}{2}\right) = 1$
$$\frac{\theta}{2} = 45°,\ (225°)$$
$$\theta = 90°$$

(d) $\sin\left(\dfrac{\theta}{2}\right) = -0.5$
$$\frac{\theta}{2} = 210,\ 330$$
$$\theta = 420,\ 660$$
No solution between $0°$ and $360°$.

(e) $\cos(2\theta - 30°) = -0.5$
$$2\theta - 30° = 120,\ 240,\ 480,\ 600$$
$$2\theta = 150,\ 270,\ 510,\ 630$$
$$\theta = 75°, 135°,\ 255°,\ 315°$$

(f) $\tan(3\theta - 20°) = 0.25$
$$3\theta - 20° = 14.04,\ 194.04,\ 374.04,$$
$$554.04,\ 734.04,\ 914.04$$
$$3\theta = 34.04,\ 214.04,\ 394.04,$$
$$574.04,\ 754.04,\ 934.04$$
$$\theta = 11.35°,\ 71.35°,\ 131.35°,$$
$$191.35°,\ 251.35°,\ 311.35°$$

(g) $\sin 2\theta = 2\cos 2\theta$

$$\frac{\sin 2\theta}{\cos 2\theta} = 2$$

$$\tan 2\theta = 2$$

$$2\theta = 63.43, \ 243.43, \ 423.43, \ 603.43$$

$$\theta = 31.72°, \ 121.72°, \ 211.72°, \ 301.72°$$

3. (a) $\cos(\theta - 100°) = 0.3126$

$$\theta - 100° = \cos^{-1}(0.3126)$$

$$= -71.78°, \ 71.78°,$$

$$288.22°, ...$$

$$\theta = 28.22°, \ 171.78°$$

(b) $\tan(2\theta + 20°) = -1$

$$2\theta + 20° = \tan^{-1}(-1)$$

$$= 135°, \ 315°, \ 495°,$$

$$675°, ...$$

$$\theta = 57.5°, \ 147.5°, \ 237.5°,$$

$$327.5°$$

(c) $\sin\left(\dfrac{2\theta}{3} + 30°\right) = -0.4325$

$$\frac{2\theta}{3} + 30° = \sin^{-1}(-0.4325)$$

$$= 205.6°, \ 334.4°, ...$$

$$\theta = 263.4°$$

Solutions to Chapter 21

Exercise 21.1

1. $C = 180 - 70 - 42 = 68°$

2. $A + B + C = 180$

$$B + C = 180 - A = 180 - 42 = 138$$

$$2C + C = 138$$

$$C = 46°$$

$$B = 2C = 92°$$

3. $A + B + C = 180$

$$A + B = 180 - C = 180 - 114 = 66$$

Since A and B are equal then $A = B = 33°$.

4. In $\triangle ABC$, $A = 50°$, $B = 72°$ and $C = 58°$:
(a) the largest angle is B and so the longest side is b, that is, AC; (b) the smallest angle is A and so the shortest side is a, that is, BC.

5. There are two solutions.
(i) Suppose $A = 40°$ and $B = C$. Then

$$A + B + C = 180°$$

$$B + C = 180 - 40 = 140$$

$$B = C = 70°$$

The angles are $40°$, $70°$, $70°$.
(ii) Suppose $A = B = 40°$. Then $C = 100°$. The angles are $40°$, $40°$, $100°$.

Exercise 21.2

1. $a^2 = c^2 - b^2 = 30^2 - 17^2 = 611$

$$a = \sqrt{611} = 24.72$$

The length of BC is 24.72 cm.

2. $d^2 = c^2 + e^2$

$$= (1.7)^2 + (1.2)^2 = 4.33$$

$$d = \sqrt{4.33} = 2.08$$

The length of CE is 2.08 cm.

3. Let the length of $b = $ AC be k cm. Then $AC/AB = b/c = k/c = \frac{2}{1}$ so $c = k/2$. The length of $c = $ AB is $k/2$ cm. Now,

$$a = \text{BC and } b^2 = a^2 + c^2$$

$$k^2 = a^2 + \left(\frac{k}{2}\right)^2 = a^2 + \frac{k^2}{4}$$

$$\frac{3}{4}k^2 = a^2$$

$$a = \sqrt{\frac{3}{4}k^2} = \frac{\sqrt{3}}{2}k$$

So AC:BC $= k : (\sqrt{3}/2)k$. This is the same as $1 : \sqrt{3}/2$ or $2 : \sqrt{3}$.

4. Let the length of $y = $ XZ be k cm. The length of YZ is then $2k$ cm. Using Pythagoras' theorem we see that

$$x^2 = y^2 + z^2$$
$$(2k)^2 = k^2 + 10^2$$
$$4k^2 = k^2 + 100$$
$$3k^2 = 100$$
$$k^2 = \frac{100}{3}$$
$$k = \sqrt{\frac{100}{3}} = 5.774$$
$$2k = 11.547$$

The length of XZ is 5.774 cm and the length of YZ is 11.547 cm.

5. $m^2 = l^2 + n^2$

$$30^2 = l^2 + 26^2$$
$$l^2 = 30^2 - 26^2 = 224$$
$$l = \sqrt{224} = 14.97$$

The length of MN is 14.97 cm.

Exercise 21.3

1. $B = 180 - 90 - 37 = 53°$

$$\tan A = \frac{BC}{AC} = \frac{a}{36}$$
$$\tan 37° = \frac{a}{36}$$
$$a = 36 \tan 37° = 27.13$$

$$\cos A = \frac{AC}{AB} = \frac{36}{c}$$
$$c = \frac{36}{\cos A} = \frac{36}{\cos 37°} = 45.08$$

The solution is
$$A = 37°   a = BC = 27.13 \text{ cm}$$
$$B = 53°   b = AC = 36 \text{ cm}$$
$$C = 90°   c = AB = 45.08 \text{ cm}$$

2. Using Pythagoras' theorem

$$e^2 = c^2 + d^2$$
$$= 21^2 + 14^2 = 637$$
$$e = \sqrt{637} = 25.24$$

$$\tan C = \frac{21}{14} = 1.5$$
$$C = \tan^{-1}(1.5) = 56.31°$$

Finally, $D = 180 - 90 - 56.31 = 33.69°$. The solution is
$$C = 56.31°   c = DE = 21 \text{ cm}$$
$$D = 33.69°   d = CE = 14 \text{ cm}$$
$$E = 90°   e = CD = 25.24 \text{ cm}$$

3. $Z = 180 - 90 - 26 = 64°$

$$\sin Y = \sin 26° = \frac{XZ}{YZ} = \frac{y}{45}$$
$$y = 45 \sin 26° = 19.73$$

and
$$\cos Y = \cos 26° = \frac{XY}{YZ} = \frac{z}{45}$$
$$z = 45 \cos 26° = 40.45$$

The solution is
$$X = 90°   x = YZ = 45 \text{ mm}$$
$$Y = 26°   y = XZ = 19.73 \text{ mm}$$
$$Z = 64°   z = XY = 40.45 \text{ mm}$$

4. Using Pythagoras' theorem,
$$j^2 = i^2 + k^2$$
$$= 27^2 + 15^2 = 954$$
$$j = \sqrt{954} = 30.89$$

$$\tan I = \frac{i}{k} = \frac{27}{15} = 1.8$$
$$I = \tan^{-1} 1.8 = 60.95°$$

and

$$\tan K = \frac{15}{27}$$
$$K = \tan^{-1}\left(\frac{15}{27}\right) = 29.05$$

The solution is
$$I = 60.95°   i = JK = 27 \text{ cm}$$
$$J = 90°   j = IK = 30.89 \text{ cm}$$
$$K = 29.05°   k = IJ = 15 \text{ cm}$$

5. $Q = 180 - 62 - 28 = 90°$ and so $\triangle PQR$ is a right-angled triangle.

$$\sin R = \frac{r}{q} = \frac{r}{22}$$
$$PQ = r = 22 \sin R = 22 \sin 28° = 10.33$$

and

$$\cos R = \frac{p}{q} = \frac{p}{22}$$
$$QR = p = 22 \cos R = 22 \cos 28° = 19.42$$

The solution is
$$P = 62° \quad p = QR = 19.42 \text{ cm}$$
$$Q = 90° \quad q = PR = 22 \text{ cm}$$
$$R = 28° \quad r = PQ = 10.33 \text{ cm}$$

6. $T = 180 - 70 - 20 = 90°$ so $\triangle RST$ is a right-angled triangle.

$$\cos R = \cos 70° = \frac{s}{t} = \frac{12}{t}$$
$$RS = t = \frac{12}{\cos 70°} = 35.09$$

$$\tan R = \tan 70° = \frac{r}{s} = \frac{r}{12}$$
$$ST = r = 12 \tan 70° = 32.97$$

The solution is
$$R = 70° \quad r = ST = 32.97 \text{ cm}$$
$$S = 20° \quad s = RT = 12 \text{ cm}$$
$$T = 90° \quad t = RS = 35.09 \text{ cm}$$

Exercise 21.4

1. (a) $\dfrac{a}{\sin A} = \dfrac{b}{\sin B} = \dfrac{c}{\sin C}$
$$\frac{a}{\sin A} = \frac{24}{\sin B} = \frac{31}{\sin 37°}$$

So

$$\sin B = \frac{24 \sin 37°}{31} = 0.4659$$

$$B = \sin^{-1}(0.4659) = 27.77°$$

Hence $A = 180 - 37 - 27.77 = 115.23°$. Finally

$$a = \frac{31 \sin A}{\sin 37°} = \frac{31 \sin 115.23°}{\sin 37°} = 46.60$$

The solution is
$$A = 115.23° \quad a = BC = 46.60 \text{ cm}$$
$$B = 27.77° \quad b = AC = 24 \text{ cm}$$
$$C = 37° \quad c = AB = 31 \text{ cm}$$

(b) Substituting the given values into the sine rule gives

$$\frac{a}{\sin A} = \frac{24}{\sin B} = \frac{19}{\sin 37°}$$

So

$$\sin B = \frac{24 \sin 37°}{19} = 0.7602$$
$$B = \sin^{-1}(0.7602) = 49.48° \text{ or } 130.52°$$

Both values are acceptable and so there are two solutions.

Case 1: $B = 49.48°$

$$A = 180 - 37 - 49.48 = 93.52°$$
$$a = \frac{19 \sin A}{\sin 37°} = \frac{19 \sin 93.52}{\sin 37°} = 31.51$$

Case 2: $B = 130.52°$

$$A = 180 - 37 - 130.52 = 12.48°$$
$$a = \frac{19 \sin A}{\sin 37°} = \frac{19 \sin 12.48°}{\sin 37°} = 6.82$$

The solutions are
$$A = 93.52° \quad a = BC = 31.51 \text{ cm}$$
$$B = 49.48° \quad b = AC = 24 \text{ cm}$$
$$C = 37° \quad c = AB = 19 \text{ cm}$$
and
$$A = 12.48° \quad a = BC = 6.82 \text{ cm}$$
$$B = 130.52° \quad b = AC = 24 \text{ cm}$$
$$C = 37° \quad c = AB = 19 \text{ cm}$$

(c) $\dfrac{17}{\sin 17°} = \dfrac{b}{\sin B} = \dfrac{c}{\sin 101°}$

$$c = \frac{17 \sin 101°}{\sin 17°} = 57.08$$

Also $B = 180 - 17 - 101 = 62°$. So

$$b = \frac{17 \sin B}{\sin 17°} = \frac{17 \sin 62°}{\sin 17°} = 51.34$$

The solution is

$A = 17°$ $a = BC = 17$ cm
$B = 62°$ $b = AC = 51.34$ cm
$C = 101°$ $c = AB = 57.08$ cm

(d) $\dfrac{8.9}{\sin 53°} = \dfrac{b}{\sin B} = \dfrac{9.6}{\sin C}$

$\sin C = \dfrac{9.6 \sin 53°}{8.9} = 0.8614$

$C = \sin^{-1}(0.8614) = 59.48°$ or $120.52°$

Both values are acceptable and so there are two solutions.

Case 1: $C = 59.48°$

$B = 180 - 53 - 59.48 = 67.52°$

$b = \dfrac{8.9 \sin B}{\sin 53°} = \dfrac{8.9 \sin 67.52°}{\sin 53°} = 10.30$

Case 2: $C = 120.52°$

$B = 180 - 53 - 120.52 = 6.48°$

$b = \dfrac{8.9 \sin B}{\sin 53°} = \dfrac{8.9 \sin 6.48°}{\sin 53°} = 1.26$

The solutions are

$A = 53°$ $a = BC = 8.9$ cm
$B = 67.52°$ $b = AC = 10.30$ cm
$C = 59.48°$ $c = AB = 9.6$ cm
and
$A = 53°$ $a = BC = 8.9$ cm
$B = 6.48°$ $b = AC = 1.26$ cm
$C = 120.52°$ $c = AB = 9.6$ cm

(e) $\dfrac{14.5}{\sin 62°} = \dfrac{b}{\sin B} = \dfrac{12.2}{\sin C}$

$\sin C = \dfrac{12.2 \sin 62°}{14.5} = 0.7429$

$C = \sin^{-1}(0.7429) = 47.98°$

Hence $B = 180 - 62 - 47.98 = 70.02°$.

$b = \dfrac{14.5 \sin B}{\sin 62°} = \dfrac{14.5 \sin 70.02°}{\sin 62°} = 15.43$

The solution is

$A = 62°$ $a = BC = 14.5$ cm
$B = 70.02°$ $b = AC = 15.43$ cm
$C = 47.98°$ $c = AB = 12.2$ cm

(f) $\dfrac{a}{\sin A} = \dfrac{11}{\sin 36°} = \dfrac{c}{\sin 50°}$

$c = \dfrac{11 \sin 50°}{\sin 36°} = 14.34$

Also $A = 180 - 36 - 50 = 94°$ so that

$a = \dfrac{11 \sin A}{\sin 36°} = \dfrac{11 \sin 94°}{\sin 36°} = 18.67$

The solution is

$A = 94°$ $a = BC = 18.67$ cm
$B = 36°$ $b = AC = 11$ cm
$C = 50°$ $c = AB = 14.34$ cm

Exercise 21.5

1. (a) $a = BC = 86$, $b = AC = 119$, $c = AB = 53$

$a^2 = b^2 + c^2 - 2bc \cos A$

$86^2 = 119^2 + 53^2 - 2(119)(53) \cos A$

$12614 \cos A = 119^2 + 53^2 - 86^2 = 9574$

$\cos A = \dfrac{9574}{12614} = 0.7590$

$= \cos^{-1}(0.7590) = 40.62°$

$b^2 = a^2 + c^2 - 2ac \cos B$

$119^2 = 86^2 + 53^2 - 2(86)(53) \cos B$

$9116 \cos B = 86^2 + 53^2 - 119^2 = -3956$

$\cos B = -\dfrac{3956}{9116} = -0.4340$

$B = \cos^{-1}(-0.4340) = 115.72°$

$C = 180 - 40.62 - 115.72 = 23.66°$. The solution is

$A = 40.62°$ $a = BC = 86$ cm
$B = 115.72°$ $b = AC = 119$ cm
$C = 23.66°$ $c = AB = 53$ cm

(b) $b = AC = 30$, $c = AB = 42$,
$A = 115°$

$$a^2 = b^2 + c^2 - 2bc \cos A$$
$$a^2 = 30^2 + 42^2 - 2(30)(42) \cos 115°$$
$$= 3728.9980$$
$$a = \sqrt{3728.998} = 61.07$$

Now using the cosine rule again,
$$b^2 = a^2 + c^2 - 2ac \cos B$$
$$30^2 = 3728.998 + 1764$$
$$\qquad - 2(61.07)(42) \cos B$$
$$5129.88 \cos B = 4592.998$$
$$\cos B = 0.8953$$
$$B = 26.45°$$
$C = 180 - 115 - 26.45 = 38.55°$. The solution is
$$A = 115° \quad a = BC = 61.07 \text{ cm}$$
$$B = 26.45° \quad b = AC = 30 \text{ cm}$$
$$C = 38.55° \quad c = AB = 42 \text{ cm}$$

(c) $a = BC = 74, b = AC = 93, C = 39°$

$$c^2 = a^2 + b^2 - 2ab \cos C$$
$$c^2 = 74^2 + 93^2 - 2(74)(93) \cos 39°$$
$$= 3428.3630$$
$$c = 58.552$$

$$a^2 = b^2 + c^2 - 2bc \cos A$$
$$5476 = 8649 + 3428.3630$$
$$\qquad - 2(93)(58.552) \cos A$$
$$10890.672 \cos A = 6601.363$$
$$\cos A = 0.6061$$
$$A = 52.69°$$

$B = 180 - 52.69 - 39 = 88.31°$. The solution is
$$A = 52.69° \quad a = BC = 74 \text{ cm}$$
$$B = 88.31° \quad b = AC = 93 \text{ cm}$$
$$C = 39° \quad c = AB = 58.55 \text{ cm}$$

(d) $a = BC = 3.6, b = AC = 2.7,$
$\quad c = AB = 1.9$

$$a^2 = b^2 + c^2 - 2bc \cos A$$
$$3.6^2 = 2.7^2 + 1.9^2 - 2(2.7)(1.9) \cos A$$
$$10.26 \cos A = -2.06$$
$$\cos A = -0.2008$$
$$A = 101.58°$$

$$b^2 = a^2 + c^2 - 2ac \cos B$$
$$2.7^2 = 3.6^2 + 1.9^2 - 2(3.6)(1.9) \cos B$$
$$13.68 \cos B = 3.6^2 + 1.9^2 - 2.7^2 = 9.28$$
$$\cos B = \frac{9.28}{13.68} = 0.6784$$
$$B = 47.28°$$

$C = 180 - 101.58 - 47.28 = 31.14°$. The solution is
$$A = 101.58° \quad a = BC = 3.6 \text{ cm}$$
$$B = 47.28° \quad b = AC = 2.7 \text{ cm}$$
$$C = 31.14° \quad c = AB = 1.9 \text{ cm}$$

(e) $a = BC = 39, c = AB = 29, B = 100°$.

$$b^2 = a^2 + c^2 - 2ac \cos B$$
$$= 39^2 + 29^2 - 2(39)(29) \cos 100°$$
$$= 2754.7921$$
$$b = 52.49$$

$$a^2 = b^2 + c^2 - 2bc \cos A$$
$$39^2 = 52.49^2 + 29^2 - 2(52.49)(29) \cos A$$
$$3044.42 \cos A = 52.49^2 + 29^2 - 39^2 = 2074.79$$
$$\cos A = \frac{2074.79}{3044.42} = 0.6815$$
$$A = 47.04°$$

$C = 180 - 100 - 47.04 = 32.96°$. The solution is
$$A = 47.04° \quad a = BC = 39 \text{ cm}$$
$$B = 100° \quad b = AC = 52.49 \text{ cm}$$
$$C = 32.96° \quad c = AB = 29 \text{ cm}$$

Solutions to Chapter 22

Exercise 22.1

1. A is 1×4; B is 4×2; C is 1×1, and D is 5×1.

2. A is square; B is neither diagonal nor square; C is square and diagonal (it is the 3×3 identity matrix); D and E are neither square nor diagonal; F is square and diagonal; G and H are square.

3. A 3×1 matrix has 3 elements, a 1×3 matrix has 3 elements. An $m \times n$ matrix has mn elements and an $n \times n$ matrix has n^2 elements.

Exercise 22.2

1. $MN = \begin{pmatrix} 13 & 46 \\ 7 & 0 \end{pmatrix}$ $NM = \begin{pmatrix} 25 & 22 \\ 1 & -12 \end{pmatrix}$

   Note $MN \neq NM$.

2. (a) $3A = \begin{pmatrix} 3 & -21 \\ -3 & 6 \end{pmatrix}$

   (b) $A + B$ not possible; (c) $A + C$ not possible;

   (d) $5D = \begin{pmatrix} 35 \\ 45 \\ 40 \\ 5 \end{pmatrix}$

   (e) BC not possible; (f) CB not possible; (g) AC not possible; (h) CA not possible; (i) $5A - 2C$ not possible;

   (j) $D - E = \begin{pmatrix} 6.5 \\ 9 \\ 8 \\ 0 \end{pmatrix}$

   (k) BA not possible;

   (l) $AB = \begin{pmatrix} 10 & 7 & 72 \\ -5 & -7 & -22 \end{pmatrix}$

(m) $A^2 = AA = \begin{pmatrix} 8 & -21 \\ -3 & 11 \end{pmatrix}$

3. (a) $IA = \begin{pmatrix} 3 & 7 \\ -1 & -2 \end{pmatrix} = A$

   (b) $AI = A$; (c) BI not possible;

   (d) $IB = \begin{pmatrix} x \\ y \end{pmatrix} = B$

   (e) CI not possible; (f) $IC = C$. Where it is possible to multiply by the identity matrix, the result is unaltered.

4. (a) $A + B = \begin{pmatrix} a+e & b+f \\ c+g & d+h \end{pmatrix}$

   (b) $A - B = \begin{pmatrix} a-e & b-f \\ c-g & d-h \end{pmatrix}$

   (c) $AB = \begin{pmatrix} ae+bg & af+bh \\ ce+dg & cf+dh \end{pmatrix}$

   (d) $BA = \begin{pmatrix} ea+fc & eb+fd \\ ga+hc & gb+hd \end{pmatrix}$

5. $\begin{pmatrix} -7x+5y \\ -2x-y \end{pmatrix}$

6. $\begin{pmatrix} 9a-5b \\ 3a+8b \end{pmatrix}$

7. (a) $BA = \begin{pmatrix} 10 & 3 & 1 \\ -2 & 9 & -5 \end{pmatrix}$

   (b) $B^2 = \begin{pmatrix} 7 & -2 \\ -3 & 10 \end{pmatrix}$

   (c) $B^2A = \begin{pmatrix} 6 & 21 & -9 \\ 34 & -9 & 13 \end{pmatrix}$

8. (a) $AB = \begin{pmatrix} 9 & 5 \\ 12 & 9 \end{pmatrix}$

(b) $BA = \begin{pmatrix} 1 & -4 \\ 1 & 17 \end{pmatrix}$

(c) $A^2 = \begin{pmatrix} -2 & 15 \\ -5 & 13 \end{pmatrix}$

(d) $B^3 = \begin{pmatrix} -6 & -1 \\ 3 & -4 \end{pmatrix}$

10. $\begin{pmatrix} 3 & 0 \\ 0 & 2 \end{pmatrix}\begin{pmatrix} a & b \\ c & d \end{pmatrix} = \begin{pmatrix} 3a & 3b \\ 2c & 2d \end{pmatrix}$

The effect of multiplying by $\begin{pmatrix} 3 & 0 \\ 0 & 2 \end{pmatrix}$ is to multiply the top row by 3 and the second row by 2.

Exercise 22.3

1. (a) -2 (b) -22 (c) 1 (d) 88
2. C and E are singular.

4. (a) $\dfrac{1}{-4}\begin{pmatrix} 10 & -4 \\ -6 & 2 \end{pmatrix} = \begin{pmatrix} \frac{-5}{2} & 1 \\ \frac{3}{2} & \frac{-1}{2} \end{pmatrix}$

(b) $\begin{pmatrix} 1 & 0 \\ 0 & 1 \end{pmatrix}$

(c) $\dfrac{1}{-8}\begin{pmatrix} 2 & -3 \\ -2 & -1 \end{pmatrix} = \begin{pmatrix} \frac{-1}{4} & \frac{3}{8} \\ \frac{1}{4} & \frac{1}{8} \end{pmatrix}$

(d) Has no inverse.

5. (a) $\begin{pmatrix} 1 & -2 \\ -4 & 9 \end{pmatrix}$

(b) $\dfrac{1}{2}\begin{pmatrix} 4 & -2 \\ -5 & 3 \end{pmatrix}$

(c) $\begin{pmatrix} -3 & -1 \\ -5 & -2 \end{pmatrix}$

(d) $\dfrac{1}{38}\begin{pmatrix} 5 & 2 \\ -9 & 4 \end{pmatrix}$

(e) $\dfrac{1}{2}\begin{pmatrix} 5 & -3 \\ -4 & 2 \end{pmatrix}$

Exercise 22.4

1. (a) Take $A = \begin{pmatrix} 1 & 1 \\ 2 & -1 \end{pmatrix}$

so that $A^{-1} = \begin{pmatrix} \frac{1}{3} & \frac{1}{3} \\ \frac{2}{3} & \frac{-1}{3} \end{pmatrix}$

Then $\begin{pmatrix} x \\ y \end{pmatrix} = \begin{pmatrix} \frac{1}{3} & \frac{1}{3} \\ \frac{2}{3} & \frac{-1}{3} \end{pmatrix}\begin{pmatrix} 17 \\ 10 \end{pmatrix} = \begin{pmatrix} 9 \\ 8 \end{pmatrix}$

so that $x = 9$ and $y = 8$.

(b) Take $A = \begin{pmatrix} 2 & -1 \\ 1 & 3 \end{pmatrix}$

so that $A^{-1} = \begin{pmatrix} \frac{3}{7} & \frac{1}{7} \\ \frac{-1}{7} & \frac{2}{7} \end{pmatrix}$

Then $\begin{pmatrix} x \\ y \end{pmatrix} = \begin{pmatrix} \frac{3}{7} & \frac{1}{7} \\ \frac{-1}{7} & \frac{2}{7} \end{pmatrix}\begin{pmatrix} 10 \\ -2 \end{pmatrix} = \begin{pmatrix} 4 \\ -2 \end{pmatrix}$

so that $x = 4$ and $y = -2$.

(c) Take $A = \begin{pmatrix} 1 & 2 \\ 3 & -1 \end{pmatrix}$

so that $A^{-1} = \begin{pmatrix} \frac{1}{7} & \frac{2}{7} \\ \frac{3}{7} & \frac{-1}{7} \end{pmatrix}$

Then $\begin{pmatrix} x \\ y \end{pmatrix} = \begin{pmatrix} \frac{1}{7} & \frac{2}{7} \\ \frac{3}{7} & \frac{-1}{7} \end{pmatrix}\begin{pmatrix} 15 \\ 10 \end{pmatrix} = \begin{pmatrix} 5 \\ 5 \end{pmatrix}$

so that $x = y = 5$.

2. The matrix

$\begin{pmatrix} 2 & -3 \\ 6 & -9 \end{pmatrix}$

is singular and so has no inverse. Hence the matrix method fails to produce any solutions. The second equation is simply a multiple of the first – that is, we really only have one equation and this is not sufficient information to produce a unique value of x and y.

3. (a)
$$\begin{pmatrix} 3 & -1 \\ 1 & 1 \end{pmatrix}\begin{pmatrix} x \\ y \end{pmatrix} = \begin{pmatrix} 3 \\ 5 \end{pmatrix}$$

$$\begin{pmatrix} x \\ y \end{pmatrix} = \begin{pmatrix} \frac{1}{4} & \frac{1}{4} \\ -\frac{1}{4} & \frac{3}{4} \end{pmatrix}\begin{pmatrix} 3 \\ 5 \end{pmatrix}$$

$$= \begin{pmatrix} 2 \\ 3 \end{pmatrix}$$

(b)
$$\begin{pmatrix} 2 & 3 \\ 1 & -2 \end{pmatrix}\begin{pmatrix} x \\ y \end{pmatrix} = \begin{pmatrix} 10 \\ -9 \end{pmatrix}$$

$$\begin{pmatrix} x \\ y \end{pmatrix} = \begin{pmatrix} \frac{2}{7} & \frac{3}{7} \\ \frac{1}{7} & -\frac{2}{7} \end{pmatrix}\begin{pmatrix} 10 \\ -9 \end{pmatrix}$$

$$= \begin{pmatrix} -1 \\ 4 \end{pmatrix}$$

(c)
$$\begin{pmatrix} 1 & \frac{1}{2} \\ 4 & -3 \end{pmatrix}\begin{pmatrix} x \\ y \end{pmatrix} = \begin{pmatrix} 2 \\ 18 \end{pmatrix}$$

$$\begin{pmatrix} x \\ y \end{pmatrix} = \begin{pmatrix} \frac{3}{5} & \frac{1}{10} \\ \frac{4}{5} & -\frac{1}{5} \end{pmatrix}\begin{pmatrix} 2 \\ 18 \end{pmatrix}$$

$$= \begin{pmatrix} 3 \\ -2 \end{pmatrix}$$

(d)
$$\begin{pmatrix} \frac{1}{2} & 3 \\ 2 & 5 \end{pmatrix}\begin{pmatrix} x \\ y \end{pmatrix} = \begin{pmatrix} 5 \\ 6 \end{pmatrix}$$

$$\begin{pmatrix} x \\ y \end{pmatrix} = \begin{pmatrix} -\frac{10}{7} & \frac{6}{7} \\ \frac{4}{7} & -\frac{1}{7} \end{pmatrix}\begin{pmatrix} 5 \\ 6 \end{pmatrix}$$

$$= \begin{pmatrix} -2 \\ 2 \end{pmatrix}$$

Solutions to Chapter 23

Exercise 23.1
1. 16 cm = 160 mm = 0.16 m
2. 356 mm = 35.6 cm = 0.356 m
3. 156 m = 0.156 km
4. $A = 14 \times 7 = 98$ cm^2
5. $A = 17.5$ m^2, cost = £87.33
6. $A = 2 \times 8 = 16$ m^2
7. $s = (a + b + c)/2 = 17$,
   $A = \sqrt{s(s-a)(s-b)(s-c)}$
   $= \sqrt{2295} = 47.9$ cm^2
8. $A = \frac{1}{2}(a+b)h = 22$ cm^2
9. $A = \pi r^2 = 1017.9$ cm^2
10. $A = 78.5$ cm^2
11. $A = 26 = \pi r^2$ and so $r^2 = 26/\pi = 8.276$,
    then $r = 2.877$ cm
12. 13430 cm^2 = 1.3430 m^2
13. $V = \pi r^2 h = 1099.6$ cm^3 = 1.0996 litres
14. $V = \frac{4}{3}\pi r^3 = 2145$ cm^3

15. $86 = \frac{4}{3}\pi r^3$, therefore
    $r^3 = (3 \times 86)/(4 \times \pi) = 20.53$,
    then $r = 2.74$ cm
16. $A = \pi r^2 \times \theta/360 = 39.27$ cm^2
17. $V = \pi r^2 h = 28.274$ m^3, now 1000 litres
    $= 1$ m^3
    and so 28.274 m^3 = 28274 litres

Exercise 23.2
1 (a) 12 kg = 12000 g
  (b) 12 kg = 12000000 mg
2. 168 mg = 0.168 g = 0.000168 kg or
   1.68×10^{-4} kg
3. 0.005 tonnes = 5 kg
4. 875 tonnes = 875000 kg = 8.75×10^5 kg
5. 3500 kg = 3.5 tonnes

Solutions to Chapter 24

Exercise 24.2

1. (a) $8x^7$ (b) $7x^6$ (c) $-x^{-2} = -1/x^2$
   (d) $-5x^{-6} = -5/x^6$ (e) $13x^{12}$
   (f) $5x^4$ (g) $-2x^{-3} = -2/x^3$
2. (a) $\frac{3}{2}x^{1/2}$ (b) $\frac{5}{2}x^{3/2}$ (c) $-\frac{1}{2}x^{-3/2}$
   (d) $(1/2)x^{-1/2} = 1/2\sqrt{x}$
   (e) $y = \sqrt{x} = x^{1/2}$ and so
   $y' = (1/2)x^{-1/2} = 1/2\sqrt{x}$; (f) $0.2x^{-0.8}$
3. $y = 8$, $y' = 0$, the line is horizontal and so its gradient is zero.

4. (a) $16x^{15}$ (b) $0.5x^{-0.5}$ (c) $-3.5x^{-4.5}$
   (d) $y = 1/x^{1/2} = x^{-1/2}$ and so
   $y' = -\frac{1}{2}x^{-3/2}$
   (e) $y = 1/\sqrt{x} = 1/x^{1/2} = x^{-1/2}$ and
   so $y' = -\frac{1}{2}x^{-3/2}$
5. $y' = 4x^3$: (a) at $x = -4$, $y' = -256$
   (b) at $x = -1$, $y' = -4$
   (c) at $x = 0$, $y' = 0$ (d) at $x = 1$, $y' = 4$
   (e) at $x = 4$, $y' = 256$. The graph is falling at the given negative x values and is rising at the given positive x values. At $x = 0$ the graph is neither rising nor falling.
6. (a) $y' = 3x^2$, $y'(-3) = 27$, $y'(0) = 0$,
   $y'(3) = 27$
   (b) $y' = 4x^3$, $y'(-2) = -32$, $y'(2) = 32$
   (c) $y' = \frac{1}{2}x^{-1/2} = 1/2\sqrt{x}$, $y'(1) = \frac{1}{2}$,
   $y'(2) = \frac{1}{2\sqrt{2}} = 0.354$; (d) $y' = -2x^{-3}$,
   $y'(-2) = \frac{1}{4}$, $y'(2) = -\frac{1}{4}$
7. (a) $2t$ (b) $-3t^{-4}$ (c) $\frac{1}{2}t^{-1/2}$ (d) $-1.5t^{-2.5}$
8. (a) $-7\sin 7x$ (b) $\frac{1}{2}\cos(x/2)$ (c) $-\frac{1}{2}\sin(x/2)$
   (d) $4e^{4x}$ (e) $-3e^{-3x}$
9. (a) $1/x$ (b) $\frac{1}{3}\cos(x/3)$ (c) $\frac{1}{2}e^{x/2}$
10. $y' = \cos x$, $y'(0) = 1$, $y'(\pi/2) = 0$,
    $y'(\pi) = -1$

11. (a) $7t^6$ (b) $3\cos 3t$ (c) $2e^{2t}$ (d) $1/t$
    (e) $-\frac{1}{3}\sin(t/3)$

Exercise 24.3

1. (a) $y' = 6x^5 - 4x^3$ (b) $y' = 6 \times (2x) = 12x$
   (c) $y' = 9(-2x^{-3}) = -18x^{-3}$ (d) $y' = \frac{1}{2}$
   (e) $y' = 6x^2 - 6x$
2. (a) $y' = 2\cos x$
   (b) $y' = 3(4\cos 4x) = 12\cos 4x$
   (c) $y' = -63\sin 9x + 12\cos 4x$
   (d) $y' = 3e^{3x} + 10e^{-2x}$
3. The gradient function is $y' = 9x^2 - 9$.
   $y'(1) = 0$, $y'(0) = -9$, $y'(-1) = 0$. Of the three given points, the curve is steepest at $x = 0$ where it is falling.
4. When x^2 is differentiated we obtain $2x$. Also, because the derivative of any constant is zero, the derivative of $x^2 + c$ will be $2x$ where c can take any value.
5. $\quad y' = 6\cos 2t - 8\sin 2t$

   $y'(2) = 6\cos 4 - 8\sin 4 = 2.1326$

6. $\quad y' = 2 - 3x^2 + 2e^{2x}$

   $y'(1) = 2 - 3 + 2e^2 = 13.7781$

7. $\quad y = (2x - 1)^2 = 4x^2 - 4x + 1$

   $y' = 8x - 4$

   $y'(0) = -4$

8. $\quad f' = 2x - 2$

   $f'(1) = 0$

9. $y' = x^2 - 1$. We see that $y' = 0$ when $x^2 - 1 = 0$, that is, when $x = -1$, 1.
10. $y' = 1 - \sin x$. When $y' = 0$ then $\sin x = 1$, and hence $x = \frac{\pi}{2}$.

Exercise 24.4

1. (a) $y' = 2x$, $y'' = 2$
   (b) $y' = 24x^7$, $y'' = 168x^6$
   (c) $y' = 45x^4$, $y'' = 180x^3$
   (d) $y' = \frac{3}{2}x^{-1/2}$, $y'' = -\frac{3}{4}x^{-3/2}$
   (e) $y' = 12x^3 + 10x$, $y'' = 36x^2 + 10$
   (f) $y' = 27x^2 - 14$, $y'' = 54x$
   (g) $y' = \frac{1}{2}x^{-1/2} - 2x^{-3/2}$,
      $y'' = -\frac{1}{4}x^{-3/2} + 3x^{-5/2}$

2. (a) $y' = 3\cos 3x - 2\sin 2x$,
      $y'' = -9\sin 3x - 4\cos 2x$
   (b) $y' = e^x - e^{-x}$, $y'' = e^x + e^{-x}$

3. $y' = 4\cos 4x$, $y'' = -16\sin 4x$, and so
   $16y + y'' = 16\sin 4x + (-16\sin 4x) = 0$ as
   required.

4. $y' = 4x^3 - 6x^2 - 72x$, $y'' = 12x^2 - 12x - 72$.
   The values of x where $y'' = 0$ are found
   from $12x^2 - 12x - 72 = 0$, that is,
   $12(x^2 - x - 6) = 12(x + 2)(x - 3) = 0$ so
   that $x = -2$ and $x = 3$.

5. (a) $y' = 6\cos 3x - 8\sin 2x$,
      $y'' = -18\sin 3x - 16\cos 2x$
   (b) $y' = k\cos kt$, $y'' = -k^2\sin kt$
   (c) $y' = -k\sin kt$, $y'' = -k^2\cos kt$
   (d) $y' = Ak\cos kt - Bk\sin kt$,
      $y'' = -Ak^2\sin kt - Bk^2\cos kt$

6. (a) $y' = 3e^{3x}$, $y'' = 9e^{3x}$
   (b) $y' = 3 + 8e^{4x}$, $y'' = 32e^{4x}$
   (c) $y' = 2e^{2x} + 2e^{-2x}$, $y'' = 4e^{2x} - 4e^{-2x}$
   (d) $y = e^{-x}$, $y' = -e^{-x}$, $y'' = e^{-x}$
   (e) $y = e^{2x} + 3e^x$, $y' = 2e^{2x} + 3e^x$,
      $y'' = 4e^{2x} + 3e^x$

7. $y' = 200 + 12x - 3x^2$, $y'' = 12 - 6x$.
   $y'' = 0$ when $x = 2$

8. $y' = 2e^x + 2\cos 2x$
     $y'' = 2e^x - 4\sin 2x$
   $y''(1) = 2e - 4\sin 2 = 1.7994$

9. $y' = 8\cos 4t + 10\sin 2t$
     $y'' = -32\sin 4t + 20\cos 2t$
   $y''(1.2) = -32\sin 4.8 + 20\cos 2.4 = 17.1294$

Exercise 24.5

1. (a) $y' = 2x$, $y'' = 2$, $(0,1)$ is a minimum;
   (b) $y' = -2x$, $y'' = -2$, $(0,0)$ is a maximum;
   (c) $y' = 6x^2 + 18x$, $y'' = 12x + 18$, $(0,0)$ is a
      minimum and $(-3,27)$ is a maximum;
   (d) $y' = -6x^2 + 54x$, $y'' = -12x + 54$, $(0,0)$
      is a minimum and $(9,729)$ is a maximum;
   (e) $y' = 3x^2 - 6x + 3$, $y'' = 6x - 6$,
      stationary point by solving
      $3x^2 - 6x + 3 = 0$, at $(1,2)$, second
      derivative test fails, inspection of gradient
      on either side of $x = 1$ reveals a point of
      inflexion.

2. $y' = \cos x$. Solving $\cos x = 0$ gives $x = \frac{\pi}{2}, \frac{3\pi}{2}$.
   The maximum and minimum points occur at
   $x = \frac{\pi}{2}, \frac{3\pi}{2}$.
   We examine y''
   $$y'' = -\sin x$$
   $$y''\left(\frac{\pi}{2}\right) = -\sin\frac{\pi}{2} < 0 \quad \text{i.e. maximum}$$
   $$y''\left(\frac{3\pi}{2}\right) = -\sin\frac{3\pi}{2} > 0 \quad \text{i.e. minimum}$$
   When $x = \frac{\pi}{2}$, $y = 1$; when $x = \frac{3\pi}{2}$, $y = -1$. In
   summary, maximum at $(\frac{\pi}{2}, 1)$, minimum at
   $(\frac{3\pi}{2}, -1)$.

3. $y' = e^x - 1$. Solving $y' = 0$ gives $x = 0$.
   Now, $y'' = e^x$ and $y''(0) = 1 > 0$. When
   $x = 0$, $y = 1$, so there is a minimum at $(0, 1)$.

4. $y' = 6x^2 - 6x - 12 = 6(x + 1)(x - 2)$.
   Solving $y' = 0$ gives $x = -1$, 2.
   $y'' = 12x - 6$, $y''(-1) < 0$ i.e. maximum
   $y''(2) > 0$, i.e. minimum
   When $x = -1$, $y = 8$; when $x = 2$, $y = -19$.
   Hence there is a maximum at $(-1, 8)$, a
   minimum at $(2, -19)$.

5. $y' = -x^2 + 1$. Solving $y' = 0$ gives
   $x = -1$, 1.
   $y'' = -2x$, $y''(-1) > 0$ i.e. minimum,
   $y''(1) < 0$ i.e. maximum
   When $x = -1$, $y = -\frac{2}{3}$, when $x = 1$, $y = \frac{2}{3}$.
   So there is a minimum at $(-1, -\frac{2}{3})$, and a
   maximum at $(1, \frac{2}{3})$.

6. $y' = 5x^4 - 5$

$= 5(x^4 - 1)$

$= 5(x^2 - 1)(x^2 + 1)$

$= 5(x - 1)(x + 1)(x^2 + 1)$

Solving $y' = 0$ gives $x = -1,\ 1$.
$y'' = 20x^3,\ y''(-1) < 0$ i.e. maximum
$y''(1) > 0$ i.e. minimum.
When $x = -1,\ y = 5$, when $x = 1,\ y = -3$.
There is a maximum at $(-1, 5)$; a minimum
at $(1, -3)$.

Solutions to Chapter 25

Exercise 25.2

1. (a) $x^2/2 + c$ (b) $x^2 + c$ (c) $x^6/6 + c$
   (d) $x^8/8 + c$ (e) $-x^{-4}/4 + c$
   (f) $2x^{1/2} + c$ (g) $x + c$ (h) $4x^{3/4}/3 + c$
2. (a) $9x + c$ (b) $\frac{1}{2}x + c$
   (c) $-7x + c$ (d) $0.5x + c$. In general,
   $\int k\,dx = kx + c$.
3. (a) $-(\cos 5x)/5 + c$ (b) $-2\cos(x/2) + c$
   (c) $2\sin(x/2) + c$ (d) $-e^{-3x}/3 + c$
   (e) $-4\cos 0.25x + c$ (f) $-2e^{-0.5x} + c$
   (g) $2e^{x/2} + c$
4. (a) $\int \cos 5x\,dx = \frac{\sin 5x}{5} + c$
   (b) $\int \sin 4x\,dx = -\frac{\cos 4x}{4} + c$
   (c) $\int e^{6t}\,dt = \frac{e^{6t}}{6} + c$
   (d) $\int \cos(\frac{t}{3})\,dt = 3\sin(\frac{t}{3}) + c$
5. (a) $\int \cos 3y\,dy = \frac{\sin 3y}{3} + c$
   (b) $\int \frac{1}{\sqrt{x}}\,dx = \int x^{-1/2}dx = 2x^{1/2} + c = 2\sqrt{x} + c$
   (c) $\int \sin(\frac{3t}{2})\,dt = -\frac{2}{3}\cos(\frac{3t}{2}) + c$
   (d) $\int \sin(-x)\,dx = \cos(-x) + c = \cos x + c$

Exercise 25.3

1. (a) $x^3/3 + x^2/2 + c$

   (b) $x^3/3 + x^2/2 + x + c$ (c) $\frac{4}{7}x^7 + c$

   (d) $5x^2/2 + 7x + c$ (e) $\ln x + c$

   (f) $\int (1/x^2)\,dx = \int x^{-2}dx = x^{-1}/(-1) + c$

   $= -1/x + c$

   (g) $3\ln x - \frac{7}{x} + c$

(h) $\frac{2}{3}x^{3/2} + c$

(i) $8x^{1/2} + c$

(j) $\int \frac{1}{\sqrt{x}}\,dx = \int x^{-1/2}dx = 2x^{1/2} + c$

(k) $\frac{3x^4}{4} + \frac{7x^3}{3} + c$

(l) $\frac{2x^5}{5} - x^3 + c$

(m) $\frac{x^6}{6} - 7x + c$

2. (a) $3e^x + c$ (b) $\frac{1}{2}e^x + c$ (c) $3e^x + 2e^{-x} + c$
   (d) $5e^{2x}/2 + c$
   (e) $-\cos x - (\sin 3x)/3 + c$
3. (a) $x^3 - 2e^x + c$ (b) $x^2/4 + x + c$
   (c) $-\frac{2}{3}\cos 3x - (8\sin 3x)/3 + c$
   (d) $-(3\cos 2x)/2 + (5\sin 3x)/3 + c$
   (e) $5\ln x + c$
4. $5\ln t - t^3/3 + c$
5. (a) $\int 3\sin 2t + 4\cos 2t\,dt$
   $= -\frac{3}{2}\cos 2t + 2\sin 2t + c$
   (b) $\int -\sin 3x - 2\cos 4x\,dx = \frac{\cos 3x}{3} - \frac{\sin 4x}{2} + c$
   (c) $\int 1 + 2\sin(\frac{x}{2})\,dx = x - 4\cos(\frac{x}{2}) + c$
   (d) $\int \frac{1}{2} - \frac{1}{3}\cos(\frac{x}{3})\,dx = \frac{x}{2} - \sin(\frac{x}{3}) + c$
6. (a) $\int e^x(1 + e^x)\,dx = \int e^x + e^{2x}\,dx$
   $= e^x + \frac{e^{2x}}{2} + c$
   (b) $\int (x + 2)(x + 3)\,dx = \int x^2 + 5x + 6\,dx$
   $= \frac{x^3}{3} + \frac{5x^2}{2} + 6x + c$
   (c) $\int 5\cos 2t - 2\sin 5t\,dt =$
   $\frac{5}{2}\sin 2t + \frac{2}{5}\cos 5t + c$
   (d) $\int \frac{2}{x}(x + 3)\,dx = \int 2 + \frac{6}{x}\,dx$
   $= 2x + 6\ln x + c$

Exercise 25.4

1. (a) $\left[\dfrac{7x^2}{2}\right]_{-1}^{1} = 0$

(b) $\left[\dfrac{7x^2}{2}\right]_{0}^{3} = \dfrac{63}{2} = 31.5$

(c) $\left[\dfrac{x^3}{3}\right]_{-1}^{1} = \dfrac{2}{3}$

(d) $\left[-\dfrac{x^3}{3}\right]_{-1}^{1} = -\dfrac{2}{3}$

(e) $\left[\dfrac{x^3}{3}\right]_{0}^{2} = \dfrac{8}{3}$

(f) $\left[\dfrac{x^3}{3} + 2x^2\right]_{0}^{3} = 9 + 18 = 27$

(g) $[\sin x]_{0}^{\pi} = 0$

(h) $[\ln x]_{1}^{2} = \ln 2 - \ln 1 = \ln 2 = 0.693$

(i) $[\cos x]_{0}^{\pi} = \cos \pi - \cos 0 = -1 - 1 = -2$

(j) $[13x]_{1}^{4} = 52 - 13 = 39$

(k) $\left[\dfrac{x^3}{3} - \cos x\right]_{0}^{\pi}$

$= \left(\dfrac{\pi^3}{3} - \cos \pi\right) - (0 - \cos 0)$

$= \dfrac{\pi^3}{3} + 1 + 1 = \dfrac{\pi^3}{3} + 2 = 12.335$

(l) $\left[\dfrac{5x^3}{3}\right]_{1}^{3} = 45 - \dfrac{5}{3} = 43.333$

(m) $\left[-\dfrac{x^2}{2}\right]_{0}^{1} = -\dfrac{1}{2}$

(n) $\left[-\dfrac{1}{x}\right]_{1}^{3} = -\dfrac{1}{3} - \left(-\dfrac{1}{1}\right) = \dfrac{2}{3}$

(o) $\left[-\dfrac{\cos 3t}{3}\right]_{0}^{1} = \left(-\dfrac{\cos 3}{3}\right) - \left(-\dfrac{\cos 0}{3}\right)$

$= 0.6633$

2. (a) $\int_{0}^{2} 3e^x + 1 \; dx = [3e^x + x]_{0}^{2} = 21.17$

(b) $\int_{0}^{1} 2e^{2x} + \sin x \; dx$

$= [e^{2x} - \cos x]_{0}^{1}$

$= (e^2 - \cos 1) - (e^0 - \cos 0)$

$= 6.8488$

(c) $\int_{1}^{2} 2 \cos 2t - 3 \sin 2t \; dt$

$= \left[\sin 2t + \dfrac{3}{2} \cos 2t\right]_{1}^{2}$

$= \left(\sin 4 + \dfrac{3}{2} \cos 4\right) - \left(\sin 2 + \dfrac{3}{2} \cos 2\right)$

$= -2.0223$

(d) $\int_{1}^{3} \dfrac{2}{x} + \dfrac{x}{2} \; dx$

$= \left[2 \ln x + \dfrac{x^2}{4}\right]_{1}^{3}$

$= \left(2 \ln 3 + \dfrac{9}{4}\right) - \left(2 \ln 1 + \dfrac{1}{4}\right)$

$= 4.1972$

3. (a) $\int_{0}^{2} x(2x + 3) dx$

$= \int_{0}^{2} 2x^2 + 3x \; dx$

$= \left[\dfrac{2x^3}{3} + \dfrac{3x^2}{2}\right]_{0}^{2}$

$= \dfrac{34}{3}$

(b) $\int_{-1}^{1} 3e^{2x} - \dfrac{3}{e^{2x}} \; dx$

$= \int_{-1}^{1} 3e^{2x} - 3e^{-2x} dx$

$= \left[\dfrac{3e^{2x}}{2} + \dfrac{3e^{-2x}}{2}\right]_{-1}^{1}$

$= 0$

(c) $\displaystyle\int_0^\pi 2\sin\!\left(\frac{t}{2}\right) - 4\cos 2t \; dt$

$= \left[-4\cos\!\left(\frac{t}{2}\right) - 2\sin 2t\right]_0^\pi$

$= \left(-4\cos\dfrac{\pi}{2} - 2\sin 2\pi\right) - (4\cos 0 - 2\sin 0)$

$= 4$

(d) $\displaystyle\int_1^2 \sin 3x - \frac{3}{x}\,dx$

$= \left[-\dfrac{\cos 3x}{3} - 3\ln x\right]_1^2$

$= \left(-\dfrac{\cos 6}{3} - 3\ln 2\right) - \left(-\dfrac{\cos 3}{3} - 3\ln 1\right)$

$= -2.7295$

4. (a) $\displaystyle\int_1^2 \frac{2}{x}\left(3 + \frac{1}{x}\right)dx$

$= \displaystyle\int_1^2 \frac{6}{x} + \frac{2}{x^2}\,dx$

$= \left[6\ln x - \dfrac{2}{x}\right]_1^2$

$= (6\ln 2 - 1) - (6\ln 1 - 2)$

$= 5.1589$

(b) $\displaystyle\int_0^1 e^{-2x}(e^x - 2e^{-x})dx$

$= \displaystyle\int_0^1 e^{-x} - 2e^{-3x}\,dx$

$= \left[-e^{-x} + \dfrac{2e^{-3x}}{3}\right]_0^1$

$= \left(-e^{-1} + \dfrac{2e^{-3}}{3}\right) - \left(-e^0 + \dfrac{2e^0}{3}\right)$

$= -0.001355$

(c) $\displaystyle\int_0^{0.5} \sin\pi t + \cos\pi t \; dt$

$= \left[\dfrac{-\cos\pi t + \sin\pi t}{\pi}\right]_0^{0.5}$

$= \dfrac{2}{\pi}$

(d) $\displaystyle\int_0^1 (2x - 3)^2 \; dx$

$= \displaystyle\int_0^1 4x^2 - 12x + 9 \; dx$

$= \left[\dfrac{4x^3}{3} - 6x^2 + 9x\right]_0^1$

$= \dfrac{13}{3}$

Exercise 25.5

1. 26
2. Evaluate $\int_{-1}^2 (4 - x^2)\,dx$ and $\int_2^4 (4 - x^2)\,dx$ separately to get 9 and $-10\frac{2}{3}$ respectively. Total area is $19\frac{2}{3}$.
3. $[e^{2t}/2]_1^4 = e^8/2 - e^2/2 = 1486.78$
4. $[\ln t]_1^5 = \ln 5 = 1.609$
5. $[\frac{2}{3}x^{3/2}]_1^2 = \frac{2}{3}2^{3/2} - \frac{2}{3} = 1.219$
6. Area $= \displaystyle\int_1^4 \frac{1}{x}\,dx = [\ln x]_1^4 = 1.3863$

7. Note that all the area is above the t axis.

Area $= \displaystyle\int_0^{0.5} \cos 2t \; dt$

$= \left[\dfrac{\sin 2t}{2}\right]_0^{0.5}$

$= \dfrac{\sin 1}{2} - \dfrac{\sin 0}{2}$

$= 0.4207$

Solutions to Chapter 26

Exercise 26.1

1. (a) Discrete (b) continuous (c) discrete
   (d) continuous (e) continuous (f) discrete
   (g) continuous (h) continuous (i) discrete
2. 1970   3
   1971   0
   1972   1
   1973   4
   1974   7
   1975   8
   1976   2
   Total 25

Exercise 26.2

1. (a) See Table 1 (b) $\frac{1}{30} = 0.033$
   (c) $\frac{2}{30} = 0.067$
   (d) $\frac{4}{30} = 0.133 = 13.3\%$
2. (a) 5499.4 hours will appear in class
   5000–5499 (b) 5499.8 hours will appear in
   class 5500–5999 (c) the class boundaries are
   4999.5–5499.5, 5499.5–5999.5, 5999.5–
   6499.5, 6499.5–6999.5 (d) all classes have
   the same width, $5499.5 - 4999.5 = 500$
   hours (e) the midpoints are 5249.5, 5749.5,
   6249.5, 6749.5.
3. See Table 2.
4. The second class boundaries are
   20.585–20.615. The class width is
   $20.615 - 20.585 = 0.03$.

Table 1

Age	Tally	Frequency
15–19	卌 l	6
20–24	lll	3
25–29	llll	4
30–34	llll	4
35–39	ll	2
40–44	ll	2
45–49	llll	4
50–54		0
55–59	ll	2
60–64	l	1
65–69		0
70–74	l	1
75–79	l	1
		30

Table 2

Mark	Frequency
0–9	0
10–19	4
20–29	3
30–39	2
40–49	6
50–59	15
60–69	10
70–79	5
80–89	11
90–99	4
	60

Exercise 26.3

1.

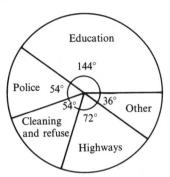

(a) Pie chart showing local council expenditure

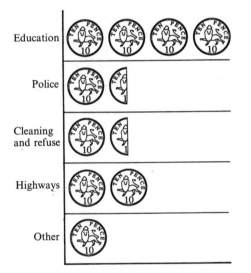

(b) Pictogram showing how each £1 spent by a local council is distributed

2.

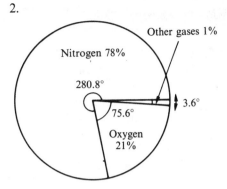

Pie chart showing composition of the atmosphere

3. (a) The class boundaries are 0–99.5, 99.5–199.5, 199.5–299.5, 299.5–399.5, 399.5–499.5 (b) a claim for 199.75 would be placed in the class 200–299

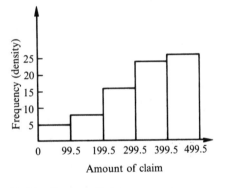

(c) Distribution of claim amounts

Solutions to Chapter 27

Exercise 27.2

1. Mean $= \bar{x} = \frac{50}{9} = 5.556$, median $= 5$, mode $= 5$.
2. Mean $= \frac{262}{7} = 37.4$. The large, atypical value of 256 distorts the mean, it may be more appropriate to use the median or mode here.
3. Mean $= \frac{498}{7} = 71.1\%$, median $= 68\%$.
4. (a) $x_1 + x_2 + x_3 + x_4$
   (b) $x_1 + x_2 + x_3 + x_4 + x_5 + x_6 + x_7$
   (c) $(x_1 - 3)^2 + (x_2 - 3)^2 + (x_3 - 3)^2$
   (d) $(2 - x_1)^2 + (2 - x_2)^2 + (2 - x_3)^2 + (2 - x_4)^2$
5. (a) $\sum_{i=3}^{6} x_i$ (b) $\sum_{i=1}^{3}(x_i - 1)$
6. Mean $= 7.57/6 = 1.26$, median $= 1.32$, no mode.
7. £32080
8.

Midpoint x_i	Frequency f_i	$f_i \times x_i$
20.57	3	61.71
20.60	6	123.60
20.63	8	165.04
20.66	3	61.98
Total	20	412.33

Mean $= \bar{x} = 412.33/20 = 20.62$ (2 d.p.).

Exercise 27.3

1. (a) $\bar{x} = 20$, variance $= 0$, standard deviation $= 0$ (b) $\bar{x} = 20$, variance $= 12.5$, standard deviation $= 3.54$ (c) $\bar{x} = 20$, variance $= 200$, standard deviation $= 14.14$. The large standard deviation of the final set reflects the widely spread values.
2. Mean $= 5$, variance $= 9.2$, standard deviation $= 3.033$.
3. Jane's mean $= 51.4$, Tony's mean $= 51.6$. Jane's standard deviation $= 12.77$, Tony's standard deviation $= 17.40$. The students have a very close mean mark but Jane has achieved a more consistent set of results.
4. (a) See Table. (b) The modal class is 45–49, this is the most common class (c) mean $= \frac{2865}{50} = 57.3$, variance $= \frac{10720.50}{50} = 214.41$, standard deviation $= 14.64$.

Class	f	Midpoint x_i	$f_i \times x_i$	$x_i - \bar{x}$	$(x_i - \bar{x})^2$	$f_i(x_i - \bar{x})^2$
20–24	2	22	44	−35.3	1246.09	2492.18
25–29	0	27	0	−30.3	918.09	0.0
30–34	1	32	32	−25.3	640.09	640.09
35–39	1	37	37	−20.3	412.09	412.09
40–44	3	42	126	−15.3	234.09	702.27
45–49	11	47	517	−10.3	106.09	1166.99
50–54	4	52	208	−5.3	28.09	112.36
55–59	3	57	171	−0.3	0.09	0.27
60–64	10	62	620	4.7	22.09	220.9
65–69	5	67	335	9.7	94.09	470.45
70–74	5	72	360	14.7	216.09	1080.45
75–79	1	77	77	19.7	388.09	388.09
80–84	3	82	246	24.7	610.09	1830.27
85–89	0	87	0	29.7	882.09	0.0
90–94	1	92	92	34.7	1204.09	1204.09
Total	50		2865			10720.50

Table for Exercise 27.3, 4(a)

Solutions to Chapter 28

Exercise 28.2

1. (a) $\frac{5}{6}$ (b) 0 (c) 1/3 (d) 0
2. $\frac{12}{13}$
3. (a) $\frac{3}{10}$ (b) $\frac{3}{10}$ (c) $\frac{7}{10}$
4. There are 36 possibilities in all. (a) $\frac{1}{36}$ (b) 0
   (c) 0 (d) $\frac{1}{36}$ (e) $\frac{26}{36}$ or $\frac{13}{18}$
5. $\frac{1}{30}$
6. (a) $\frac{1}{7}$ (b) $\frac{12}{35}$ (c) $\frac{2}{7}$ (d) $\frac{18}{35}$ (e) $\frac{6}{7}$ (f) $\frac{23}{35}$
7. If H represents Head and T represents Tail
   then the possible outcomes are: TTT, TTH,
   THT, THH, HTT, HTH, HHT, HHH
   Note that all the outcomes have equal
   probability.

(a) number of outcomes with two heads and
one tail is three. So
P (two heads and one tail)$=\frac{3}{8}$
(b) Four outcomes have at least two heads.
P (at least two heads)$=\frac{1}{2}$
(c) 'No heads' is equivalent to the outcome
TTT.
P (no heads)$=\frac{1}{8}$

Exercise 28.3

1. (a) $\frac{143}{150}$ (b) $\frac{7}{150}$
2. 0.95
3. 89760
4. Number reached in less than one hour is
   $0.87 \times 17,300 = 15051$.
5. P (component works) $= \frac{48\,700}{50\,000} = 0.974$
   Number in batch working well
   $= 0.974 \times 3000 = 2922$.
   Number not working well $= 78$.

Exercise 28.4

1. $\frac{3}{6} \times \frac{1}{2} = \frac{1}{4}$
2. $\frac{1}{52} \times \frac{1}{52} = \frac{1}{2704} = 0.000370$
3. $\left(\frac{1}{2}\right)^8 = 0.0039$
4. $\left(\frac{1}{6}\right)^4 = 0.000772$
5. (a) 0.5 (b) 0.25
6. (a) P (all work well) $= (0.96)^4 = 0.8493$
   (b) P (none work well) $= (0.04)^4 = 2.56 \times 10^{-6}$

7. (a) P (passing three modules) $= (0.91)^3$
      $= 0.7536$.
   (b) Suppose a student passes two modules;
       the possible outcomes are PPF, PFP,
       FPP. Each outcome has a probability of
       $(0.91)^2(0.09)$. Hence total probability is
       $3(0.91)^2(0.09) = 0.2236$.
   (c) Passing one module, the possible
       outcomes are FFP, FPF, PFF. Each
       outcome has probability $(0.09)^2(0.91)$.
       Hence total probability is
       $3(0.09)^2(0.91) = 0.0221$.
   (d) Passing no modules is equivalent to
       failing all three. This has a probability
       of $(0.09)^3 = 7.29 \times 10^{-4}$.

Index